From Tool Use to Site Function

Use-wear analysis in some Final Upper Palaeolithic sites in the Basque country

Juan José Ibáñez Estévez
Jesús Emilio González Urquijo

BAR International Series 658
1996

Published in 2019 by
BAR Publishing, Oxford

BAR International Series 658

From Tool Use to Site Function

ISBN 9780860548478 paperback
ISBN 9781407349671 e-book

DOI https://doi.org/10.30861/9780860548478

A catalogue record for this book is available from the British Library

This book is available at www.barpublishing.com

BAR Publishing is the trading name of British Archaeological Reports (Oxford) Ltd. British Archaeological Reports was first incorporated in 1974 to publish the BAR Series, International and British. In 1992 Hadrian Books Ltd became part of the BAR group. This volume was originally published by Tempvs Reparatvm in conjunction with British Archaeological Reports (Oxford) Ltd / Hadrian Books Ltd, the Series principal publisher, in 1996. This present volume is published by BAR Publishing, 2019.

BAR
PUBLISHING

BAR titles are available from:

BAR Publishing
122 Banbury Rd, Oxford, OX2 7BP, UK
EMAIL info@barpublishing.com
PHONE +44 (0)1865 310431
FAX +44 (0)1865 316916
www.barpublishing.com

ACKNOWLEDGEMENTS

We would like to thank Dr. Juan María Apellániz, director of the doctoral theses which led to this publication, for his invaluable support during the work on these theses. Dr. Amelia Baldeón and Eduardo Berganza, directors of the excavations at Berniollo and Santa Catalina, kindly turned over to us the archaeological material for study and provided us with unpublished data on the sites. Dr. Hugues. Plisson and Dr. Patricia Anderson offered valuable commentary on our work, the fruit of their long years of experience in the field of use-wear analysis of lithic tools. Dr. Anderson also corrected the specialized vocabulary of the English version. To all we wish to express our thanks and friendship, at the same time assuming entire responsability for any error which may appear in this publication.

The work which we have carried out has been partially financed by the Ministerio de Educación y Ciencia, of the Spanish Government, project PB87-0186. During the time of its development we have been beneficiaries of pre-doctoral and post-doctoral grants from the Basque Government. The translation of the text from Spanish to English, by Elizabeth Willcox, was financed by the Departamento de Cultura de la Diputación Foral de Bizkaia

INDEX

INTRODUCTION 5

CHAPTER 1. **METHODOLOGY OF FUNCTIONAL ANALYSIS**. 7
1.1. THE EXPERIMENTAL PROGRAMME 8
1.2. THE DESCRIPTION OF THE TRACES. 11
1.2.1. Chipping. 11
1.2.2. Polish. 11
1.2.3. Striations and linear traces 13
1.2.4. Rounding. 15
1.2.5. The relationship between the traces. 15
1.3. THE CRITERIA. 15
1.4. THE SYSTEM OF FUNCTIONAL INFERENCE. 16

CHAPTER 2. **THE SITES STUDIED AND THE OVERALL RESULTS**. 19
2.1. THE SITES. 19
2.2. THE SAMPLING. 21
2.3. ANALYSIS OF ARCHAEOLOGICAL DATA. 23
2.4. THE OVERALL RESULTS OF FUNCTIONAL ANALYSIS. 23

CHAPTER 3. **THE FUNCTION OF THE DIFFERENT TYPOLOGICAL GROUPS**. 25
3.1. END-SCRAPERS. 25
3.2. BURINS. 31
3.3. BORERS. 34
3.4. RETOUCHED FLAKES AND BLADES. 34
3.5. TRUNCATED TOOLS. 35
3.6. THE POINTS. 39
3.7. BACKED BLADELETS. 45
3.8. OTHER TYPES OF RETOUCHED TOOLS. 47
3.9. UNRETOUCHED TOOLS. 48
3.9.1. Unretouched flakes. 48
3.9.2. Unretouched bladelets. 48
3.9.3. Unretouched blades. 48
3.9.4. Cores. 50

CHAPTER 4. **GENERAL CHARACTERISTICS OF THE UTILISATION OF LITHIC TOOLS.** **51**
4.1. DIFFERENTIAL USE OF THE RAW MATERIALS AT BERNIOLLO. 55
4.2. DIFFERENTIAL USE OF TOOLS ACCORDING TO THE TYPE OF BLANK. 58
4.3. THE USE OF RETOUCHED AND UNRETOUCHED TOOLS. 59
4.4. THE CHOICE OF MORPHOTYPES FOR USE. 60
4.5. CHARACTERISTICS OF TOOL USE IN SANTA CATALINA AND BERNIOLLO. 60

CHAPTER 5. **THE TECHNICAL PROCESS.** **63**
5.1. THE TECHNICAL PROCESS AS AN INSTRUMENT FOR INTERPRETATION. 63
5.2. THE TECHNICAL PROCESS ON THE DIFFERENT RAW MATERIALS. 66
5.3. HIDE. 66
5.3.1. The technical process on hide 66
5.3.2. The working of hide in Berniollo and Santa Catalina. 71
5.4. BONE MATERIALS. 75
5.4.1. The technical process on bone materials. 75
5.4.1.1. Raw materials. 75
5.4.1.2. The blank. 76
5.4.1.3. Shaping. 77
5.4.1.4. Finishing and maintenance. 78
5.4.2. Bone tools in Santa Catalina. 79
5.4.3. The work on bone and antler in Berniollo and Santa Catalina. 81
5.5. WOOD. 83
5.5.1. The technical process on wood. 84
5.5.2. Wood working in Berniollo and Santa Catalina. 87
5.6. MEAT. 89
5.6.1. The technical process related to meat. 89
5.6.2. Butchering in Berniollo and Santa Catalina. 90
5.7. PLANTS. 90
5.7.1. The technical process on plants. 90
5.7.2. The working of plants in Berniollo and Santa Catalina. 91

CHAPTER 6. **SETTLEMENT ECONOMY.** **93**
6.1. THE STRATEGY FOR THE EXPLOITATION OF THE ENVIRONMENT. 93
6.2. ACCES TO INFORMATION ON SETTLEMENT ECONOMY: POSSIBILITIES AND LIMITATIONS OF THIS STUDY. 96
6.3. THE AZILIAN LEVEL OF SANTA CATALINA 98
6.4. THE EPIPALAEOLITHIC SITE OF BERNIOLLO. 100
6.5. THE MAGDALENIAN LEVEL OF SANTA CATALINA. 101

CONCLUSION. **103**

APPENDIX 1. **CRITERIA FOR FUNCTIONAL INFERENCE.** **105**

APPENDIX 2. **THE EXPERIMENTS CARRIED OUT.** **109**

APPENDIX 3. **PHOTOGRAPHS OF EXPERIMENTAL USE TRACES.** **115**

APPENDIX 4. **DIAGRAMS OF EXPERIMENTAL USE TRACES.** **133**

APPENDIX 5. **RESULTS OF THE USE-WEAR ANALYSIS OF THE TOOLS.** **161**

APPENDIX 6. **PHOTOGRAPHS OF ARCHAEOLOGICAL USE TRACES.** **173**

BIBLIOGRAPHY. **185**

RESUMEN. SPANISH SUMMARY **195**

LABURPENA. BASQUE SUMMARY **199**

INTRODUCTION

Interest in how stone tools were used is not recent. It developed in the 19th century at the same time that these objects were recognized as tools of human manufacture. The first responses to the questions posed emerged from comparison with other similar tools whose function was known, from ethnographic analogy. However, these interpretations were hypothetical, little more than suppositions. The archaeologists of the time lacked a method which would allow them to verify these supposed uses, so they quickly abandoned their attempts. During the first half of the 20th century, the general orientation of the discipline in the West did not demand a precise answer to the question of the tools' use. Archaeologists were preoccupied with establishing chrono-cultural sequences based on the morphology of the objects; function was relatively superflous information.

In another milieu, S.A. Semenov, a Soviet researcher, was interested in the technological evolution of past societies, within a Marxist perspective. From 1930 he began to develop a method of analysis which would allow him to understand the use of prehistoric tools and the technical activities which could be deduced from their use. This method was based on the analysis of use-wear traces.

The publication in the West of the methodological contributions of S. Semenov (1964) encouraged the development of functional studies from the middle of the 1970's. The "New Archaeology" had introduced a new interest in the technology and economy of prehistoric groups. Within this approach, the function of the tools offered important information.

The work of R. Tringham *et al.* (1974) and of L.H. Keeley (Keeley and Newcomer, 1977; Keeley, 1980) in the West and that of Schelinski (see Plisson, 1988) in the USSR established the first bases for use-wear analysis. The work which continued that of these pioneers (Anderson 1981, Moss 1983, Mansur-Franchomme 1983, Vaughan 1985, Plisson 1985, etc.) had a strong methodological content. It attempted to refine and widen the possibilities of a method which is relatively new.

Soon, use-wear analyses began to provide information concerning the behaviour of prehistoric groups. Information concerning the function of stone tools has contributed to some of the more relevant questions in archaeology.

The work of L.H. Keeley and N. Toth and of C. Sussman helped in determining the human origin of stone tools and the conception of tools among the oldest human groups (Keeley and Toth, 1981; Sussman, 1987). The discussion on the causes of the variability of Mousterian "facies" has brought a new approach after the work on tool use by P.C. Anderson (1981) and S. Beyries (1987).

The organization of hunter-gatherer groups at the end of the Palaeolithic is a theme which has been treated intensively. We are concerned with it in this publication. The function of the settlement, the spatial distribution of activities in the settlements and the management of the tools have been the subject of numerous studies based on use-wear information (Van Noten *et al.*, 1978; Cahen *et al.*, 1980; Audouze *et al.*, 1981; Plisson, 1985; Donahue, 1985). All these have contributed to a better understanding of the economic strategies of the hunter-gatherer groups of this period.

The study of tool use in the Neolithic has revealed participation in activities which were often more specialized, intense and complex than in the final Upper Palaeolithic (Caspar, 1988; Gassin, 1993; Juel-Jensen, 1994). The results of functional analyses have helped to construct the models which explain the domestication of cereals in the Near East (Anderson 1992). Other aspects as diverse as the preparation of wood intended for building (Caspar and Cahen, 1986) and the the existence of specialist workshops (Calley and Grace, 1988; Anderson *et al.*, 1989) have also been demonstrated.

In general, use-wear analyses have contributed to a better understanding of lithic tools, to placing them in a framework (that of their use) where they acquire a relevant meaning: the relationship between form and function, the systems of hafting, resharpening and maintenance and the management of the assemblage of tools. Use-wear analyses have provided information on working techniques for perishable materials, such as hide, meat, wood and other plants. They represent the few data which are within our reach in most archaeological contexts.

As may be observed, use-wear analysis has been applied often to resolve questions of a technological or economic nature. Tool use is more obviously related to these aspects. However, in recent years, social and ideological questions have taken a more important place on the agenda of use-wear analysts. Studies have been carried out aimed at understanding status and social differentiation among populations: the type and intensity of work in various family units (Hayden, 1993) or the activities reflected in the tools found with inhumations (Gibaja, 1994). New studies attempt to understand the spatial distribution of activities in domestic spaces in the first sedentary societies, such as those of L. Astruc at Kirokitia (Cyprus) and ours in Mureybet and Halula (Syria). It is to be hoped that this type of approach continues to interest use-wear analysts in the immediate future.

The present work begins, in chapter 1, with a summary presentation of the methodology of use-wear analysis which we use. This is an abstract of a specific publication on the subject (González and Ibáñez, 1994a), in which we set out the principal lines of our methodological approach and specify the criteria upon which we base the examination of the function of stone tools. Chapter 2 deals with the sites from which the tools analysed in the study come, as well as with the overall results of use-wear analysis in each site. Chapter 3 deals with the association between form and function in the stone tools, showing the modes of use of the retouched tools (end-scrapers, burins, points and backed bladelets, etc.) and the unretouched tools. Then in chapter 4, use strategies for the stone tools are taken up, to attempt an understanding of the factors which determined why a tool with certain characteristics was chosen for each task. These characteristics which define the tool are the type of flint in which it was made, the type of blank (blade, flake, or bladelet) , whether or not it was retouched, and if so, the morphotype which resulted. Chapter 4 deals with the stone tools as objects subjected to technical behaviour, and chapter 5 considers the tools as agents which participate in the transformation of various materials: antler, hide, wood, etc. In chapter 5 we try to reconstruct the types of tasks that were carried out with stone tools. Finally, in chapter 6, our intention is to demonstrate how the study of stone tool use is valuable in understanding the function of the settlement and, in general, the economic strategies of prehistoric groups. For this, the information on the use strategy of the tools (chapter 5) and on the characteristics of the tasks carried out in each settlement (chapter 6) is applied. In this way we seek to explain the information concerning tool function on three levels: the first refers to the action carried out with the tool (the motion and material worked), the second corresponds to the technical process or task in which the action takes part, and the third refers to the assemblage of economic activities which took place in the settlement.

CHAPTER 1

METHODOLOGY OF FUNCTIONAL ANALYSIS

When the active zone of the tool comes in contact with the material which it transforms, an ensemble of alterations are produced on this zone. These alterations depend for the most part on the type of action carried out and the characteristics of the material worked. The analysis of these alterations allows, up to a certain point, the reconstruction of the kinematics of the tool and the nature of the material with which it was in contact. To be able to accomplish the functional identification it is necessary to have "rules of inference", which relate each type of work to the alterations it causes in the tool. This is achieved through experimentation, reproducing the work which could have occurred in prehistory and analyzing the traces which it causes.

Since S.A. Semenov published his work, great advances have been made in the methodology of the functional analysis of stone tools. The work of Semenov revealed the relation between the microscopic traces on the tools and the characteristics of the work in which they participated, which constitutes the basis of functional interpretations. The studies of L.H. Keeley (1980) on the polish produced by use and of R. Tringham *et al.* (1974) on chipping more clearly defined the types of observable traces and the interpretive possibilities of each type of trace. In the last 15 years, other researchers have perfected the first approaches (Anderson, 1981; Odell, 1981; Moss, 1983; Mansur-Franchomme, 1983; Fischer *et al.*, 1984; Vaughan, 1985; Plisson, 1985; Shea, 1988; van Gijn, 1989; Gassin, 1993; Yamada, 1993, among others), adding to the number of materials worked and consolidating the criteria of interpretation.

If the first methodological advances centred on the study of flint tools, recently the fields of application of functional analysis have been extended to a great variety of rocks (Plisson, 1985; Sussman, 1985; Binder and Gassin, 1988; Knutsson, 1989; Hurcombe, 1992; Rodríguez, 1993; Clemente, 1995). The method of functional analysis is today supported by the thousands of experiments carried out by dozens of independent specialists, who have arrived at fundamentally similar conclusions.

At the present time there exists a degree of acceptable consensus among the specialists concerning the diagnostic possibilities and limitations of the different types of traces. The information which use-wear analysis has provided concerning the behaviour of prehistoric groups, which we have briefly dealt with in the introduction, is proof that the discipline is entering maturity.

Our method of analysis is intended to bring together the contributions made by the various researchers, but at the same time to improve some aspects of the methodology which to us present certain deficiencies. The experimentation carried out to date has led to the establishment of a series of qualitative criteria for functional interpretation. Thus we know, for example, that the working of hard materials produces chipping of larger size than the working of soft materials, or that the polish produced by scraping wood is smoother than that produced by scraping hide. However, these criteria are merely qualitative, and may be influenced by the subjectivity of the particular researcher.

The identification of a tool's use implies taking into account a large number of criteria in a coordinated way. It is necessary that the criteria be integrated coherently, within a system of analysis in which the argument is clearly presented, leading the analyst to understand the use of the prehistoric tool. Therefore, the functional inference is made up of a complex system of decisions. Except for some tests (Grace *et al.*, 1988; Grace, 1993), the system of inference has not been explained.

Therefore, the analyst of use traces has a whole set of qualitative criteria for functional identification, which are not expressed in an explicit system. In this state of methodological development it is possible to contribute interesting information on the use of prehistoric tools. However, the method has various limitations. Not being expressed in quantitative form, the criteria present a certain level of sujectivity, which has an effect on the degree of reliability of the interpretations. The capacity of interpretation of the analyst depends upon his personal experience, his visual memory, and his capacity for logical argument. Thus the credibility of the results is based more on his personal prestige than on the critical knowledge of the methods of analysis put into practice. The qualitative and inorganic character of the criteria makes the development of the discipline more difficult, since the training of new researchers implies new experimental programs which for the most part repeat those already carried out, and also makes the gradual refinement and perfecting of the system of interpretation very difficult.

Consequently, our efforts concerning the methodology of interpretation are directed towards advancing two of the aspects mentioned: quantitative expression of the criteria of interpretation and organization of the criteria in such a way as to constitute a decision-making system. In the present study we will avoid repeating those aspects of the methodology of use-wear analysis which have been sufficiently dealt with by other researchers. Because of this we will focus upon those aspects of our work which may prove to be original, including ourselves among those researchers who have attempted to improve the method of use-wear analysis, a method which is not finished and which still needs further work. This chapter on methodology summarizes the approaches which are presented in greater detail in another monograph published on the subject (González and Ibáñez, 1994a).

1.1. THE EXPERIMENTAL PROGRAMME

The proposed objectives are achieved based upon the results obtained from an experimental programme which at present consists of 405 tools (Appendix 2).

We have controlled four main variables in defining the experiments: the action, the material worked, the time duration, and how the tool was held (in the bare hand or by its haft). These are the major variables which can be inferred from study of use-wear traces on archaeological material. The effect of other variables was also examined, such as the type of flint, the angle and the morphology of the active edge, which also influence the formation of traces.

The action is defined by subvariables such as:
- the morphology of the active zone: linear (figure 1.1., n. 1), pointed (figure 1.2., n. 2), or bevelled (i.e. burin, figure 1.2., n. 3),
- the way in which force was applied: percussion (figure 1.1., n. 6) or pressure (figure 1.2., n. 1)
- the position of the edge in relation to the movement: longitudinal (figure 1.1., n. 2) or transverse (figure 1.1., n. 3).
- the working angle: high (figure 1.1., n. 4) or low (figure 1.1., n. 3),
- the edge angle: high (figure 1.1., n.4) or low (figure 1.2., n.1)

The activities carried out are defined as a combination of all these subvariables.

The materials worked in the basic experimental programme were those which could be expected in Palaeolithic contexts. Each of the materials was worked in different states (dry, soaked, with additives, cooked, etc....), and Appendix 2 presents the details of all these experiments.

The time of work is an extremely relevant factor. Part of the experiments were carried out over an extended and constant period, an hour, in order to isolate this variable. Another group of experiments was intended to illustrate the influence of time in the formation and appearance of traces; in these experiments, each of the tools was cleaned and observed every 1, 2, 4, 7, 10, 15, 20... minutes during the work until the tool presented developed traces.

The majority of the tools were held in the bare hand, or the hand protected with leaves or leather. A minority were hafted, especially the end-scrapers and the bladelets used for working hide, some of the tools used for butchering and the points used as projectiles. Wooden shafts and bone diaphyses were used as handles for those lithic tools attached by means of resin and wax, tar, tendons or leather straps.

The flint used in the experiments comes mostly from outcrops situated in the Basque country, in Kurtzia (Vizcaya) from the Upper Senonian and in Albardon and Cucho (Alava), from the Upper Cretaceous (Elorza and Bustillo, 1989; Elorza and García-Garmilla, 1993).

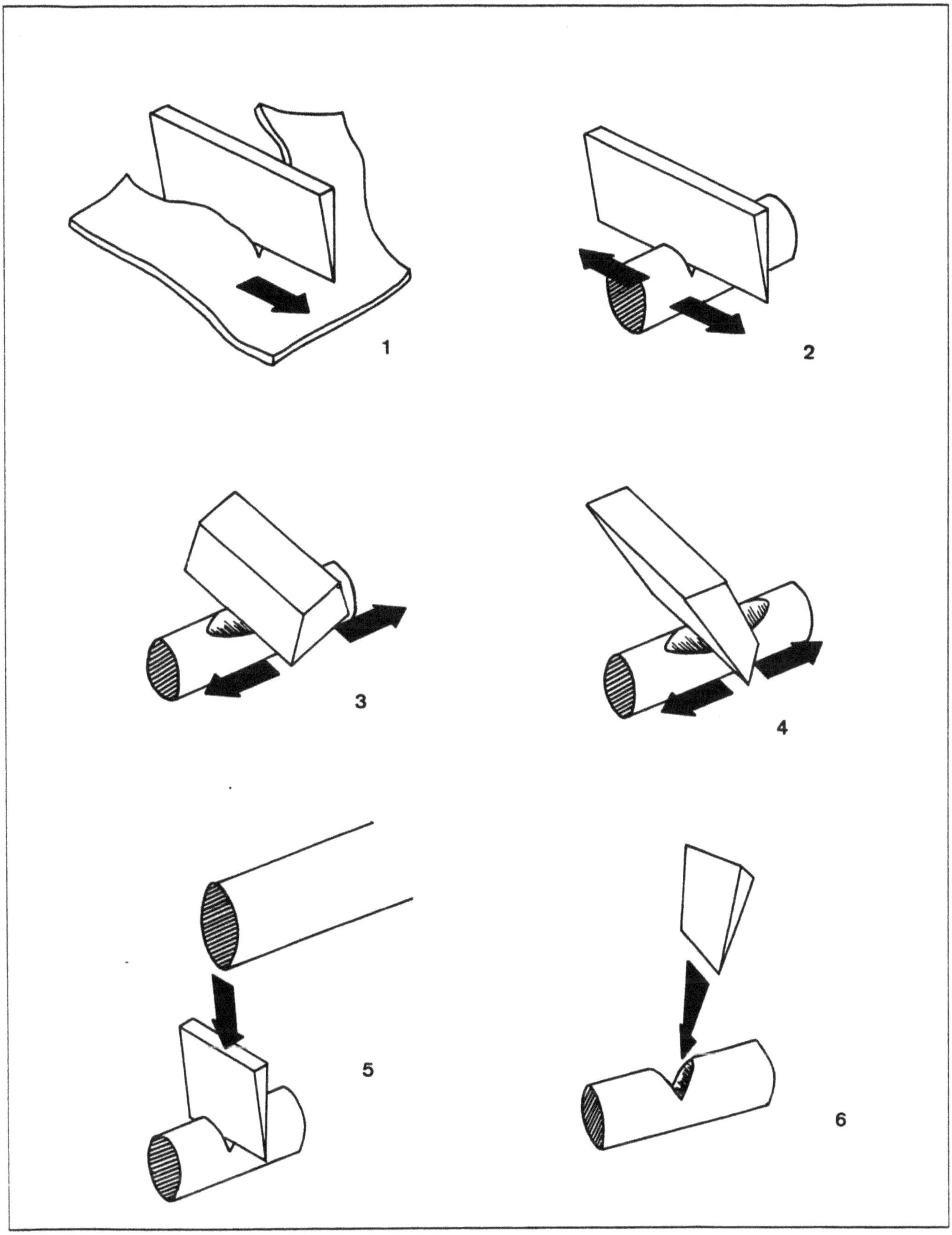

Figure 1.1. Description of the actions carried out with lithic tools.

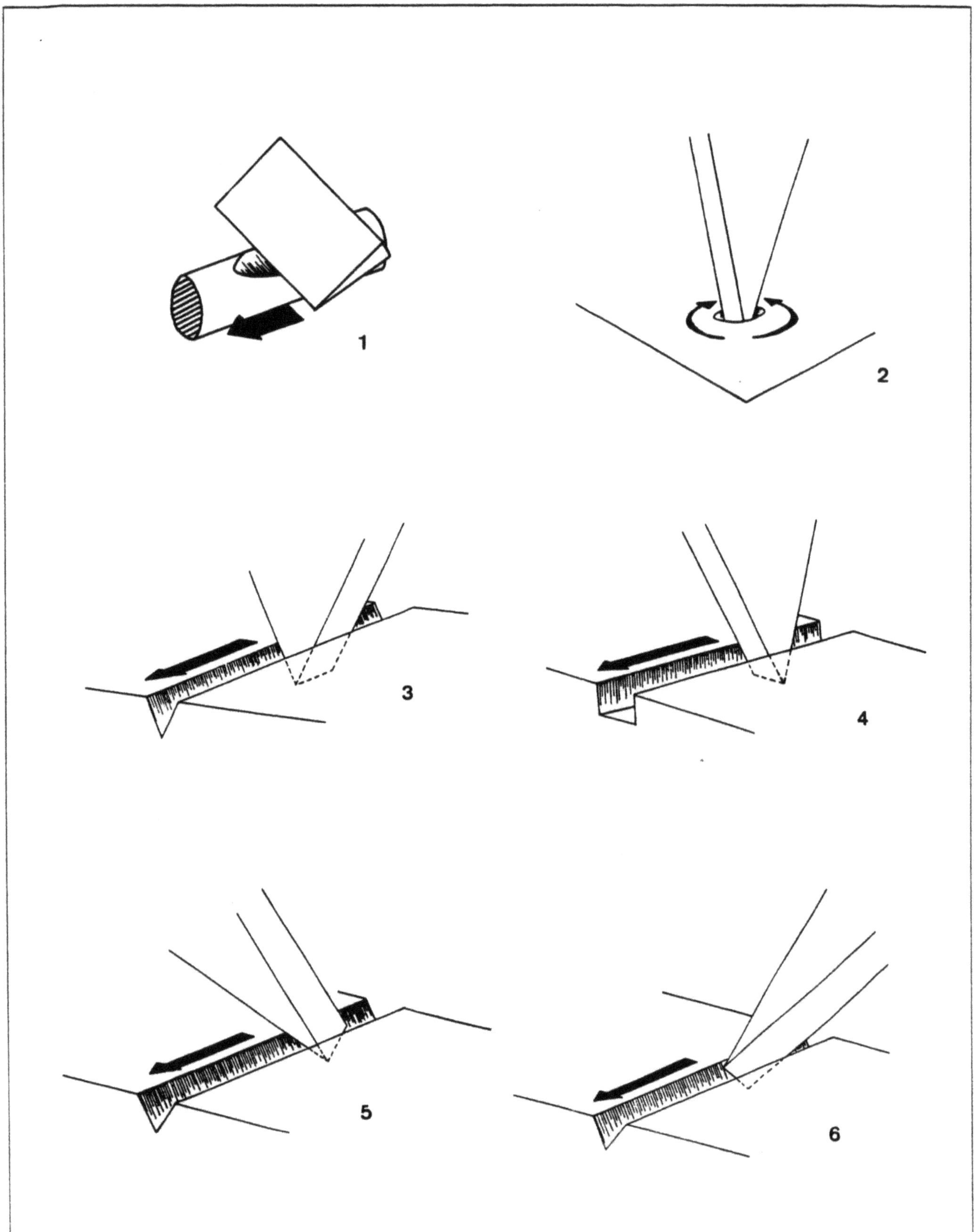

Figure 1.2. Description of the actions carried out with lithic tools.

1.2. THE DESCRIPTION OF THE TRACES

We recorded the traces produced by use in such a way that they could be used later in establishing specific functional criteria. To do this, we organized the traces into particular attributes (i.e. polish topography) and categories (smooth, undulating, irregular). The categories were quantified when the nature of the traces and the characteristics of the means of observation made this possible.

1.2.1. **Chipping**

The resistence presented by the material worked causes pressure on the edge of the tool, which, once past a certain threshold, causes formation of chipped areas. The main factors which determine the characteristics of this chipping are the hardness of the material worked, the force applied during the work, the position and movement of the tool and the resistence of the active zone, conditioned by the edge angle and its profile.

The degree of reliability offered by chipping as an element of functional identification has been the subject of arduous discussion, constituting the central theme of the debate between the so-called "high-power" school and the "low-power" school, at the beginning of the 1980's (Keeley, 1980; Odell and Odell-Vereecken, 1980). According to P. Vaughan (1985), one of the principal analysts who criticized the use of chipping in functional inference, this type of trace does not offer sufficient reliability either for understanding the movement of the tool or the hardness of the material worked.

To verify the interpretative potential of this type of trace, we carried out an exhaustive study of chipping on 100 of the tools in our experimental programme. We examined the active edges of these tools with an Olympus SZH binocular microscope, at between 3 and 30X, drawing the chipped areas with a camara lucida at 10 magnifications. The following attributes were derived from the more than 4,300 micro-scars analysed: quantity, distribution along the edge, disposition in relation to the other micro-scars, morphology, termination, length and width and location of use polish in relation to them. It was possible to quantify number and length, treated as continuous variables, to half a tenth of a millimetre. For the other attributes -disposition, morphology, termination- discrete categories were established, shown by drawings (see Appendix 4).

1.2.2. **Polish**

The rubbing of the tool against the worked material alters the surface of the active zone, producing a progressive regularization of the microtopography, and consequently an increase in reflectivity (figure 3).

The mechanisms involved in the formation of this trace have been the subject of intense discussion throughout the 1980's, between those who maintain that the polish is the product of the abrasion of the siliceous surface (Diamond, 1979; Kamminga, 1979; Meeks *et al.*, 1982; Levi-Sala, 1988; Yamada, 1993) and those who hold that the siliceous surface becomes amorphous, with a subsequent redeposition (Anderson, 1981; Mansur-Franchomme, 1983). In any case, the importance of this trace resides in the fact that its appearance depends, for the most part, on the characteristics of the contact material.

Although polish from use had already been observed by other researchers (Curwen, 1930; Bordes, 1950; Semenov, 1964), it is especially important as an element of use-wear analysis in the work of L.H. Keeley (1980). He considers that there exists a direct relationship between the appearance of the polished surface and the material worked, speaking of "antler polish", "wood polish", etc. His work was continued by a series of researchers who both confirmed to a large extent his results, and increased the number of materials worked and types of polish known. These authors also pointed out some interpretative limitations of this trace. In the first stages of formation, the polish does not present characteristics allowing distinction between different worked materials. Moreover, they point out that the differences in the appearance of polish from the working of different materials can overlap in some cases, causing certain difficulties in distinguishing between the polish produced by working bone and that produced by working antler, between that from antler and that from wood, between that from wood and that from plants, etc (Keeley, 1980, 83). This phenomenon is shown by P. Vaughan (1985, 28): the types of polish generated by working different materials, after a certain amount of time, present distinguishable characteristics, although there is a certain level of overlap.

R. Grace (1989) maintained that any association between appearance of polish and material worked is ficticious, therefore any appearance of polish can be caused by any contact material. He proposes a model of development of polish as a "continuum" which depends on the material worked and the duration of use. Thus, different materials could produce the same type of polish after different lengths of time. Working of bone for 5 minutes, of antler for 10 minutes, wood for 14 and hide for 27 could create polish with the same characteristics (Grace, 1989, 44).

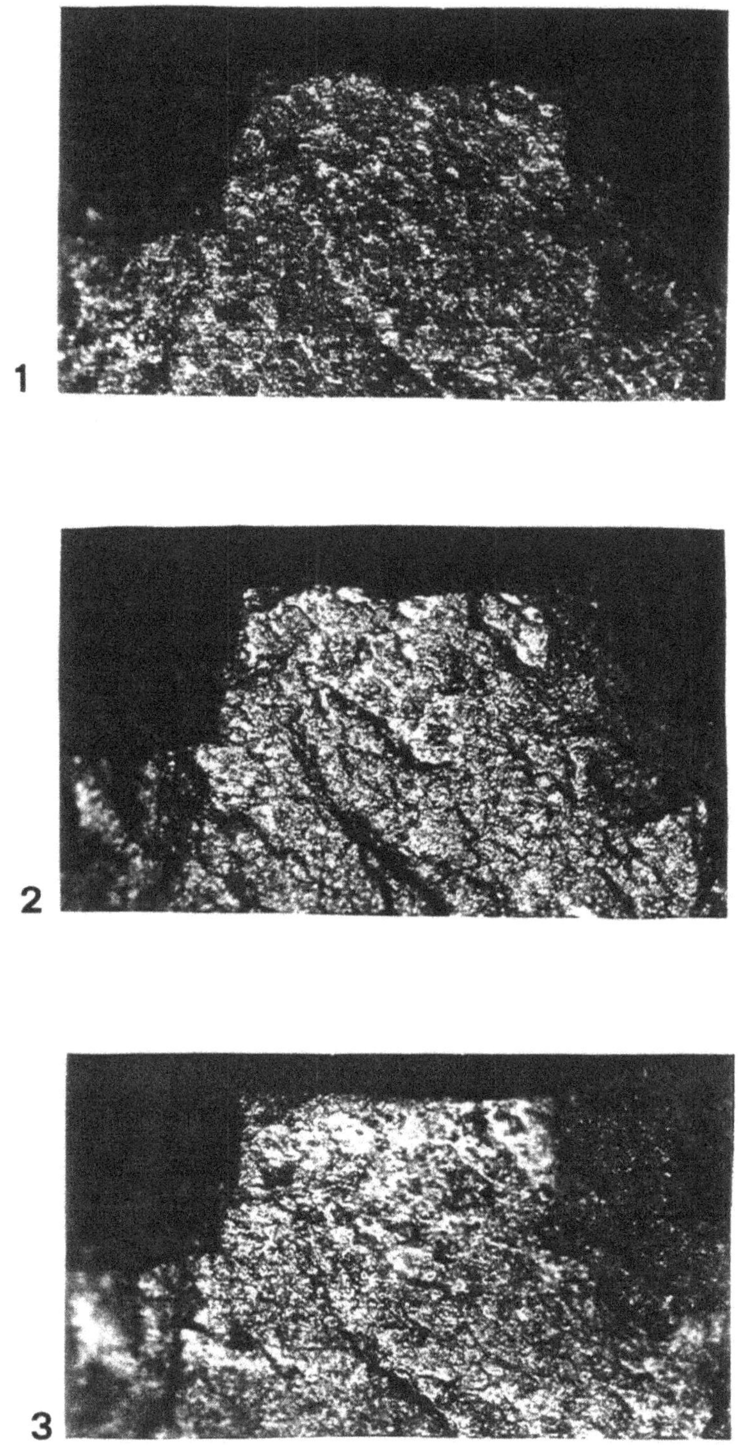

Figure 1.3. Process of polish development. 1. Polish after scraping wood 3 minutes; 2. polish after 7 minutes; polish after 15 minutes.

In Grace's opinion, as the duration of the work on archaeological objects is not known, we cannot infer the material worked. However, the prediction implicit in Grace's model does not hold, as we have proved experimentally, that for example, the working of wood for 14 minutes produces polish which is clearly distinguishable from that produced by working hide for 27.

To examine these extremes in our experimental program we have considered different characteristics of the polish: the distribution, the longitudinal and transverse extension and the appearance of the polished surface. For the appearance of polish we have introduced three descriptive elements: microtopography, pattern, and network (see Appendix 3).

We consider microtopography to be the degree of regularization of the polished surface. This can be irregular, undulating or smooth, including the intermediate stages between each one of the three categories. By pattern we mean the proportion of the surface which is polished and the degree of the linking of polished areas. In the process of the development of polish both factors evolve at the same time, so that as the proportion of the surface altered increases, a process of linking of the isolated polished areas is produced (see figure 1.3.). The pattern (figure 1.4.) can be open (isolated points of polish), semi-closed (starting to become linked, the polished surface occupies less than 50% of the contact-use zone), closed (the polished zones are linked, occupying more than 50% of the altered area), and compact (almost all the surface is polished). The network, which is considered when the pattern is semi-closed, refers to the width of each area of polish, as measured between two unaltered areas. The network can be wide (average width more than 15 microns), moderate (between 10 and 15 microns), narrow (between 5 and 10 microns) and very narrow (less than 5 microns).

We should understand that polish resulting from use is a product of the dynamic process of alteration of the surface, by which the microtopography tends to be regularized and polished areas tend to link and to present a wider network as the time of contact or the intensity of the friction between the tool and the worked material increases.

During the work, not all of the points of the active zone of the tool are involved in the transformation of the worked material with the same intensity. The zones closest to the edge or the most projecting points receive more friction than those farther from the edge or more protected from the rubbing. Therefore, on the same active edge appear zones with different stages of development of the polish. The zones that, although altered, are more protected from friction reflect the appeareance of the type of polish in the first stages of development. Other zones, more affected by friction, show the successive stages of polish development, and a progresively more closed pattern of wear and more regular topography. Because of this, in the same active zone we can observe the different stages of evolution of a polish, from the first stages to the maximum development reached for that object. It is the analysis of the concurrence of these progressive stages of development that allows inference of the worked material.

The idea that polish seen on an active zone of a tool represents only one point in the evolution of a type of polish through time of use is implicit in the model of R. Grace. On the contrary, as we have seen, observation of an active zone shows different stages of evolution of a polish. That is why it must be emphasized that the inference of the material worked by a tool is not based on the analysis of the appeareance of the polish in isolated points of the micro-surface, but on the analysis of polish over the whole active zone.

Of course, if the maximum degree of polish development that is reached in an active zone is too low, the worked material cannot be inferred.

If we accept the dynamic nature of the polish, reflected in each active zone, we can conclude that different stages of pattern, from more open to more closed, can be observed in the same active zone. Because of this we can speak of an "association of patterns", expressing the different types of pattern that are present on the same edge[1].

1.2.3 Striations and linear traces

Striations are linear depressions with clear limits which appear within the polished areas. Other linear traces are **polish in linear disposition** (appendix 3, photo 7) bands of polish on the surface of the flint having a clear direction, and the **linear components of polish** (appendix 3, photo 8) which are superficial linear depressions with indistinct margins, which cannot be seen as individual units.

Striations are caused by scratching the surface of the polished zone by some abrasive agent, originating from the worked material itself, from the chipping of the edge or accidental contamination (soil, grit, etc.). These traces are important for determining the kinematics of the activity.

For striations, the attributes to be considered are length, width, orientation, type (ex. classification of M.E. Mansur-Franchomme, 1983), the distance from the edge

---

1. For a more detailed presentation of polish description atributes, see González and Ibáñez, 1994a.

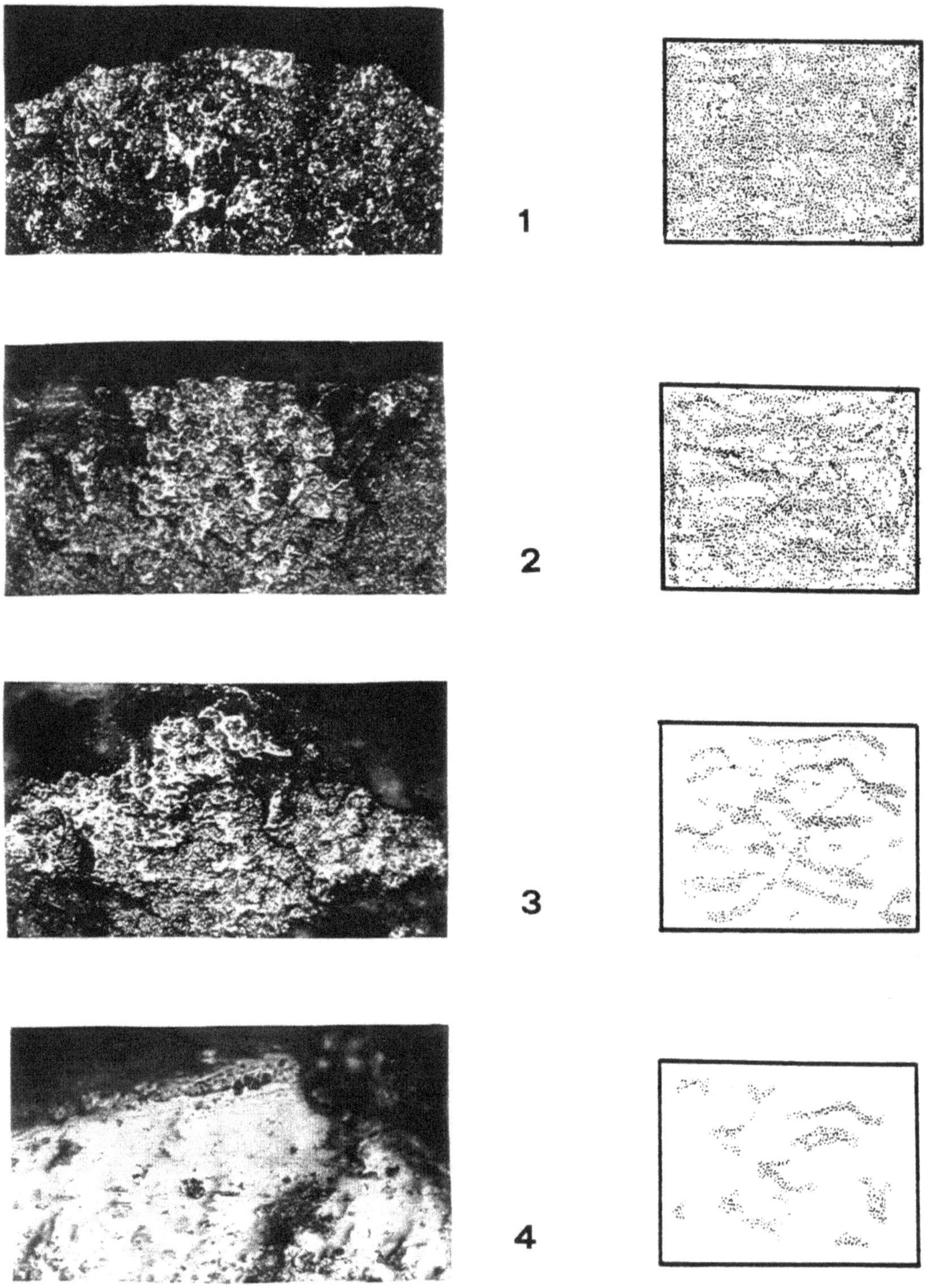

Figure 1.4. Pattern of use-polish: 1. open; 2. semi-closed; 3. closed; 4. compact.

and the quantity. For polish with linear disposition, the same attributes except type are used. For the linear components of polish, the orientation is considered.

1.2.4. **Rounding**

The rounding of the active edges and ridges is produced by the loss of material from erosive contact with the worked materials. It is related to the type of material worked. In some cases it can even be macroscopic, visible to the naked eye. We have recorded the degree, in relative terms given the difficulty of precise measurement, and its position on both edge aspects (dorsal and ventral).

1.2..5. The relationship between the traces.

In this case we are referring to a topographic location, where the traces are situated in relationship to one another and to the active edge. In each microscar we recorded the relationship to polish, taking into account whether this occupied the sides, the upper part, the interior of the microscar or various zones at the same time.

1.3. **THE CRITERIA**

To establish the criteria for functional interpretation it is necessary to contrast the hypotheses proposed concerning the relationship between the characteristics of the traces and of the work.

For this we have used the Chi^2 significance test. In the calculation we have attempted to best isolate the variable which we want to know, holding constant the other variables which might influence the result. For example, if we want to know whether the hardness of the material worked influenced the size of the microscarring, we have to eliminate the influence of the type of action and the angle of the edge, since these two variables also have an influence on the size of the microscars. For this, we have to calculate the relationship between the hardness of the material worked and the size of the microscars for each type of action and for each category of edge angle separately.

The first step is to establish the existence of a significant relationship between a variable of use and an attribute of the traces. The next is to find the criterion based on this attribute which best distinguishes between specific categories within the variable. For this we turn to our experimental programme, testing which of the possible limits provides a better level of certainty.

An example will clarify our procedure. We have proved that, with active edges greater than 60 degrees, the termination of microscarring is hinge/step much more often when harder materials are worked than when softer materials are worked (table 1.1.). It is necessary now to look for a criterion of functional interpretation which makes good use of this relationship. A limit must be found which differentiates, which can predict, when an edge with these characteristics has worked a hard or soft material. We have proved that in this case the most useful limit is found in 60% of hinge/step terminations. Seven of the eight pieces used on hard materials show a high proportion of these, and seven of the eight pieces which worked soft materials had a far lower proportion. That is, the application of this criterion would have produced 14 certain determinations and two erroneous ones. For this particular attribute, no other limit explored produces better results.

With this procedure we have obtained two explicit criteria: the edges of more than 60 degrees which present more than 60% of hinge/step terminations have worked hard materials; the edges with the same characteristics with less than 60% of hinge/step terminations have worked soft materials. Moreover, we know how far we can have confidence in this criterion, whose application results in a 12% margin of error.

Table 1.1. Termination of the microscars in pressure (non-percussion) actions with linear active zones. Edges with angle greater than 60 degrees.

	Hard materials	Soft materials
Feather	129 *(21,5%)*	96 *(62,5%)*
Hinge/step	468 *(78,5%)*	58 *(37,5%)*

In some cases, the overlap is great enough to establish different limits, one for each category of the variable, leaving some of the tools undetermined, in particular those which produce results between the two limits. For example, to distinguish between longitudinal and transverse actions we use the extension of the polished areas on each of the aspects of the edge. According to the position of the polish, symmetry indicates longitudinal movement and disymmetry indicates transverse movement. If a tool presents disproportion in extension of polish between the two aspects in a ratio greater than 70% to 30%, this indicates a transverse movement; if it is 60% to 40% or 50% to 50%, then it is a longitudinal movement Those tools which have an intermediate proportion (i.e. 65% to 35%) cannot be attributed to either movement, because the possibility of error is too high. The criteria could be correctly applied to 16 of the 25 pieces considered. However, movement for 9 of the 25 (36%) could not be determined with this criterion.

A third type of criterion is established when a limit allows the identification of tools which fulfill a certain condition with either category of the variable, but does not confirm anything concerning the tools which do not fulfill the condition. For example, in longitudinal movements, tools which have more than 20 microscars per centimetre of edge can be attributed to cutting bone or harder materials. Whereas, nothing can be confirmed concerning the tools which do not fulfill this condition, among which some had also worked bone materials.

Knowledge of the level of certainty, percentage of error and the degree of indeterminacy establishes the confidence of each criterion.

This kind of procedure has allowed us to establish a series of quantitative criteria on chipped areas occurring from use, on striations, and on the extension of polish. We emphasize, however, that these criteria are not universally applicable. The experimental programme was carried out using fine-grained flint from the outcrops of Albardón and Kurtzia, types of flint common on the sites which we have studied. These raw materials present fracture characteristics not necessarily shared with coarser-grained flint or other types of rock (Akoshima, 1987). Moreover, the chipped areas have been observed using a binocular microscope at 10X, with low-angle lighting; using another observation technique may produce different results from ours. This method can be used to verify how directly applicable these criteria are to other contexts.

The appearance of polish on the surface is more difficult to describe and quantify. Observation took place with microscopes using incident lighting between 100X and 400X. In these conditions, the depth of field is very small, which limits the size of the area which is well-focused; also, the description of surfaces in visual terms is always complex. In the literature on use-wear analysis, informal and intuitive terms are common for expressing the appearence of these surfaces. A feeling that there exists a certain lack of definition is shared by many colleagues and has produced different attempts to resolve the problem. Some have sought to reinforce a certain rigour in the description of the attributes and categories employed (cf. Plisson, 1985; Yamada and Sawada, 1993). Others have attempted to objectivize and quantify the characteristics of polish. Thus, attempts have been made to measure the brilliance of the surface (Keeley, 1980), to apply techniques of interferometry (Dumont, 1982, 1988), analytical tribology (Beyries *et al.* 1988) and image analysis (Grace *et al.* 1985; Knutsson, 1988; Grace 1989; Vila and Gallart, 1991) and surface analysis through the atomic force microscope (Kimball, et al, 1995).

In our case we have attempted two complementary approaches; the first, with the goal of making more objective the terms used to characterize polish; the second, to attempt to quantify some of the more characteristic features using digitalized image analysis. Image analysis of polished surfaces has begun to give promising results, but this research is in progress, so that in the present work we will limit ourselves to presenting results on polish using only qualitative criteria.

1.4. . THE SYSTEM OF FUNCTIONAL INFERENCE

The data obtained from use traces make up a fundamental part of the system of functional inference. However, use traces are not the only factor which should be considered in the identification of the function of a tool. The archaeological context in which the tools appeared should be taken into account, as it can suggest hypotheses of use, as well as limiting the range of activities which could have occurred (for example, pottery was not scraped in the Palaeolithic). The inherent characteristics of a tool (mass, form, type of active edge) provide substantial information on the range of uses for which the tool is efficient, that is, concerning its functional potential. Finally, it is necessary to situate the hypothesized activity in the context of a logical working process.

Besides the traces produced by use, the archaeological tool goes through a series of alterations which can interfere in the examination of the use traces. Among these alterations are those related to the making of the tools, those caused by soil movement and those produced by the activities of archaeologists. Their effect on use traces has been checked by specific experimental programmes (Gutiérrez *et al.*, 1988; Ibáñez *et al.*, 1987). Thus, chemical and mechanical alterations in the sediment, the action of fire (Clemente, in press) or poor storage conditions can eliminate the evidence of use or, at least, make the analysis of the traces more difficult.

It is necessary to take into account also that some uses do not produce traces, or the traces are too faint for reliable interpretation. Short time of use and working soft materials produce traces so tenuous that in many cases they are difficult to identify and interpret.

Despite these limitations, we can confirm that use traces, together with the data from the archaeological context studied and the potential of the tool, constitute the key to understanding the function of the tool. Concerning traces, for many years the study of chipped areas with the binocular microscope was in opposition to the analysis of polish with the incident-light microscope. To our way of thinking and in the opinion of the majority of functional analysts today, such a dilemma is false, since these approaches are complementary. The characteristics of chipped areas provide an understanding of the activity and the relative hardness of the material worked. However, this approach cannot be applied to retouched edges, as it is not possible to distinguish use-derived chipping from the smallest components of retouch. Polish from use provides identification of the activity and the material worked. But it must be taken into account that when the period of use is brief the polish is weak and the characteristic differences between every type of polish are not yet developed. If we take into account the possibilities and limitations of each type of trace, we have to conclude that all of them have to be studied in a complementary way. It is the concurrence of the data provided by the different types of traces which insures the reliability of the interpretation (Keeley, 1980, 83).

We have established the criteria of interpretation based on traces and we know that other data support or restrict the functional interpretation. The diagnosis should be based on the integration and the organization of the assemblage of criteria at our disposal. The question now remains of how to structure the system of inference. The option which seems the most convenient to us is to organize the deduction process in the form of a flow chart, deciding upon the features of the work performed by the tool step by step, from the most general and therefore the simplest interpretation, to the most specific, which are generally more complicated. This technique has the virtue of allowing a partial interpretation if the criteria are not sufficient to understand more specific functional details.

The procedure to follow in the interpretation is presented in the diagram in Appendix 1. The first step consists of locating the use-traces on the tool, in order to identify the active zone. This can consist of an edge (linear), a point (pointed), or a bevel of burin type (bevelled). The next step in the interpretation is to decide whether the action is one of percussion or pressure. Percussion can be direct (chopper), indirect (wedge) or thrown (projectile).

Linear active zones with pressure actions can be cutting (longitudinal) or scraping (transverse). For pointed or beveled zones used with pressure actions, these could be for incising (engraving) or rotation (boring). Boring can be carried out in one direction or both directions.

Once the activity has begun to be defined we can proceed to the identification of the material worked, starting with examination of its relative hardness and go on to the inference of the particular material (hide, bone, antler...), when this is possible.

We have devised a series of criteria which will allow us to decide which is the most probable choice in each of the steps of the process of inference. Thus, using linear active zones as an example, the presence of more than 20% of chipped areas with a length greater than 2 mm. and a width greater than 1.2 mm. will indicate, together with other criteria, that the tool was used with a percussion action. The different criteria which we use in each step of the inference are explained in Appendix 1.

We have thus established an explicit system of functional inference. Of course, this system must not be understood as a closed system which dominates the analyst's capacity for decision, but rather on the contrary, facilitates the process of functional identification, since it allows for a systematic comparison of all the criteria for understanding the activity of the tool and the material which it worked.

CHAPTER 2

THE SITES STUDIED AND THE OVERALL RESULTS

2.1. THE SITES

The site of Santa Catalina is situated next to the lighthouse of the same name, in Lekeitio, on the east coast of Vizcaya, in northern Spain. It is a cave site whose entrance is in a rocky limestone cliff, which constitutes the present coastline.

Within a project of investigation of the Epipalaeolithic in Vizcaya in 1982, Eduardo Berganza began excavation. Since then there have been 9 excavation campaigns and the field work is continuing.

The excavation is located in one of the exterior chambers and occupies nine square metres, in a square of 3 by 3 metres. So far 2 levels of occupation have been found. The first of these corresponds to an Azilian assemblage with flat harpoons, with an unpublished radiocarbon date from the eighth millennium B.C. (Berganza, pers. comm.). The second level is from the late Magdalenian, with a group of dates which range between the tenth and the eleventh millenniium B.C.

The lithic material is fine-grained black flint. The state of preservation of this material is optimum, since few natural alterations have affected the objects; at the same time special care has been taken to preserve the material during the excavation and in storage.

The open-air site of Berniollo is situated in the middle of the Bayas river basin, in a cultivated field located on the right bank of the river, in Subijana-Morillas (Alava).

Excavation has taken place since 1984, directed by Doctor Amelia Baldeon. Two campaigns have been carried out, in 1984 and 1985. During the excavation two habitation areas occupying different zones of the site were found, one corresponding to the Neolithic and the other to the Epipalaeolithic, which we studied. The archaeological site consists of a single level of between 10 and 15 cms. thick, producing numerous pebbles and abundant remains of lithic industry, among which end-scrapers, burins, points and backed bladelets are prominent.

The assemblage has been dated to 9940+490 B.P. (Mariezkurrena, 1990). In this site, the state of preservation of the archaeological materials is not as good as it is at Santa Catalina. The bone remains are scarce, and in most cases comprise only the parts of the skeleton most resistant to deterioration, such as teeth. The processes of alteration have also affected the flint tools to a certain extent, some causing patination in differing degrees, which in some cases has impeded the functional analysis of the objects.

The lithic tools were made in three different types of flint, one black (type 1), one brown (type 2) and a third, also brown, with Lesengang rings (type 3).

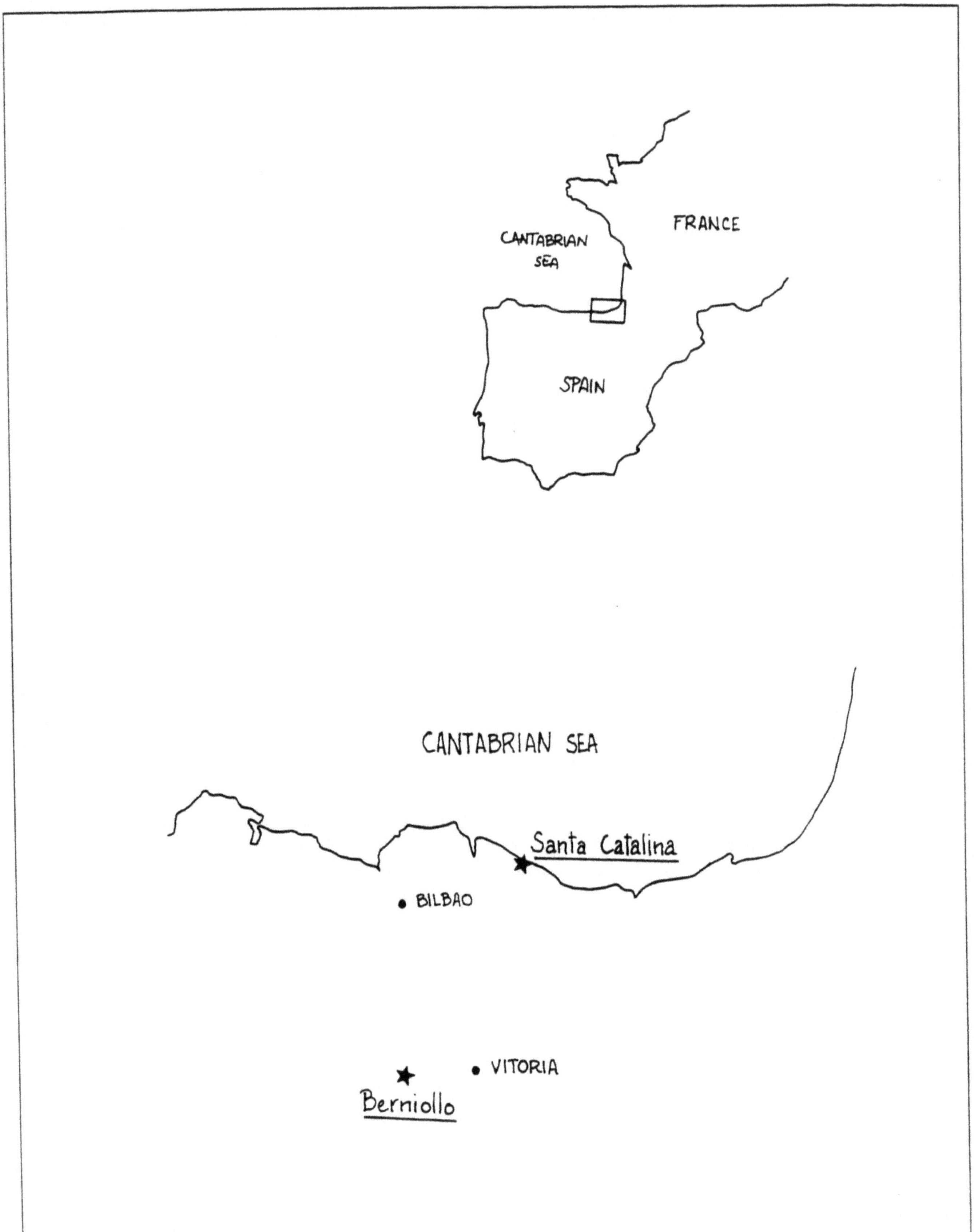

Figure 2.1. The sites studied.

2.2 THE SAMPLING.

The great amount of time necessary to carry out functional analysis of an object, which includes cleaning and detailed examination with a binocular microscope and a metallographic microscope of the edges and surfaces, limits the total number of objects which can be studied. Because of this, any functional study of assemblages containing many objects must rely on sampling.

The criteria for sampling vary depending on the type of information which is desired (Moss, 1983, 106). Thus, understanding of the areas of activity on a site cannot lead to the same criteria of selection as a study which strives to comprehend the relations between form and function.

However, there are a series of general criteria based upon the method of analysis and the functional potentials of the lithic objects which can guide the sampling. Patination of objects modifies the surface of the flint, making analysis difficult and even impossible. Therefore, it is advisable to reject the more patinated objects (Keeley, 1980; Moss, 1983, 106; Vaughan, 1986, 185; Beyries, 1987, 26; Sussman, 1988, 21). In principle this selection does not distort the final results, unless tools of a particular functional category are altered differentially from others.

Another selection criterion commonly used is the functional potential of the object, those with slight probability of use being rejected. The determination of the probability of use becomes clear in the case of the retouched pieces, since the secondary shaping is related to use of the tool. For the unretouched pieces, a series of criteria can guide the selection. The, first is size, since it is awkward and inefficient to work with small pieces, unless they are going to be fixed into handles, as in the case of backed bladelets. Because of this, for unretouched tools, pieces less than 2 cms. are rejected (cf. Vaughan, 1985, 184).

The edges of the piece should also present certain characteristics. E.H. Moss (1983, 106) proposes the selection of unretouched pieces when they have edge profiles which are not sinuous and are larger than two cms., or a pointed zone.

Other authors go by the traces of use which can be observed microscopically or with a magnifying glass, to sample for objects which were most probably used. P. Vaughan (1986) proposes that objects with edge damage be selected. H. Plisson (1985, 271-272) selects pieces which are chipped, rounded, retouched or which simply present characteristics which make them suitable for use.

Our investigation will be oriented principally towards the examination of use in the different typological groups and towards exploring the possibilities of understanding prehistoric economic strategies through the study of use-traces. Therefore, we need to have information on the function of each of the morphological types of tools, as well as an overall view of the activities which were carried out on the site.

To fulfill the first objective we have chosen a broad sample of the main typological groups present in the three assemblages. However, we have not selected the same proportion of pieces within each of the typological groups, and have increased the sample from the groups for which a greater variability of use was expected.

The intent of having an overall view of the activities carried out in each of the archaeological assemblages has led us to incorporate in the sample studied a group of unretouched tools: blades, bladelets and flakes, larger than two centimetres.

In the case of Berniollo, we were especially interested in understanding the use criteria for the pieces made in the three types of flint. Therefore, we have reinforced the sample of the tools made in flint less common on the site, as opposed to the more common flint.

We have not used as criteria for selection the presence of chipped areas on the edges or other morphological characteristics of the objects. It is evident that the assemblage of tools with chips due to use is going to offer a much greater use ratio than those which do not present substantial chipped areas; however, we consider that such a practice could introduce inconvenient biases in the final result which is sought. Indeed, chipped areas are produced more easily on edges with acute angles, as well as during the working of hard materials. Therefore, if we choose the tools with this type of trace, we will favour activities on hard materials over soft materials and work with acute angles over those achieved with higher angles, causing the final result to be distorted.

On the other hand, we consider that some of the information most interesting to the description of use strategies of lithic tools in a given assemblage concerns the proportion of use of the unretouched blanks. If we choose among unretouched tools only those pieces with evidence of use, we will not be able to obtain this type of information. Consequently, we have not considered any criterion concerning the presence of chipped or rounded areas in the objects chosen.

Finally, in the sampling process the total quantity of tools is taken into account, as is logical. For example, in the Magdalenian level of Santa Catalina were found, among other tools, 26 truncated pieces and 1864 unretouched flakes of more than 2 cm. Many of the features of the function which unretouched flakes fulfill (percentage of used pieces, selection by size, by edge angles or by type of flint, worked materials, etc.) are recognizable to a great extent in a small sample, of 2% or 3%, given the large number of pieces. However, to establish a viable conclusion on the use of truncated pieces the percentage should be much higher and include nearly all of the pieces.

With all these criteria we have achieved a sample whose composition is reflected in tables 2.1, 2.2, and 2.3, where the number of tools analysed and the total number of pieces found on the site are shown.

Table 2.1. Objects analysed in relation to total objects in the Azilian level at Santa Catalina.

	Objects analysed	Total objects
End-scrapers	17	19
Borers	1	3
Burins	11	17
Backed points	14	14
Truncated pieces	1	2
Notched and denticulated tools	2	2
Retouched blades and flakes	10	10
Backed bladelets	34	76
Other retouched bladelets	5	5
Unretouched flakes	29	679
Unretouched blades	20	260
Unretouched bladelets	23	659

Table 2.2. Objects analysed in relation to total objects in the Magdalenian level at Santa Catalina.

	Objects analysed	Total objects
End-scrapers	52	53
Borers	5	9
Burins	23	44
Backed points	29	29
Truncated pieces	20	26
Notched and denticulated tools	8	13
Retouched blades and flakes	9	21
Backed bladelets	35	177
Other retouched bladelets	4	20
Backed blades	5	15
Composite tools	4	4
Diverse	9	9
Unretouched flakes	39	1864
Unretouched blades	44	1408
Unretouched bladelets	14	1512

Table 2.3. Objects analysed in relation to total objects at the site of Berniollo.

	Objects analysed	Total objects
End-scrapers	36	49
Borers	1	3
Burins	12	30
Backed points	12	14
Truncated pieces	5	20
Wedges	1	22
Retouched blades and flakess	11	48
Backed bladelets	21	87
Other retouched bladelets	6	9
Unretouched flakes	64	1883
Unretouched blades	34	547
Unretouched bladelets	14	495

2.3. ANALYSIS OF THE ARCHAEOLOGICAL DATA

The dorsal and ventral surfaces of each of the objects analysed was drawn, as well as any burin facet present. Afterward the objects were washed in a solution of 25% acetic acid for 15 minutes, and then in a solution of 25% potassium hydroxide, also for 15 minutes. Then followed observation of the edges and surfaces of the object with a binocular microscope Olympus SZH and a metallographic microscope Olympus BHT.

The traces were located on the drawing. For the functional identification the criteria described in the section on methodology and Appendix 1 were followed.

2.4. THE OVERALL RESULTS OF FUNCTIONAL ANALYSIS

Although this has been described previously, it is necessary to mention the limitations of the functional study of lithic material in order to understand all the economic activities which are produced in a particular context. In the first place, it must be considered that numerous tasks had to have been carried out without flint tools or without any kind of tool. Moreover, although some uses, such as bone- or wood-working, are easily discerned, others leave much fainter traces on the active edges, such as the cutting of meat or fresh hide, and therefore may not be detected in analysis (Gassin, 1993; van Gijn, in press).

Therefore, we cannot attempt to reconstruct the proportion in absolute terms of each of the tasks carried out. However, if we assume that the different archaeological assemblages are affected equally by these basic limitations, it is possible to establish comparisons between them, so that we can observe variations in the proportion of the activities discerned in each of the contexts studied. This is valid for comparison of the activities carried out in the Azilian and Magdalenian levels at Santa Catalina, as the type of flint used in both is the same and its state of preservation is similar. However, when these two assemblages are compared to the site of Berniollo, which has different types of flint and poorer preservation than at Santa Catalina, it is necessary to take certain precautions.

On the one hand, we have carried out experiments with the three types of flint present at Berniollo and have not found any substantial differences in the degree of development of the traces, either between the three types or compared with the flint of Santa Catalina. On the other hand, concerning the state of preservation of the lithic materials, we have to assume that the greater degree of alteration of the materials of Berniollo would imply that in this assemblage activities which leave fainter traces are under-represented, such as cutting meat and fresh hide and butchering in general.

In general terms, we can show that in the Azilian level it is activities concerning fresh hide and butchering that predominate, whereas in the Magdalenian level it is the working of antler and hide. In Berniollo the main activities concern the working of wood, bone material and hide (tables 2.4., 2.5. and 2.6.).

Table 2.4. Activities and material worked in the Azilian level of Santa Catalina

	Engraving	Indeterminate	Cutting	Percussion	Boring	Scraping
Hunting				7		
Butchering			28			
Hide/meat			1			
Fresh hide			6			22
Hide	1		3		2	8
Dry hide	1		4			6
Plants						1
Wood	1		1			9
Antler	4		1		1	16
Antler/Bone	1					6
Bone						9
Soft material			2			1
Hard material	1		1		2	9
Indeterminate	4	3	2			5

Table 2.5. Activities and materials worked in the Magdalenian level of Santa Catalina.

	Engraving	Indeterminate	Cutting	Percussion	Boring	Scraping
Hunting				11		
Butchering			39			
Hide/meat			2			
Fresh hide			8			23
Hide	1	1	5		2	20
Dry hide		2	16		1	24
Plants			1			4
Wood	3		5	2	2	16
Antler	8		13		3	82
Antler/Bone	5					21
Bone	6		5		2	12
Soft material		1	3			5
Hard material	2					5
Indeterminate	1	7	4	2	1	20

Table 2.6. Activities and materials worked in Berniollo.

	Engraving	Indeterminate	Cutting	Percussion	Boring	Scraping
Hunting				8		
Butchering			18			
Fresh hide			8			7
Hide			2			17
Dry hide			2			24
Plants			3			2
Wood/plants			2			
Wood	2		10		2	42
Antler	9		4		1	23
Antler/Bone	4		1			19
Bone						3
Soft material		1	2			4
Hard material	1		8	1		12
Indeterminate		7	15	3	2	21

CHAPTER 3

THE FUNCTION OF THE DIFFERENT TYPOLOGICAL GROUPS

In this chapter we will discuss the results of the analysis of use traces for the different types of tools analysed, retouched as well as unretouched.

3.1 END-SCRAPERS

In our study end-scrapers were the tools studied most extensively. In all 117 tools were examined, upon which we found 273 active zones. In this assemblage there were 17 end-scrapers with 31 active zones from the Azilian level of Santa Catalina, 56 others with 136 active zones from the Magdalenian level on the same site. The remaining 39, with 106 active zones, came from the Epipalaeolithic settlement of Berniollo.

In all the cases studied the retouched front of the end-scraper was moved transversely on the material worked. In the majority of cases the end-scraper was used at a high angle to the material, that is, with the ventral surface facing the direction of movement, although a third of the tools were used to scrape at a low working angle, with the dorsal surface leading.

Table 3.1. Materials worked with the front of end-scrapers.

	Berniollo	S.C. Azilian	S.C. Magdalenian	Total
Wood	4 (9.8%)	1 (6.7%)	2 (3.8%)	7 (6.4%)
Bone	1 (2.4%)	0	1 (1.9%)	2 (1.8%)
Antler	1 (2.4%)	0	8 (15.1%)	9 (8.3%)
Antler/Bone	4 (9.8%)	0	4 (7.5%)	8 (7.4%)
Antler/Wood	0	1 (6.7%)	0	1 (0.9%)
Hide	10 (24.4%)	2 (13.3%)	6 (11.3%)	18 (16.5%)
Fresh hide	1 (2.4%)	7 (46.7%)	10 (18.9%)	18 (16.5%)
Dry hide	9 (22%)	2 (13.3%)	12 (22.6%)	23 (21.1%)
Indeterminate	11 (26.8%)	2 (13.3%)	10 (18.9%)	23 (21.1%)

The material most often worked with the fronts of the scrapers on the sites studied is hide (table 3.1.). If we eliminate from the total those tools which, because of the scarcity of traces, the material worked cannot be determined, the scrapers which worked hide increases to 68.6% of the total number of end-scrapers with defined uses. In the late Magdalenian level of Santa Catalina 65.1% of scrapers were used on hide, rising in the overlying Azilian level to 84.6%, while the Epipalaeolithic of Berniollo is closer to the lower level of Santa Catalina with 66.6% used on hide. [2]

In the working of hide with scrapers, taking into account only those cases in which the state of the hide could be determined, the Magdalenian level of Santa Catalina produced a higher proportion of dry hide worked as opposed to fresh hide. In the Azilian level of Santa Catalina, the working of fresh hide was clearly dominant over dry hide, the reverse of Berniollo, where dry hide was worked more regularly than fresh.

The second material worked with the fronts of the end-scraper is antler. Again, the differences between the three lithic assemblages is very marked. While in Berniollo the proportion is clearly marginal and in the Azilian level of Santa Catalina there is not a single scraper which performed this activity, in the Magdalenian level it acquires greater importance, involving at least 18% of the scrapers on which the material worked by the front of the tool can be determined.

The working of bone appears in very low proportion among the scrapers, as much in the Magdalenian level of Santa Catalina as in the Epipalaeolithic site of Berniollo, while we have not found traces of this material on the fronts of end-scrapers from the Azilian level.

In some cases the traces of working bone and antler cannot be differentiated, so that the functional interpretation can go no further. This occurs for four end-scraper fronts from the Magdalenian level of Santa Catalina and on another four from Berniollo; together with previous data this reinforces the interpretations concerning the moderate importance of working bone materials in Berniollo and the greater degree to which they are worked with end-scrapers in the Magdalenian of Santa Catalina, while in the Azilian level there was no working of bone or antler with end-scraper fronts.

The working of wood also appears to rarely occur using end-scrapers. Only on the site of Berniollo does it exceed 10% of the fronts used.

In general the data observed coincide with those from other assemblages of end-scrapers of the Upper Palaeolithic and the first Epipalaeolithic in Europe, indicating the fundamental first use of these tools for scraping hide and their more marginal use on other materials, such as antler, bone, or wood (van Noten *et al.*, 1978; Moss, 1983; Vaughan, 1985; Plisson, 1985; Symens, 1986; Winiarska-Kabacinska, 1988; Keeley, 1988; Philibert, 1993; Roodríguez, 1993).

The analysis of the formal characteristics of the end-scrapers in relation to the material worked with the front has produced some results. End-scrapers which worked dry hide are larger and have a more curved front than those which worked fresh hide, wood or bone materials. Probably the particularity of the end-scrapers which work dry hide is due to the characteristics of the task for which they are made. The scraping of dry hide wears down the edge of the tool intensively and demands much force (Gusinde, 1930/1982, 194; Gallagher, 1977). The long end-scrapers allow an easier application of force, the hafting of the tool and also the possibility of intensive resharpening. The curved front favours more contact between the tool and a flexible material such as hide.

The end-scrapers were used not only on the retouched front, but also a high proportion of them have use-traces on the lateral edges and in the distal zone (table 3.2.). The analysis of the type of activities carried out in these zones shows that in the majority of cases there is a concurrence between the material modified with the front and with the other zones of the tool. This leads us to think that the different active zones of the end-scraper were used in a complementary way within the same task. With these zones scraping was carried out, substituting the front as the active zone, or cutting was performed, complementing the action carried out by the front with a new one. The distal zones were used for scraping, while the lateral zones were intended for scraping as well as cutting. Cutting is more common for the lateral edges of the end scrapers made from blade blanks.

We can conclude from this that the retouched front should not be considered the only active zone of the end-scraper, but that the prehistoric individual took advantage of other effective zones which the edges of the tool naturally present, using the zone most suitable to each activity. As for cutting, straight long edges with acute angles are most efficient, the lateral edges of the end-scrapers were chosen for this, especially if they were made on blades. If the work to be done was scraping, thicker edges were used, such as in proximal zones (opposite to the front), as well as the lateral zones, which had a suitable edge angle.

This practical behaviour does not imply that the prehistoric individual considered all the zones of the

[2] In the calculations cited from here on, concerning the proportion of a worked materials, the cases in which the material worked could not be determined will not be taken into account.

end-scraper to be equally useful. Of course the retouched part is favoured, as can be seen in most of the tools, because more traces occur in that zone in comparison with the unmodified ones. Also the front of the. end-scraper offers the enormous advantage of being able to be resharpened, causing this zone to be used preferentially.

Observation of the average number of active zones per tool suggests that the degree of use of the end-scrapers in Berniollo (2.7) was greater than in the Magdalenian level of Santa Catalina (2.4). The lowest index corresponds to the Azilian level of Santa Catalina (1.8)

The average number of active zones per tool seems to indicate differences in the degree of use of end-scrapers on blades or on flakes in each of the assemblages (table 3.3). In Berniollo, the end-scrapers on flakes show a greater number of active zones per piece, contrary to what happens in the Azilian level, where the end-scrapers on flakes seem to be exploited less intensively. In the Magdalenian level, the average use of the two types of end-scrapers is equal.

The degree of development of polish supports this information, showing greater intensity of use of the end-scrapers on flakes compared to those on blades in Berniollo, whereas the reverse occurs in the Azilian level at Santa Catalina.

Within the assemblage of Berniollo we can see a differential exploitation of the end-scrapers depending on the type of flint. The average number of active zones per piece is high for flint 2 (2.8) and flint 1 (3.1), while the veined flint 3 reflects the lowest average number of active zones (2.2).

The analysis of the proportion of retouched active zones with traces of resharpening is higher for the assemblages of Berniollo (19%) and the Magdalenian of Santa Catalina (31%), which indicates the end-scrapers were sharpened more often in these two archaeological contexts than in the Azilian level (7%). This information correlates with that regarding the intensity of use, the intensity of work with end-scrapers in the two above assemblages being greater in comparison to the Azilian level of Santa Catalina.

As regards the relation between resharpening and the materials worked, there is a higher incidence of resharpening on those end-scrapers which worked dry hide, which is logical, considering the abrasive nature of this material.

Table 3.2. Materials worked and activities carried out with lateral edges and proximal zones of end-scrapers.

	Cutting	Scraping	Engraving
Wood	2	16	2
Bone	2	1	3
Antler	3	19	0
Antler/Bone	2	6	1
Antler/Wood	1	0	0
Hide	4	14	0
Fresh hide	10	15	0
Dry hide	15	26	0

Table 3.3. Average number of active zones acording to the blank.

	Flake	Blade
Berniollo	2.9	2.4
S.C. Azilian	1.7	2.3
S.C. Magdalenian	2.5	2.5

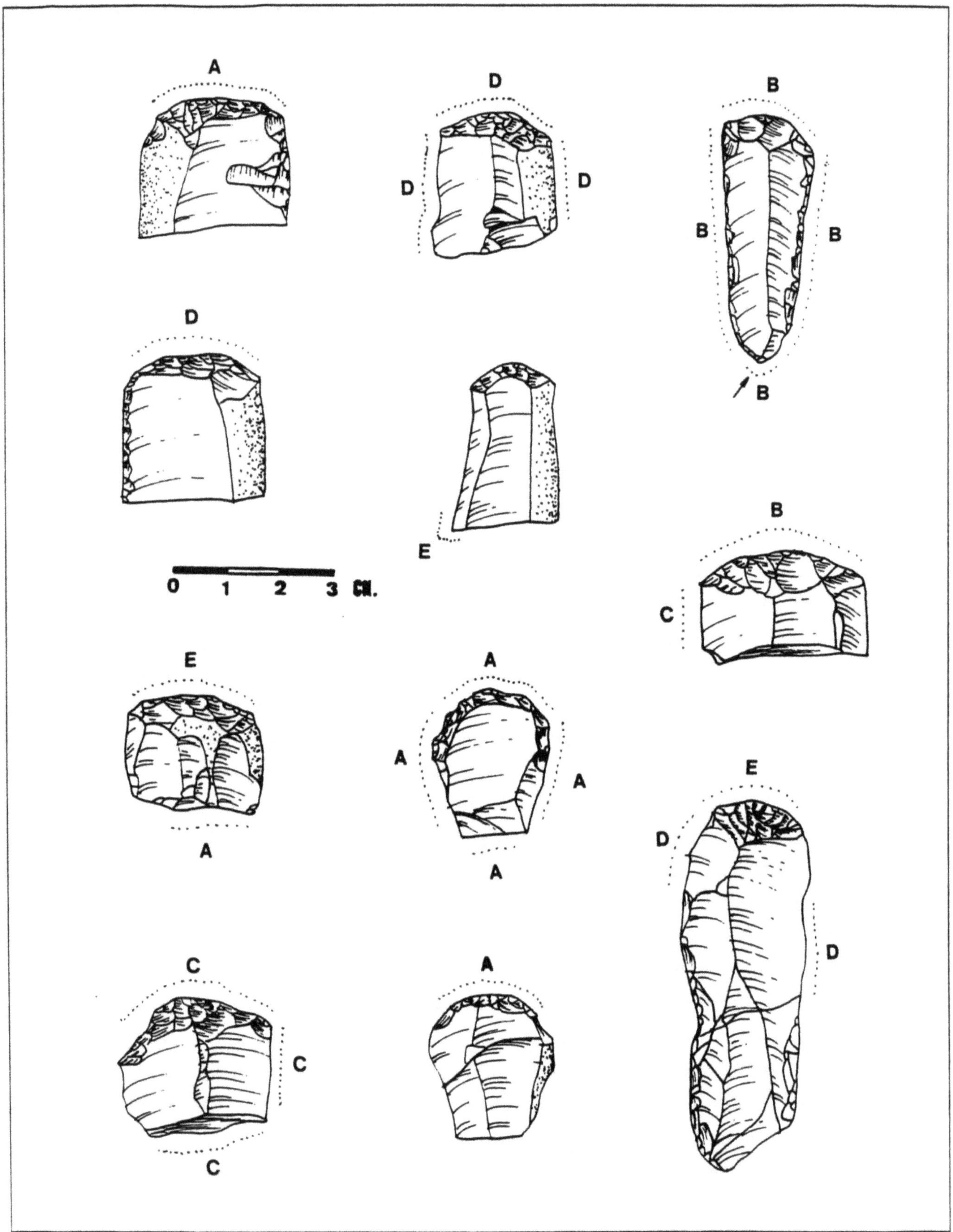

Figure 3.1.. The use of end-scrapers from flint 3. A: Scraping hide; B: Scraping bone/antler; C: Scraping wood; D: Cutting hard material; E: Indeterm.

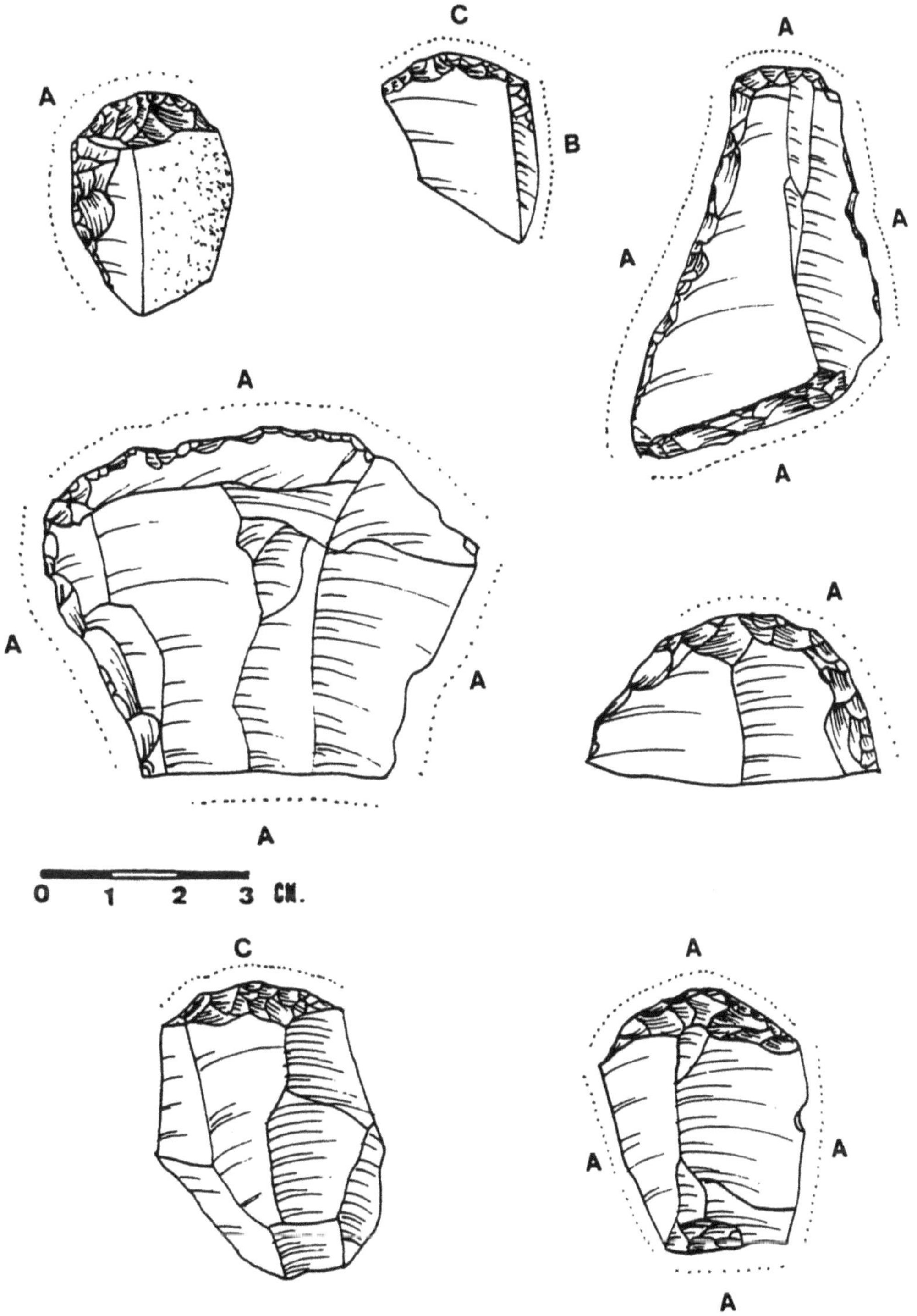

Figure 3.2.. The use of end-scrapers from flint 1 and 2 at Berniollo. A: Scraping dry hide; B: Cutting antler or wood; C: Indeterm.

Graph. 3.1. Length and width of of flint 3 end-scrapers at Berniollo.

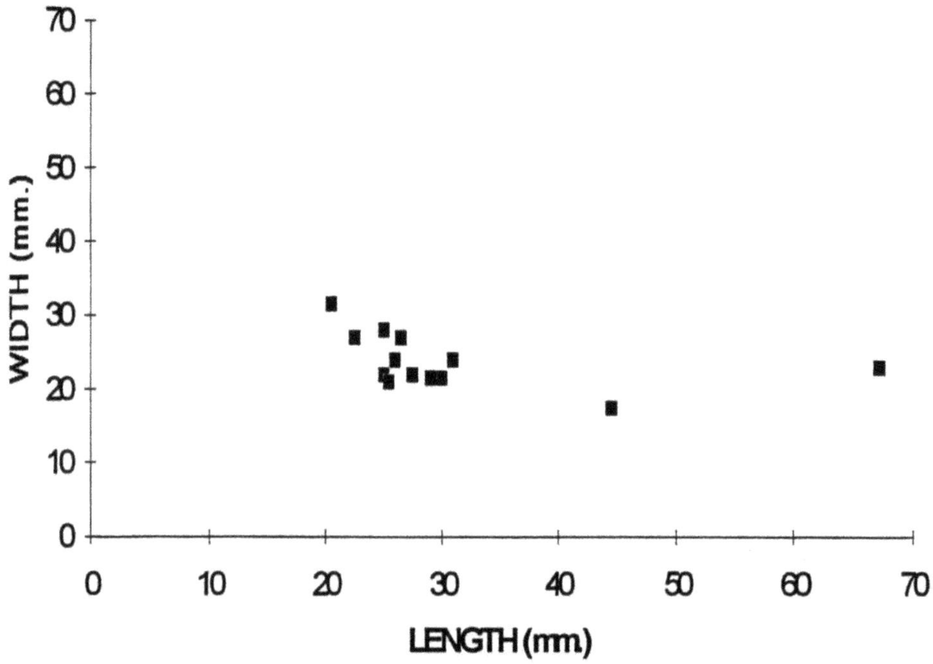

Graph 3.2. Length and width of flint 1 and 2 end-scrapers at Berniollo

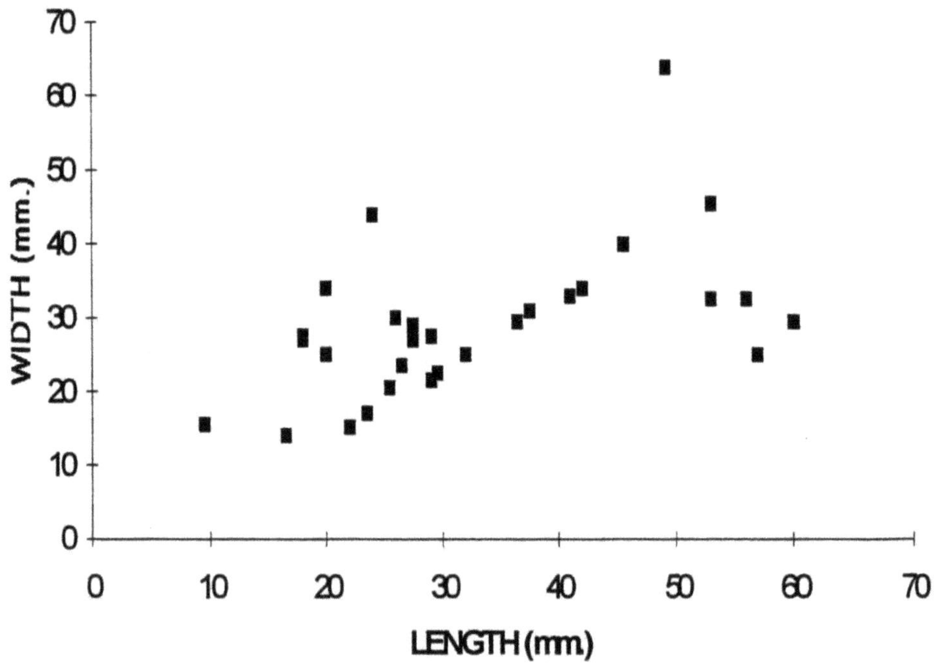

Moreover, there are differences in the intensity with which the end-scrapers of Berniollo were resharpened depending upon the lithic raw material in which they were made. The lower index occurs for end-scrapers of flint 1 (10%), while that for flint 3 (30%) is clearly higher. The differences in the degree of resharpening between the three types of flint is not due to the nature of the materials worked with each type, since we have not found an association between the type of flint and the material worked, but rather to a different technical behaviour for each type of lithic raw material.

It is not surprising that the tools in flint 2 show more intensive resharpening than those in flint 1, since the degree of use of end-scrapers in flint 1 is greater. This is shown by the number of active zones per tool and the degree of development of the polish. It is striking, however, that the tools in flint 3, which were used with less intensity than those in flint 2, were resharpened more often. The characteristics of use of the tools according to the type of lithic raw material will be dealt with later. It is enough to say here that the intensity of use is not the only factor which conditions the degree of resharpening of the tools: aspects related to the use-strategies of a particular type of tool are also influential.

Finally, functional study has allowed us to obtain some information on the hafting of the tools. Various authors have demonstrated the existence of traces of hafting in different archaeological assemblages, specifically in the case of end-scrapers (Anderson and Helmer, 1987; Symens, 1986, 219; WIniarska-Kabacinska, 1988; Caspar and Cahen, 1987,193; Vila 1987, 121).

For our part, we have found it difficult to detect microscopic traces of the fixing of a tool into a handle, so, for the end-scrapers, we have attempted to understand this through other criteria, such as: the fact that an end-scraper is made on a blade blank, its fracturing, the absence of use-traces on the sides as well as on the proximal end of the tools, and the standardization of the size of the scraper. When such attributes converge in an assemblage of end-scrapers, then these were likely to have been usually hafted.

The above criteria do not appear to be correlated in particular sites or levels, or in particular worked materials; therefore, we cannot demonstrate that any of the tools in these assemblages were preferentially hafted. These criteria are usually present, however, for end-scrapers made in flint 3 from Berniollo. As the group of end-scrapers of flint 3 belongs to the same archaeological assemblage, we can analyse the degree of standardisation of the group with respect to the rest of the end-scrapers in the assemblage.

We see that end-scrapers made in flint 3 are quite homogenous in size (Graph 3.1) in comparison to the rest of the end-scrapers (Graph 3.2) (see also figures 3.1. and 3.2.).

According to this information, the end-scrapers of flint 3 from Berniollo were used hafted in a greater proportion than the general assemblage of end-scrapers.

3.2. BURINS

We selected 46 burins for functional analysis, of which 12 are from Berniollo, 11 from the Azilian level of Santa Catalina and 23 from the Magdalenian level of the latter site. In such an assemblage we found 132 active zones, which can give an idea of the intensity with which these tools were used.

The burins of Berniollo and Santa Catalina are primarily used for the working of bone material, of which antler is the material worked most regularly. These data coincide with those from other assemblages of burins of Upper Palaeolithic and Epipaleolithic sites in Europe, such as Verberie, Rascaño and El Juyo (Keeley, 1988; Symens, 1986), Meer (van Noten *et al.*, 1978), Laminak II (González and Ibáñez, 1994b), Star Carr (Dumont, 1983) and Vaenget Nord (Juel Jensen and Brinch Petersen, 1985), where burins were used for working bone materials.

The activities carried out are varied: scraping, cutting and engraving having been detected. There is a clear relationship between the morphological features of the active zone and the activity carried out. The dihedral angle and the trihedral point are used for engraving and the facets for scraping, while the natural edges not affected by the burin facet were used for cutting or scraping.

There is a certain relationship between the morphology of the burin and the activity which occurs with the tool. The dihedral burins are more often associated with the engraving of bone material than the truncated burins. The same can be said of the burins made on blades in respect to those made on flakes. For burins made on flakes and truncated burins, the proportion of engraving activity is lower and the variety of materials worked is comparatively higher. So dihedral burins on blade were the most specialized type of burin for engraving bone material.

The average number of active zones for burin used varies in the three assemblages. In Berniollo there seems to be a more intensive average use of each burin, with 3.5 active zones per tool, dropping to 2.9 in the Azilian level. In the Magdalenian of Santa Catalina, burins average 3 active zones.

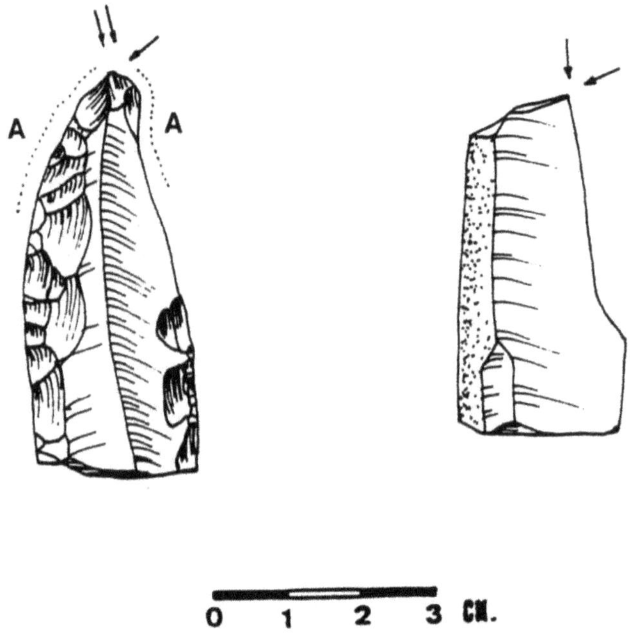

Figure 3.3. The use of burins from flint 3 at Berniollo. A: Scraping antler/bone.

If we consider the number of active zones in relation to the blank from which the burin is made, in overall terms we observe a similar intensity of use of the tools made on flakes in relation to those made on blades. However, the diversity between the assemblages is significant for this attribute. In Berniollo the burins made from flake blanks present a greater number of active zones per piece than those made from blade blanks. On the other hand, in the Azilian level of Santa Catalina the results are inverted, the burins made from blades appearing to be more intensely used. In the Magdalenian of Santa Catalina both kinds of burin blank show a similar average number of active zones (table 3.4.).

Table 3.4. Average number of active zones per burin.

	Flake	Blade
Berniollo	3.8	2.5
S.C. Azilian	2.5	3.3
S.C. Magdalenian	3	3.3

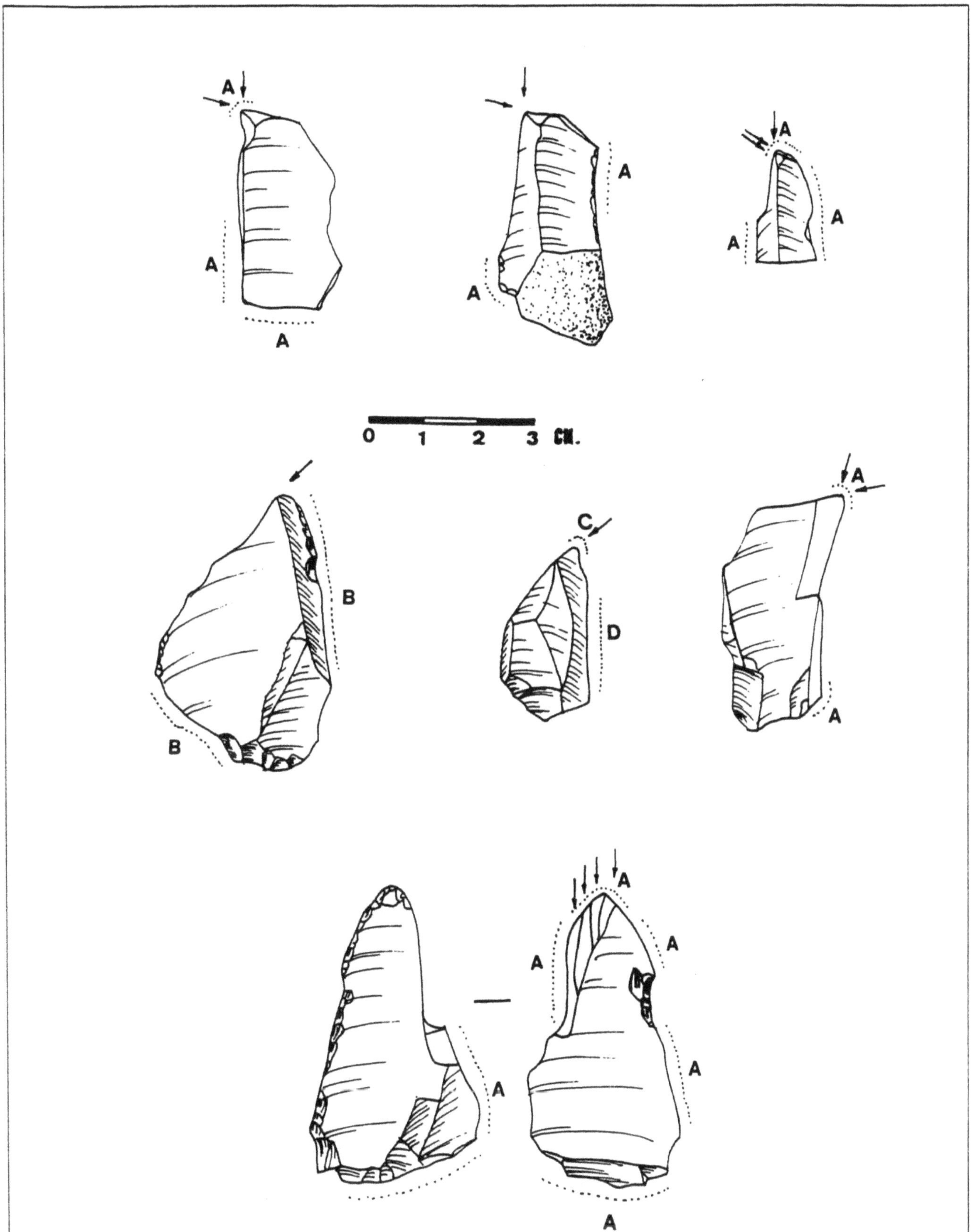

Figure 3.4. The use of burins from flint 1 and 2 at Berniollo. A: Engraving (bevel) or scraping antler; B: Cutting unknown material; C: Engraving hide; D: Cutting soft material.

We note that these tendencies of tool-use according to assemblage and type of blanks are identical to those which we observed for the end-scrapers. This suggests that these factors are related not so much to the function of each particular type, but rather to factors which are more generally related to the use-strategies of the stone tools.

The relation between the number of active zones and the type of flint in which the piece was made, which we have established for the assemblage of Berniollo, shows less use of the burins made in flint 3. The two burins in type 1 flint have a very high average number of active zones, 4.5, an intensive use which is also shown, although to a lesser degree, by the 8 burins of flint 2 with an average of 3.3 zones. On the other hand, the intensity of use of type 3 burins is less: of the two analysed one presented no traces and on the other only two active zones were found.

In general terms the development of the polish, another of the indexes which reflect the intensity of use of the tools, gives information complementary to the average number of active zones per piece. Polish due to use present on the burins of Berniollo and the Magdalenian level of Santa Catalina shows a greater proportion of high development than in the Azilian level. At the same time, in Berniollo we observe a lower proportion of traces which are little developed.

To summarize the data on the intensity of the use of burins, we have observed that these pieces were used with greater intensity in Berniollo than in the two levels of Santa Catalina. In the Epipalaeolithic site of Berniollo, moreover, burins made from flake blanks were used with greater intensity as compared to those made on blade blanks, and those made from flints 1 and 2 more intensively than those made in flint 3. In the Azilian level the strategy of use in regard to the blank was the reverse, since the burins made from blade blanks were used more intensively.

In the majority of cases, the burins present traces of only one worked material (85.8%), although having more than one active zone (78.5%) and carry out on this material at least two types of different activities (57.1%). This leads us to think that the burins were considered to be tools suitable for complex tasks on bone materials, for which it was necessary to have a tool which offered the possibility of carrying out different actions; this necessity was covered by the burin, due to the diversity of active zones which it offers.

3.3. BORERS

We studied 7 borers, one from Berniollo, another from the Azilian level of Santa Catalina and 5 from the Magdalenian level of the same site. This type of tool is very rare in these assemblages. In fact, of the 7 tools selected, probably only two can be considered authentic borers, that is, tools having a point made by retouch. In the other four examples the "retouching" is in fact chipping produced by use.

The activities carried out with the pointed zones of the borers were, in all cases except one, of boring, carried out on different materials. Bone was bored with two of the tools, while one tool worked on antler and another on wood. On two pieces the traces are more ambiguous; for the first, the material worked was probably hide, for the other it could have been antler or wood.

Despite the small number of tools analysed, the proportion of materials worked by them seems to correlate with the results of the analysis of borers in other investigations of lithic assemblages from the late Upper Palaeolithic. These show a predominance of work on hard animal materials, antler and bone, and to a lesser extent on wood and hide (Van Noten *et al.*, 1978, 90; Cahen *et al.*, 1979, 666; Audouze *et al.*, 1981, 139; Plisson, 1985, 196-197).

On the sites which we are studying, the proportion of borers among the retouched tools is very low. During the course of this functional analysis we will be able to prove that boring actions were carried out with a wide range of tools, such as burins, pointed backed bladelets, oblique truncated pieces, denticulated pieces, and retouched and unretouched flakes and blades. Therefore, when it was necessary to bore a material, a typological "borer" was not necessarily chosen; rather a wide spectrum of tools was used.

3.4. RETOUCHED FLAKES AND BLADES

The functional study of Berniollo and Santa Catalina has included a series of retouched flakes and blades which could be classified in the section of side-scrapers according to Laplace's classification or in types 65 and 66, flake or blade with retouch on one edge or in both, according to the list of Sonneville-Bordes. In total, the analysis was carried out on 28 tools; 9 correspond to the site of Berniollo, 10 to the Azilian level of Santa Catalina and 9 others to the Magdalenian.

Observation of the activities carried out and the materials worked with these types of tools reflects the great functional heterogeneity of the group. In fact, we can determine the work of these tools on materials as different as antler, bone, meat, wood, fresh hide and dry hide, that is to say, practically the whole spectrum of materials which one would expect to have been worked on the sites of the period (table 3.5.).

For the actions carried out with these tools an equilibrium is observable between cutting and scraping. Chopping and boring were also detected, though less so. With respect to the worked material, no particular activity stands out.

This information seems to show that this type of tool is a product of the adaptation of a blank to a wide range of necessities. Thus, this assemblage of tools does not form a defined functional group, as do the burins, which present great uniformity of use in activity as well as in material worked, or the end-scrapers, used exclusively for scraping.

Concerning the relationship between the type of blank and the activity, blades are preferred for cutting, while on the flakes the proportion of longitudinal and transverse actions is very similar.

In regard to the edge angle, for cutting activities more acute edges were used than for scraping. This is also true for other morphotypological groups (see below).

If we compare information concerning the use of retouched and unretouched flakes and blades (*infra*), we see that dry hide and antler/bone are more often worked with retouched tools. This is probably due to the speed with which both materials abrade active edges, which need to be resharpened by retouch. Apart from that, the variety of uses of retouched vs. unretouched flakes and blades is very similar. This leads us to think that, in this morphological group, retouch is more the result of resharpening of the tool than the form given to the blank before use. Probably these tools were first used without any retouch, then resharpened by retouch when the edge became too blunt for use.

We can find differences between the assemblages studied in the intensity of use of these tools. Again, the assemblage of Berniollo shows the highest average of active zones per tool (2.3) the Magdalenian level is in an intermediate position (1.7) and the Azilian has the lowest average (1.4).

Table 3.5. Materials worked with retouched flakes and blades.

	Cutting	Scraping	Boring	Chopping	Indeterminate
Wood	4 (9.8%)	5 (12.2%)	0	0	0
Wood/Antler	0	1 (2.4%)	0	0	0
Antler/Bone	0	3 (7.3%)	0	0	0
Antler	0	3 (7.3%)	1 (2.4%)	0	0
Butchering	5 (12.2%)	0	0	0	0
Fresh hide	1 (2.4%)	1 (2.4%)	0	0	0
Dry hide	2 (4.9%)	0	0	0	0
Indeterminate	5 (12.2%)	8 (19.6%)	0	1 (2.4%)	1 (2.4%)

3.5. TRUNCATED TOOLS

The study of truncated flakes and blades from Santa Catalina and Berniollo has involved the analysis of 24 tools: 5 from the site of Berniollo, 1 from the Azilian level of Santa Catalina and 18 from the Magdalenian level.

Taken as a whole the group of truncated tools shows high functional diversity, as they were used for working antler, wood, hide and butchering. Within this assemblage of tools we can establish a primary distinction between those which present traces of use in the truncated end (13 tools) and those which in this area show no indications of use (9 tools).

The truncated tools with used retouched ends were used on hard materials, mainly for working antler (table 3.6.). The truncated ends were used for scraping, while the rest of the active zones of the tool were used for scraping or for cutting.

The reason for the truncation of the tool is not difficult to explain. It was sought to condition an active zone with a right edge angle for scraping hard materials. To work on these materials is not necessary to use active zones with a curved front, as end-scrapers have, as the truncated tools

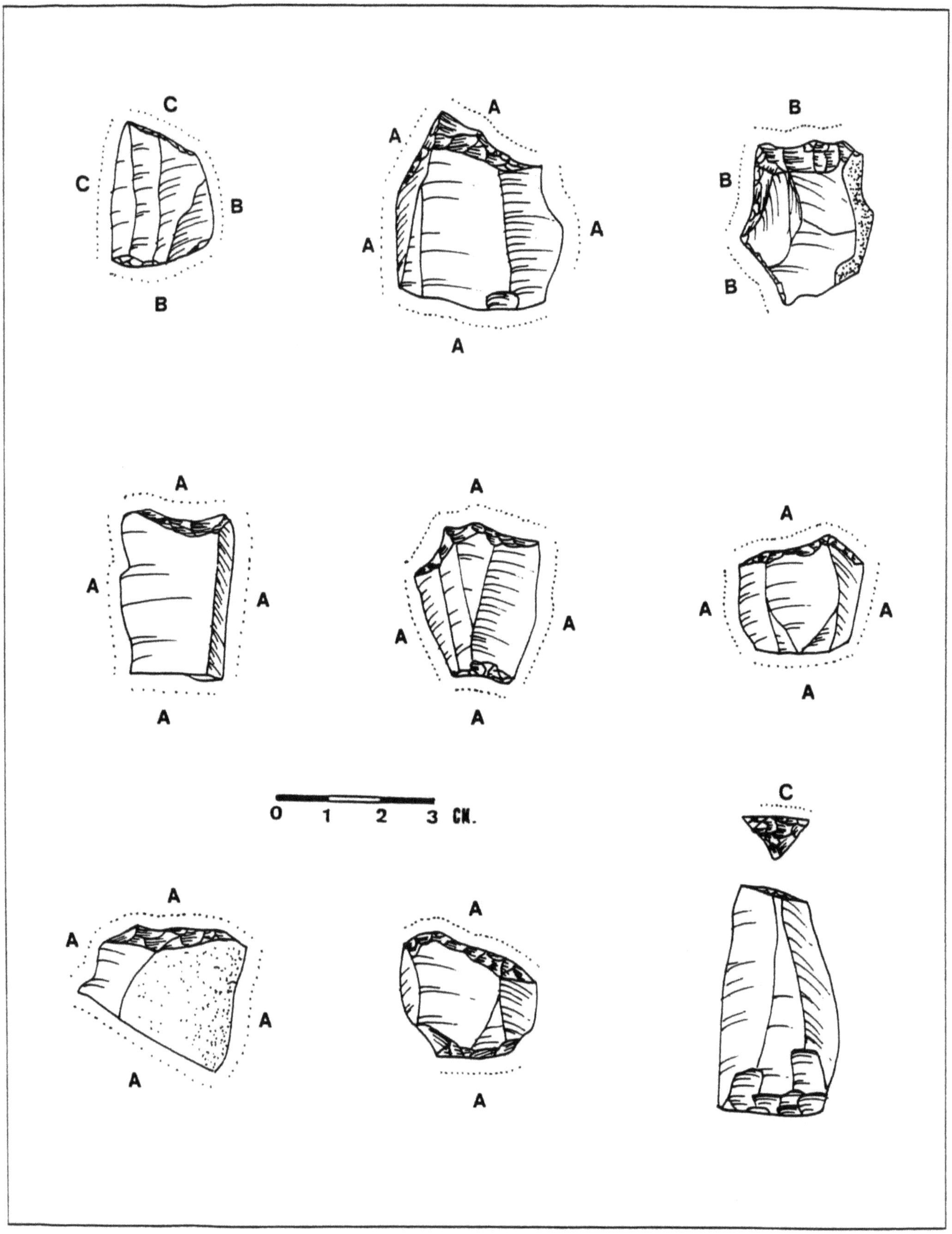

Figure 3.5. The use of truncated tools type "a". A: Scraping antler; B: Scraping wood; C: Scraping antler or wood.

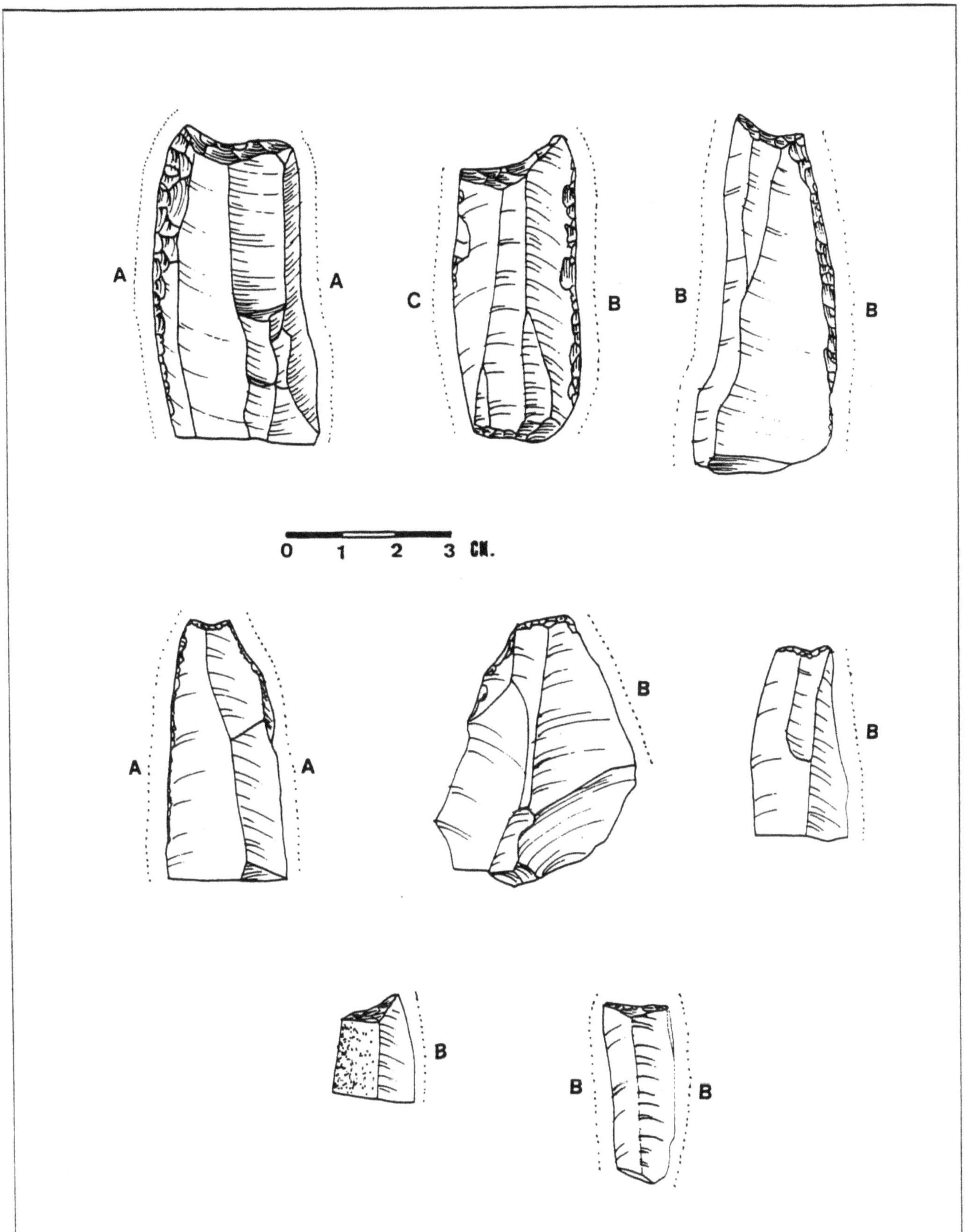

Figure 3.6. The use of truncated tools type "b". A: Cutting or scraping hide; B: butchering; C: cutting unknown material.

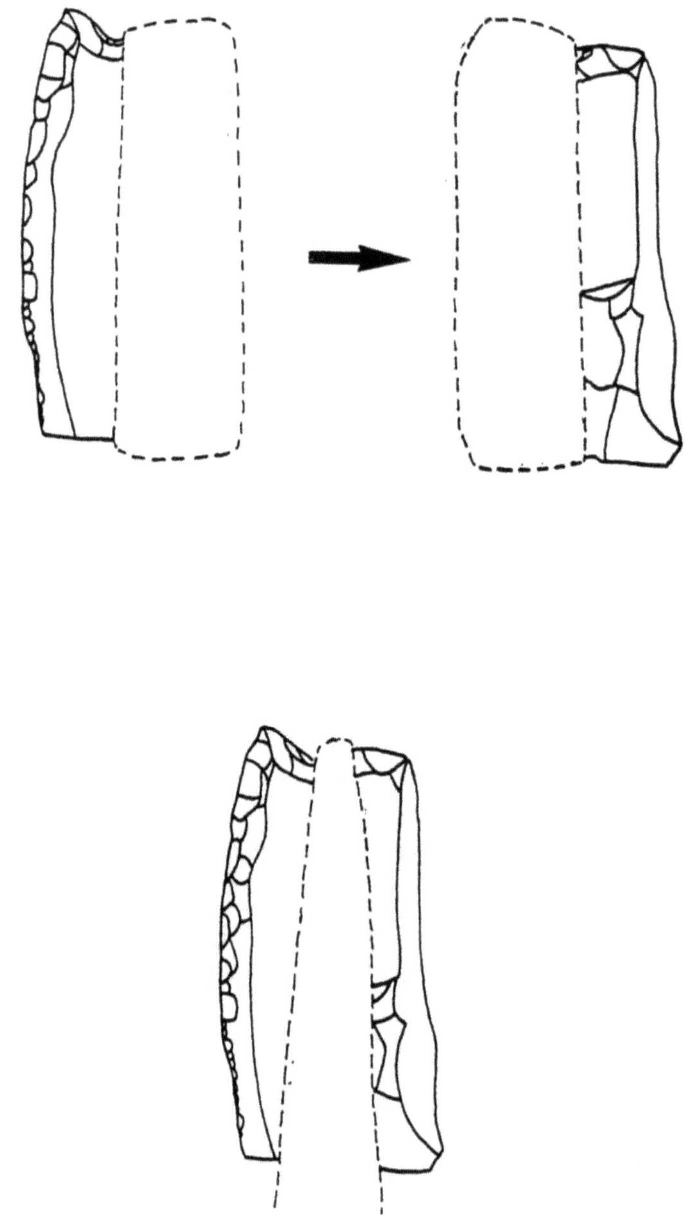

Figure 3.7. Two hypothesis of hafting of truncated tools type "b".

carry out adequately the function of scraping. In relation to this, it is worth remembering that end-scrapers used on hard materials have fronts that are less curved than those which worked hide.

The truncated ends of the tools which have no traces on the retouched zone were intended for hafting (Plisson, 1985, 306 and 1987, 86) (see figure 3.7.).

Lateral edges in this assemblage of tools were used for butchering or cutting hide (table 3.7.). This type of work carried out with truncated tools has its parallel on the Italian site of Paglicci, where these tools were used for cutting meat and to work hide (Donahue, 1985, 38). In Habitation 1 of Pincevent, although the truncated blades were used for various tasks, they were best for cutting meat (Plisson, 1985, 190-192 and 1987, 86).

Two predominant functional groups are thus established for the truncated tool, that in which the retouched end was used for scraping (group a), mainly hard materials, and that in which the tools were used as knives, once hafted, for butchering or cutting hide (group b).

A series of formal characteristics defines each of the two groups. In the group "a", the truncated tools intended for scraping, are included those with a straight, oblique, concave and double truncation, while in group "b" the type of truncation represented is predominantly concave. For group "a" flakes were chosen as much as blades, though flakes dominate. Tools from group "b" were made on blades. Both groups differ also in their measurements, the tools of group "b" tending to be elongated and more stylized than those of "a" (see figures 3.5. and 3.6.).

Thus, within the group of truncated tools there are different functional groups. On the one hand, we find truncated tools intended for scraping hard materials, and on the other the knives. For these knives, blades were chosen, a concave truncation was made, and once hafted they were used for butchering or cutting hide.

Table 3.6. Activities and materials worked with truncated tools having traces on the truncated end.

	Cutting	Scraping	Boring	Total
Wood	1 *(2,6%)*	4 *(10,5%)*	0	5 *(13,1%)*
Antler	6 *(15,8%)*	16 *(42,1%)*	1 *(2,6%)*	23 *(60,8%)*
Antler or wood	0	3 *(7,9%)*	1 *(2,6%)*	4 *(10,5%)*
Fresh hide	1 *(1,6%)*	2 *(5,3%)*	0	3 *(7,8%)*
Indeterminate	0	2 *(5,3%)*	1 *(2,6%)*	3 *(7,8%)*

Table 3.7. Activities and materials worked with truncated tools without traces on the truncated end

	Cutting	Scraping	Indeterminate	Total
Butchering	9 *(60%)*	-	-	9 *(60%)*
Antler	0	1 *(6,7%)*	0	1 *(6,7%)*
Hide	2 *(13,3%)*	1 *(6,7%)*	1 *(6,7%)*	4 *(26,6%)*
ndeterminate	1 *(6,7%)*	0	0	1 *(6,7%)*

3.6. THE POINTS

Backed points are abundant in the three assemblages which we are analysing, as is usual on the sites of this period. Among these, we chose for microscope examination 12 pieces from Berniollo, 14 from the Azilian and 29 from the Magdalenian of Santa Catalina.

Within the assemblage of backed points we have differentiated the microgravette type from the Azilian point type. We classify in the first type the narrow points, which are straight-backed and not very thick. Azilian points are those which have curved backs and are relatively thick. Points which do not correspond to these forms have not been included in these two categories. Among them are some double backed points, one with a truncated base and another with a basal uni-lateral notch which forms a kind of small tang.

The functional results of the backed points allows us to establish three groups of points, according to their use-traces:

1. Points used to work soft animal materials: to cut, scrape or bore hide, to cut meat, or for butchering (17%).

2. Points with traces of impact, resulting from their use as projectiles (38%).

3. Points without distinguishable use-traces (45%).

According to these results, the typological assemblage of "backed points" can consist of pieces which correspond to two functional groups, knives for working soft animal materials and projectile points, or perhaps projectile points which were used occasionally for another type of work, cutting soft animal material, when set into shafts (*infra*). However, among the 55 points analysed we have found only one case where the traces of impact are associated with traces of another activity, which was butchering. Therefore, we consider that this possible duality of function in the same point was not very common in the assemblages studied.

These groups are not evenly distributed throughout the three assemblages, since in Berniollo no point with traces other than those of impact have been found. Moreover, in the Azilian level of Santa Catalina the proportion of tools which worked soft animal materials is low in comparison with the results of the Magdalenian level. We shall first deal with the points which show traces of working soft animal materials.

Four of the points from the Magdalenian level were used for scraping, boring or cutting hide, in two of the cases in a dry state. Two of the tools show butchering traces. On the unretouched edges of four points of the Magdalenien level and one point from the Azilian level we detect some traces which could have also been attributed to the butchering of animals, although with less certainty. In the site of Andernach, almost 50% of the backed points with use-traces were used as knives for cutting meat or hide (Plisson, 1985, 276).

The small size of this type of tool and the presence of the back in the zone opposite to the active edge leads us to believe that most of these tools had to be hafted. Thus, although the perforation of hide could be carried out easily using one of these tools without a handle, other actions such as the cutting of hide and meat or butchering would be difficult to execute using a tool of such a small size. Therefore, we assume that the tools used to carry out these activities were hafted, either alone or with other backed tools, in the form of composite knives.

The activity inferred for another broad group of points, that of projectile points, offers no surprise, since this was the supposed activity for this typological group from an early date (Peyrony, 1933; Bordes, 1952; Brezillon, 1983).

However, interestingly, there is the clear association between the morphology of these tools and the work carried out (table 3.8), at Santa Catalina, where the points have a dual function.

Table 3.8. Work carried out whith backed points according to their morphology.

	Microgravette	Azilian point
Without use traces	8 *(40%)*	7 *(41.1%)*
With impact traces	10 *(50%)*	3 *(17.6%)*
Butchering or hide	2 *(10%)*	5 *(29.4%)*
Butchering+impact	0	1 *(5.9%)*

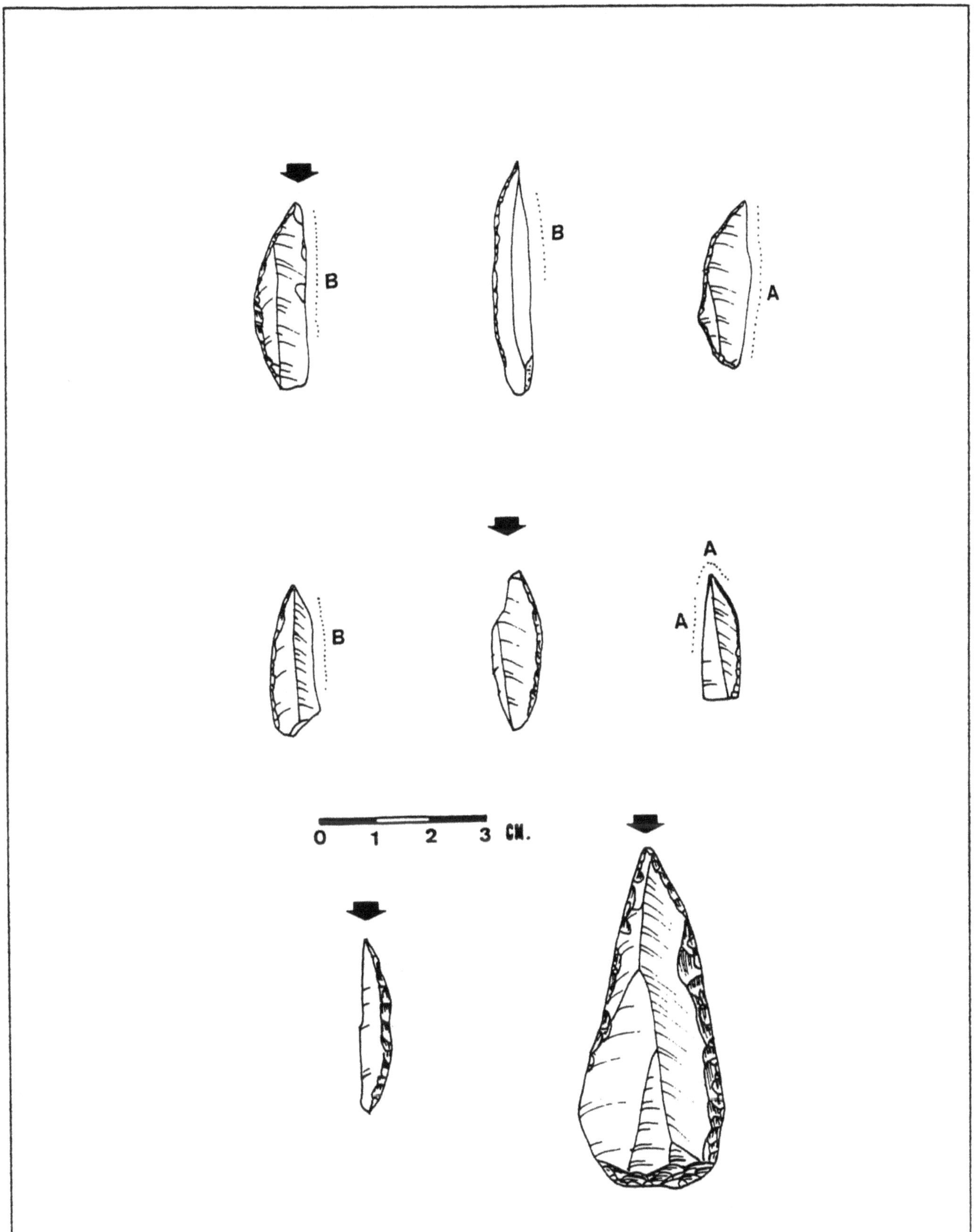

Figure 3.8. The use of some "Azilian" points and one pointed blade at Berniollo and Santa Catalina.: A: Cutting or boring hide; B butchering.

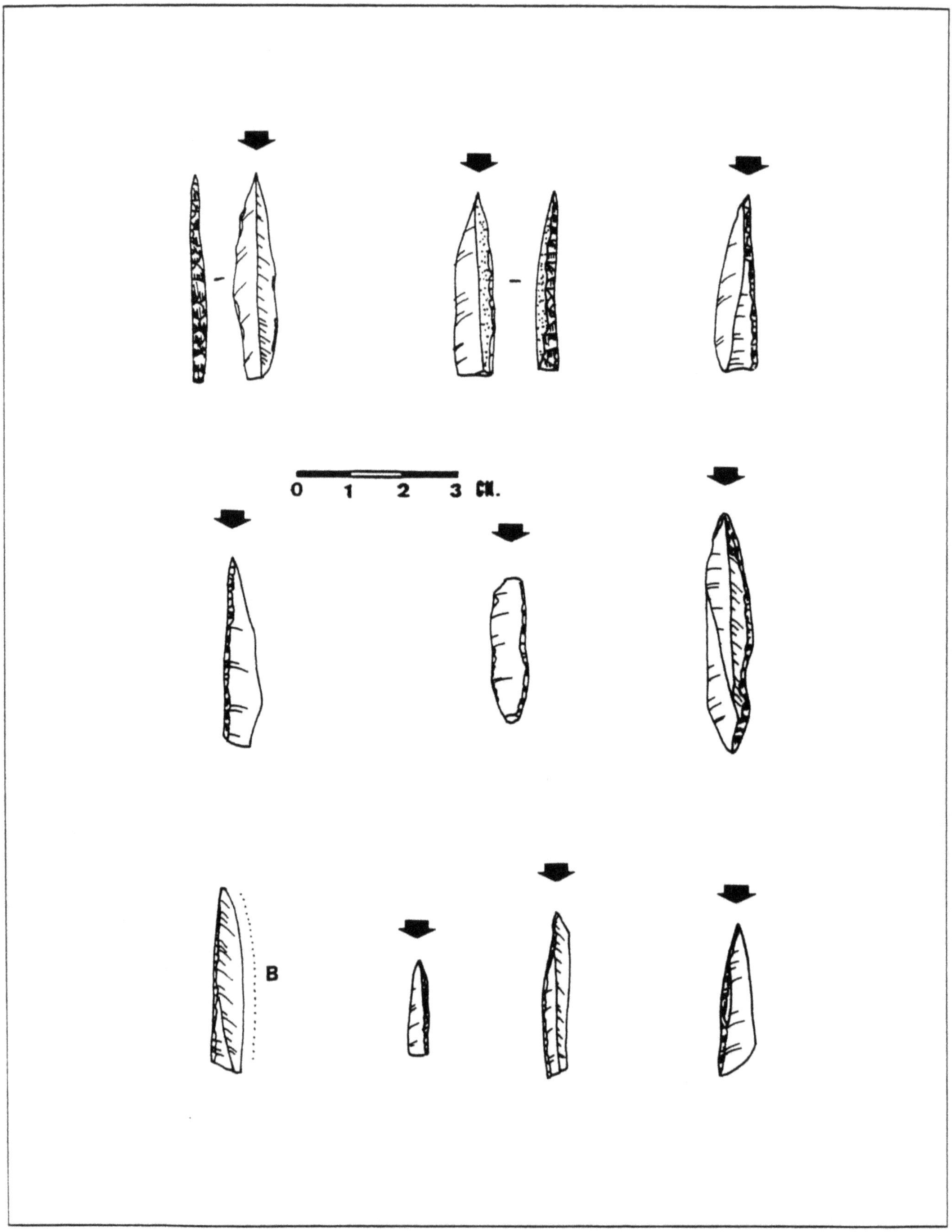

Figure 3.9. The use of "Micrograveтte" points at Berniollo and Santa Catalina. A: butchering

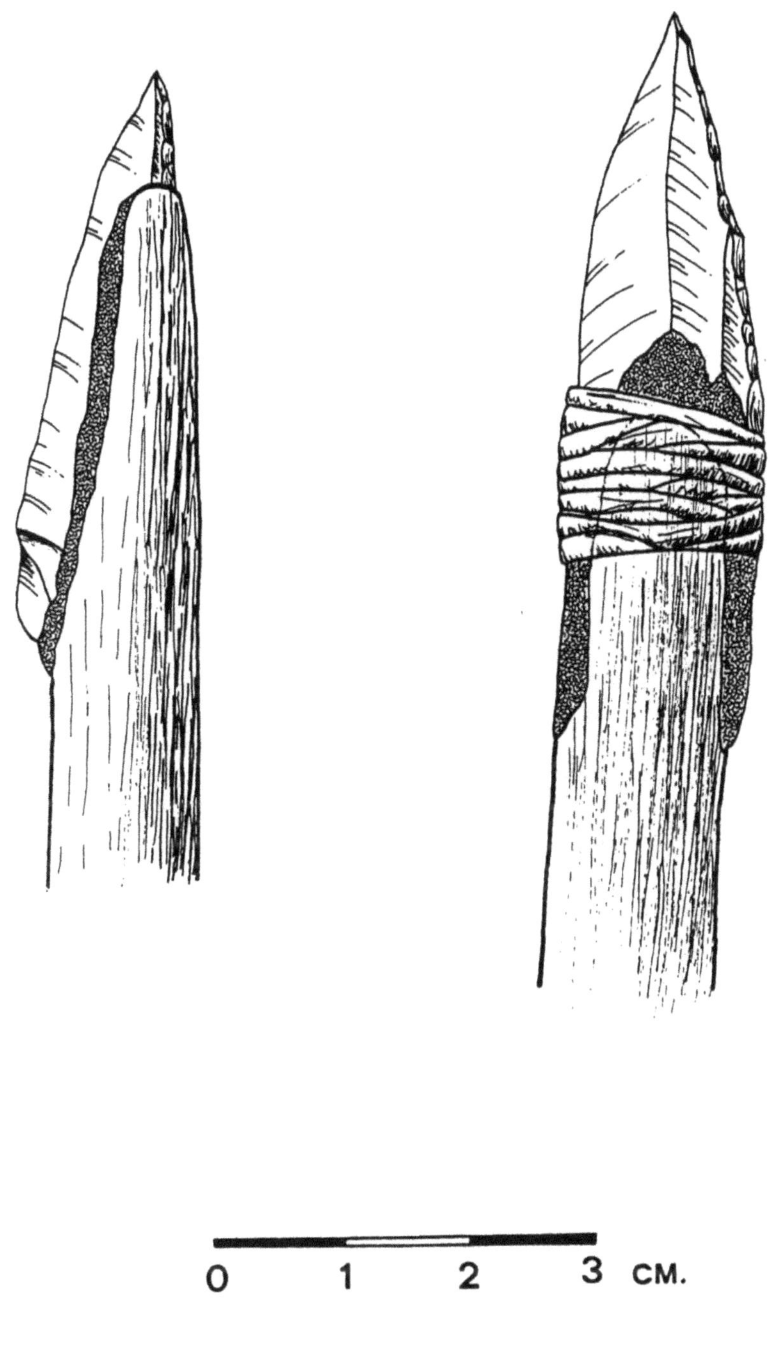

Figure 3.10. Hypothetical hafting of backed points.

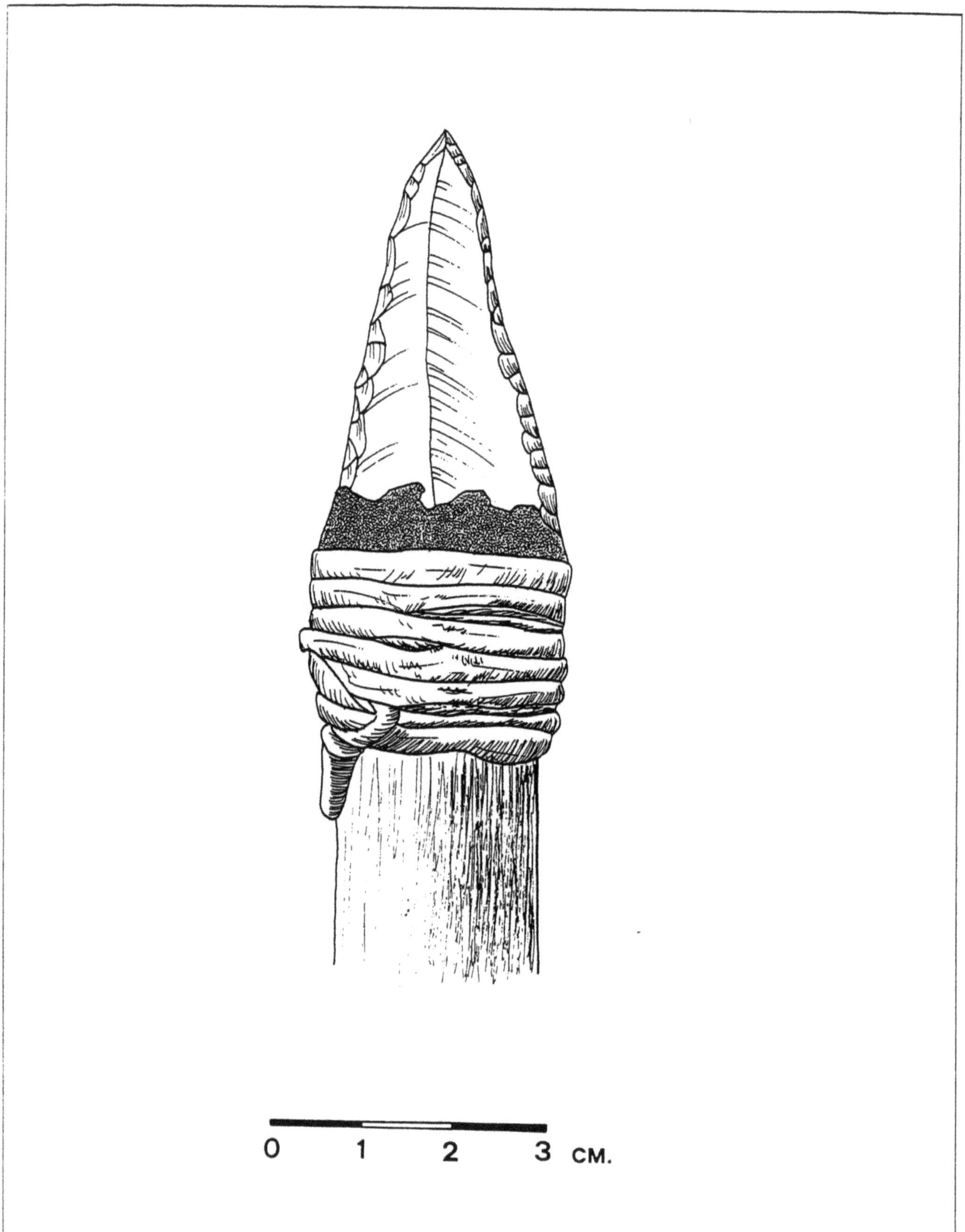

Figure 3.11. Hypothetical hafting of pointed blades.

It can be observed that microgravettes are preferentially used as projectile points (figure 3.9.), while Azilian points tend to be used to work soft animal materials (figure 3.8.). Thus, for projectile points, straight, relatively narrow bladelets were chosen, retouched to a symmetrical point (figure 3.10.). On the other hand, the Azilian points tended to be used for animal materials, probably forming parts of composite knives. The question arises as to whether the making of points of "Azilian" or "Microgravette" morphology was incidental, favoured by the form of the blank, or whether it was intentional. It will be necessary to have a larger sample of analysed Azilian points to be able to answer this question.

The third group of points, which show no trace, could represent projectile points or tools for working soft animal materials, whose use did not produce enough traces to be recognizable during analysis.

Other than backed points, we have analysed two pointed blades. On the site of Berniollo there are two examples of retouched blades measuring between 56 and 62 mm. in length and between 21 and 28 mm. in maximum width. Functional analysis has indicated that the tools were used as projectile points. The traces which allowed the identification consist of a slight polish on the retouched side as well as on the ventral surface and a series of very pronounced striations, which run parallel to the lateral edges of the point. Chipped areas are also present on the points of both tools. The band of polish and striations which is found all along the edges stops at the wider zone of the tool so that the area opposite the point has no traces. We consider it probable that this disposition of the traces indicates the point to which the handle reached.

The points have a width of between 27 and 28 mm. in the basal zone and a thickness of 8 mm. Because of this, the shaft in which they were set would have had to have been a relatively large diameter, at least 15 mm. Moreover, a tool intended to be thrown or held in the hand had to have a long shaft (figure 3.11.).

It will be necessary to verify whether this type of projectile element is common on Magdalenian and Epipalaeolithic sites. If so, we will be able to examine the significance of the presence of two different types of projectile elements, backed points and pointed blades, within the same industrial assemblage.

3.7. BACKED BLADELETS

The functional analysis of the backed bladelets includes the study of 92 tools, of which 22 are from the site of Berniollo, 34 from the Azilian level and 36 from the Magdalenian level of Santa Catalina.

In these results for the 92 backed bladelets studied, the high proportion of tools which have no use-traces is striking, *circa* 50%.

Within the assemblage of tools having traces of use is a group of bladelets used for cutting meat and butchering (61.2%). On some of these (16,3%) are traces which indicate with certainty a butchering activity, because they include traces of contact of the tool with cartilage and bone.

In another assemblage of backed bladelets (44.9%) more ambiguous traces are present, also identified as those of butchering. In this case, the level of certainty in the functional interpretation is lower.

The second most common group of backed bladelets, is those with no trace at all. It is very probable that some of the backed bladelets used for cutting meat or fresh hide did not develop recognizable traces. Because of this, the assemblage of bladelets without traces could form part of the same functional group as the backed bladelets with traces of butchering.

However, if we compare the measurements of length and width of the backed bladelets with traces of butchering to those which have no traces, we observe that in the three assemblages the backed bladelets with traces are larger (Table 3.9.). It is probable that this difference in size indicates that, even if some of the backed bladelets without traces could have been butchering knives, the majority of them belonged to a different functional group (figure 3.12.). This is clear for the two levels at Santa Catalina and to a lesser extent in Berniollo.

Table 3.9. Length and width of the backed bladelets.

	Berniollo	S.C. Azilian	S.C. Magdalenian
Butchering traces	22.7x5.8mm.	21x7.6mm.	40x6.5mm.
Without traces	19x6.3mm.	14.8x5.1mm.	23.6x4.6mm.

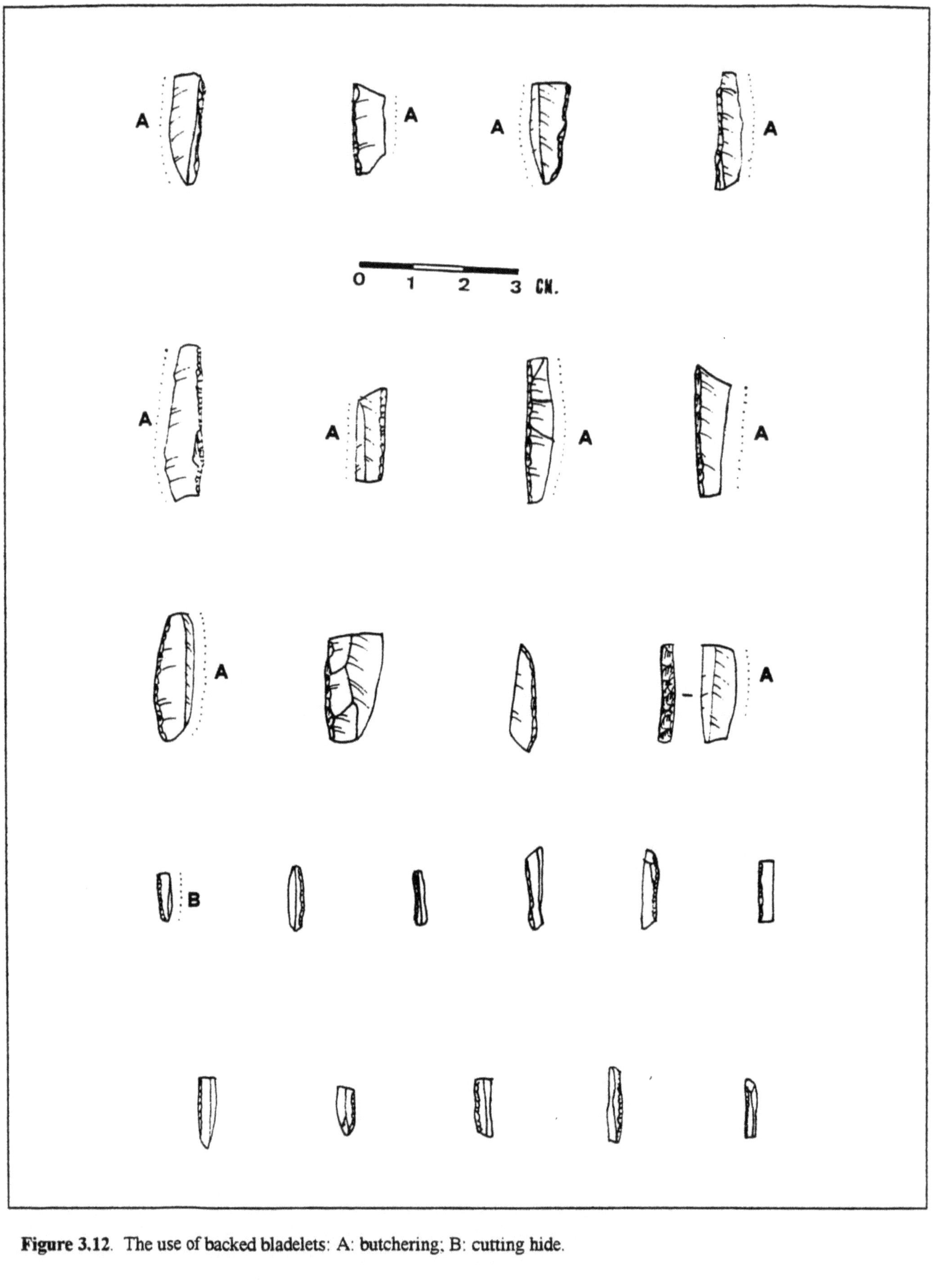

Figure 3.12. The use of backed bladelets: A: butchering; B: cutting hide.

When it comes to identifying the function of most of the bladelets without traces, the systematic transformation of a material other than meat or perhaps fresh hide has to be rejected, as this produces identifiable traces. However, if we reject the possibility that the assemblage of bladelets without traces could have been used in butchering, the only activity that could be attributed to this group is that of projectile elements, since this type of activity rarely produces traces.

Other information could back up the fact that some of the backed bladelets were used as projectile elements. The particular distribution of the tools in the site of Berniollo (González, 1993) shows a strong association between the bladelets and the backed points, whose use as projectiles has been established. This association is especially strong with small backed bladelets. This would suggest that a number of the backed bladelets, especially the small ones, could have had a complementary function to that of the points, forming part of the projectiles as barbs. This hypothesis suggests that if the points and the bladelets formed part of the projectile, then both groups would have been abandoned in the same zones.

On the other hand, the problem acquires a higher degree of complexity if we take into account that the line which separates the backed bladelets used as butchering knives and the backed bladelets possibly used as projectile elements (barbs parallel to the shaft) was more vague than we have indicated above. Therefore, in some cases projectile elements could have been used as knives (Moss, 1983, 139; 1987, 173-175; Caspar, 1985).

According to the data which we have, some backed bladelets were used with certainty as elements in composite-knives, the bladelets being fitted into shafts either alone or in groups. Moreover, it is probable that another group were used as barbs on projectiles. The use of a part of the assemblage of backed bladelets as knives and of another part as barbs is mentioned for some Upper Palaeolithic and Epipalaeolithic sites, such as Pincevent (Moss and Newcomer, 1982, 292), Bergumermeer (Odell, 1977, 262) and Verberie (Symens, 1986). The larger backed bladelets would tend to have been used as knives, while the smaller pieces were set into projectile shafts.

Moreover, the backed tools were occasionally used for a variety of functions. Three were used for boring, the material worked being identified as wood in two cases, while in the other case it could only be determined that it had bored a hard material. At least five tools were used with a transverse motion on various materials (fresh hide, wood, ...).

Thus, the larger backed bladelets comprised knives, with which a wide range of tasks were carried out, especially working soft animal materials. This type of use would be similar to that practised among the Selk'nam with a type of hafted knife, which though used for a wide range of tasks, is especially meant for cutting meat and soft leather (Gusinde, 1932/1982, 228).

Worthy of separate mention are three tools which were classified as backed bladelets, but which show clear traces of an original use as projectile points. In our experiments we have verified that in some cases the fractures of impact which the backed points suffer lead to the loss of the tip, so that the points acquire the form of backed bladelets. In these cases, the impact traces on one extremity of the tool reveals that it was originally a point.

3.8. OTHER TYPES OF RETOUCHED TOOLS.

In this section we will consider those tools which are little represented in the assemblages studied.

We have studied five tools with denticulated retouch which had various functions. The retouched zone was used to scrape bone materials in two cases. One blade fragment was used for butchering tasks, and the two others for working hide.

On none of the five tools with notches which we have studied is the notch used as an active zone. We do not have the data to elucidate the function of the notched zones, which could perhaps be related to the prehension of the tool. As in the previous group, the activities carried out with these pieces are varied: butchering tasks, working of dry hide and bone.

Functional study has not led to a detailed definition of the use to which an assemblage of 8 denticulated bladelets could have been put. On six of them there are no use-traces, while on the others the traces are so faint that a conclusion cannot be made as to the activity or the material worked.

The traces on the five backed blades studied indicate that these tools were used on soft animal material. Of the three tools whose use can be reconstructed with certainty, one was used for butchering and the other two for working hide.

As regards the truncated bladelets, one of them served as a projectile point, another for cutting a soft material, while in another case the truncated end served as the active zone, to scrape an unknown hard material.

On a Dufour type bladelet we found traces which indicate that it was used for cutting hide. The one microburin which was analysed had no use-traces.

3.10. THE UNRETOUCHED TOOLS

3.9.1. Unretouched flakes.

In this study we analysed 108 flakes. Of these, 29 are from the Azilian level of Santa Catalina, 39 from the Magdalenian level and 64 from Berniollo. At Berniollo, 25 were type 1 flint, 27 type 2 and 11 type 3 [3]

As concerns the criteria for choosing flakes for use at Berniollo and Santa Catalina, we observe that flakes were not selected randomly. In Berniollo and in the Magdalenian level of Santa Catalina there was certainly a selection of larger blanks, as the average size of the flakes with traces is larger than those which do not present use-traces. The measurements of the flakes from the Azilian level at Santa Catalina reflect a less-restricted choice in regard to the length of the pieces.

The analysis of the intensity of use of unretouched flakes shows that Berniollo was the site where flakes were the least used, although, when one of these tools was chosen for an activity, it was exploited with great intensity, demonstrated by the high index of active zones. In the other two sites, flakes were used with more regularity, although they were used less intensively than in Berniollo.

Taking the three occupations as whole, the flakes were used for a wide range of activities, among which are scraping, cutting, boring, engraving and chopping, which are carried out on different materials. Among these, scraping actions are dominant. The second activity carried out most often was cutting, while other activities were much less represented.

For the material worked, the results obtained differ from one assemblage to another. In the Magdalenian level at Santa Catalina, the flakes were used on the materials most worked by all tools in the occupation, bone materials, for which nearly half of the active zones were employed; a high percentage were used on hide and lesser quantities on meat and wood.

In the Azilian, materials most worked at Santa Catalina were fresh hide and meat, while work of bone materials took second place. Neither wood nor any other plant material was modified using these tools.

In Berniollo there is a sufficiently clear relationship between retouched flakes and the material worked to affirm that the unretouched flakes were preferentially used to work wood. Although a certain number of active zones were applied to antler and bone, the working of wood is represented by 65% of the used edges for which the worked material could be determined.

3.9.2 **The unretouched bladelets**

A total of 52 unretouched bladelets were examined. 15 of these are from the site of Berniollo, another 14 from the Magdalenian level and 23 from the Azilian level at Santa Catalina.

The unretouched bladelets were little used in the three assemblages, only in 11% of cases. In the three assemblages the criterion for selection of bladelets available was size. The larger bladelets were chosen, whose size is borderline between blades and bladelets.

In the Magdalenian level of Santa Catalina, three of the four tools were used for cutting meat (in one case, either meat or hide), the other bladelet was used to briefly scrape antler or bone. It is probable that the angle of the active edge influenced the choice of each particular bladelet as those for cutting have a narrow range of about 40 degrees, while those used for scraping have angles of about 80 degrees.

In Berniollo the natural pointed end of two of the bladelets was used, one to bore wood and the other as a projectile point. The unretouched pointed bladelet used as a projectile has a morphology and size very similar to the backed points on the site.

Only in the Azilian level of Santa Catalina was a bladelet among those studied used on an undetermined material.

The low proportion of unretouched bladelets which were used and the variety of activities carried out with these tools leads us to believe that these pieces were not systematically, but only occasionally used as tools. The functional meaning of these pieces was more as blanks with which to make bladelets and backed points through retouch. However, occasionally they were used without shaping for a wide range of tasks, which were probably not very specialized and did not involve intensive use.

3.9.3 **The unretouched blades**

There are a total of 99 unretouched blades analysed, of which 35 are from Berniollo, 20 from the Azilian level of Santa Catalina and 44 from the Magdalenian level. Of those from Berniollo 8 were made from flint 1, 17 in flint 2 and 10 in flint 3.

[3] One of the flakes was not attributed to any of the three types of flint.

Among the selection criteria for the blades to be used is size. In Berniollo it is clear that those of greater size were chosen, as in the two levels at Santa Catalina, although in these the degree of selection is less marked.

The analysis of the active edge angles on blades that were chosen for use establishes that there is a bimodal distribution: the majority of the edges used are between 20 and 50 degrees, with the smaller mode representing edge angles greater than 80 degrees. This last group corresponds to edges created by use of breaks on blades.

This bimodality in the distributions originates principally in the type of actions carried out with the edges. Longitudinal actions are carried out with acute angles, while transverse actions are made using wider angles, although depending upon the material they may be also practised with acute angles.

It must be emphasized that although a bimodal distribution is appreciable in the three assemblages, it is less pronounced in the Azilian assemblage. We interpret this fact as a reflection of less care in the selection of the active zones to be used, since the edges chosen are distributed more evenly within the range of potential angles.

A definite relation exists between the fact that a blade was fractured and its utilisation. In Berniollo the whole blades were used in a limited proportion (17%), whereas this proportion rises when the blade has a fracture (47%), and even more when its two extremities are fractured (75%). However, this does not occur in the two levels of Santa Catalina.

Therefore, the data from Berniollo indicate that whole blades tended to be kept without being used, blades already fractured were chosen or a whole blade was fractured, then used.

In regard to the degree of use of the blades, each assemblage studied presents certain peculiarities. The Azilian assemblage of Santa Catalina stands out, with more than 50% of blades used. The blades from Berniollo were used in a lesser proportion, but when one of these blanks was chosen for use it was exploited more than in the other two assemblages, which reminds us of the intensive use made of the flakes on this site. In the Magdalenian occupation of Santa Catalina only 35% of the blades were used, with an intensity similar to that of the Azilian assemblage.

If we analyse the angles of the active edges corresponding to the whole assemblage of unretouched flakes and blades, we observe that the distribution of the angles is again bimodal: the greater mode situated between 25 and 40 degrees and another between 70 and 85 degrees, in Berniollo and the Azilian level, and between 85 and 100 degrees in the Magdalenian (Graphic 3.3.). In the Azilian level, the bimodality is less pronounced.

Graphic 3.3. The edge angle of the active zones of unretouched tools.

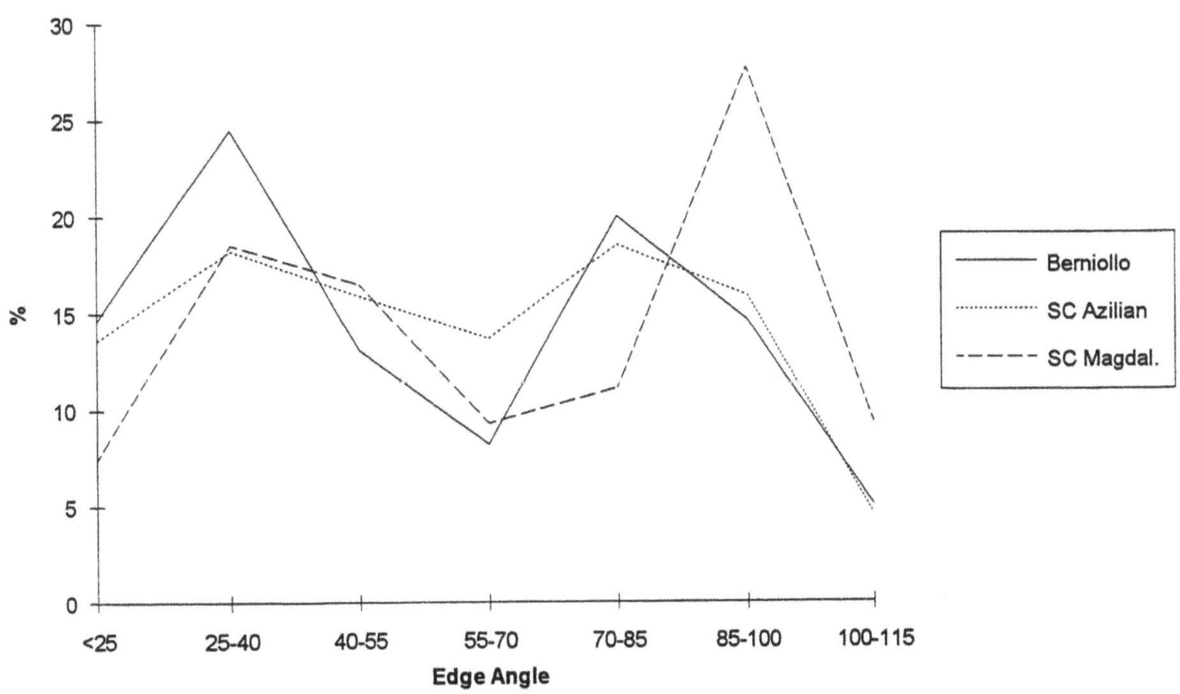

At Berniollo, unretouched blades made from flint 2 were used intensively, those made from type 1 less, and those from type 3 were hardly used.

Concerning the activities carried out with unretouched blades and flakes in Berniollo and Santa Catalina, these tools were used as much for cutting as for scraping. However, cutting was principally carried out with blades and scraping with flakes. In regard to the material worked, the unretouched blades were used with a certain intensity to cut soft materials, mainly meat and hide. The use of blades for cutting meat and hide have been observed in many Upper Palaeolithic and Epipalaeolithic sites, such as Meer (Van Noten *et al.*, 1978, 88; Cahen *et al.*, 1980, 244), Pincevent (Plisson, 1985, 173-176; Plisson, 1987, 76), Verberie (Symens, 1986, 215), Paglicci (Donahue, 1985, 40), Andernach (Plisson, 1985, 256 and 272) and La Tourasse (Plisson, 1985, 313).

In the Azilian level of Santa Catalina about two thirds of the blades with traces were used for butchering tasks. The remaining tasks carried out with blades were practised less: scraping fresh hide, scraping antler, plants, etc. For these tasks high edge angles produced by fractures were used .

In the Magdalenian level butchering and meat-cutting were the most common uses of blades. However, the unretouched blades were used in this level for more varied tasks than in the Azilian. 25% of the active zones have traces from scraping bone material; also present are traces of working wood, dry hide and probably fresh hide. For working on antler and bone, the edges produced by fracture were always chosen, with angles higher than 80 degrees. On the other hand, the soft animal materials were cut using acute angles, corresponding to unretouched edges of blades.

In the occupation of Berniollo the use found for unretouched blades is different. Although butchering and meat-cutting is present in a high proportion, here wood-working is the most frequent use of blades, for cutting and for scraping in equal proportion.

3.9.4. **The cores.**

We have analysed 16 cores, of which 8 from Berniollo, 2 from the Azilian level of Santa Catalina and 6 from the Magdalenian level.

The morphological distinction between cores and retouched tools, is difficult in certain cases, since for some small pieces it is not easy to discern whether they were bladelet cores or authentic tools. Thus, the morphology of some pieces can correspond equally to bladelet cores and to burins or carinated end-scrapers.

Of the 16 cores studied, use-traces were found on only four, which indicates that the use of cores as tools was somewhat sporadic. One core was used for working wood, another two for hide and the last for working a bone material. Three of the cores with use-traces have an ambiguous morphology, which could be interpreted as that of a core, even of a burin.

The fact that the cores were very rarely used is reinforced by the fact that the four core resharpening tablets (*tablettes de réaffutage*) which we have analysed do not have use traces.

CHAPTER 4

GENERAL CHARACTERISTICS OF THE UTILISATION OF LITHIC TOOLS

The technology of hunter-gatherer peoples has been generally described as simple, expedient and transportable (Lee, 1979, 119; Keeley, 1983, 252).

In regard to lithic tools, the fulfilling of economic needs entails the taking of a long chain of decisions which ends with the discard of the tool. This chain includes the procurement of the lithic raw material, the making of the tool, its use and maintenance. In this way a complex cycle of production and use is established, which has been dealt with by some authors (Schiffer 1976, 46; Collins 1975; Driskell, 1986; Binder and Perlès, 1990; Perlès, 1992; González, 1993). This cycle is a complex process, made up of a chain of coordinated decisions with a purpose, such as the achievement of certain types of work using lithic tools.

When the lithic raw material is procured, it is followed by the manufacture of blanks. Once the flakes and blades are obtained by means of knapping, the exploitation of the tools begins. This process may consist of a first stage of secondary shaping, in which the blank can be fractured, retouched or hafted. Of course, these options are not exclusive, but usually go together.

Afterward a logistic phase may occur, in which the pieces may be stored or transported to another place. These operations may be complementary and recurring, generating complex cycles of storage/transport.

The third stage of the process concerns the use of the object, which may be applied to different tasks, activities and worked materials, and be subject to different intensities of use and degrees of resharpening. Finally, the tool is discarded.

Not all the knapped blanks necessarily have to pass through each of the mentioned stages. The flakes or blades may be discarded directly, when they are abandoned in knapping waste heaps, or may be used unretouched or stored or transported to another place. These operations may also take place in reverse order and a discarded tool may return to use, which is a recycling of the piece, or a worn implement may be stored or transported with the intention of reusing it.

This process can vary within the same community according to the characteristics of the implements, thus establishing a "differential usage of the tools". The features which justify a differential usage can be encountered in each step in the process of manufacture. In the first stage different raw materials may be selected, which may give rise to a "differential usage of raw materials"; in the second stage flakes, blades and bladelets are created, establishing a "differential usage of blanks"; and in the secondary shaping there may be established a differential use of the retouched and unretouched tools, and a "differential usage of morphotypes".

The reconstruction of the cycle of manufacture and use of stone tools in prehistoric communities must involve a wide range of different approaches. Studies concerning the location and description of lithic raw materials and the technological analyses concern the first two stages. The exploitation of the tools is described through typological and functional approaches and by the spatial distribution of the lithic remains within the settlement (González, 1993).

Our functional study has a bearing on the final part of this chain of decisions which makes up the cycle of production and use, that is, in the stages of utilisation and maintenance of the tools. Functional study has contributed certain facts in this respect. The information which we have concerning the assemblages of lithic tools from Berniollo and Santa Catalina is expressed in:

-the intensity of use
-the degree of resharpening of the tool
-the hafting
-the type of work carried out

Moreover, we have information concerning the knapping techniques in Berniollo and the spatial distribution of the lithic remains on this site (González, 1993).

We are aware that a precise and detailed examination of the cycle of manufacture and use of lithic tools in the contexts studied should be carried out through an interdisciplinary study. However, we consider that a functional approach provides lines of explanation for the whole process, offering hypotheses for conducting future investigations, especially as we have access to a larger body of information for the site of Berniollo.

The first characteristic which defines the cycle of manufacture and use of lithic tools on the sites studied concerns the limited importance of the actions of re-use and recycling. A good part of the tools which we have analysed presented traces of more than one activity, although in almost all cases it was clear that the active zones of a tool had worked on the same material, a factor which has already been pointed out in other functional studies (Mansur-Francomme, 1983; Celerier and Moss, 1983, 99; Coqueugniot, 1983, 167). Consequently, it is probable that the different active zones of a tool were employed in the same task, acting in a complementary way. If there had existed a significant operation of re-use and recycling of the tool, as occurs in the Neolithic periods (Gassin, 1993), there would be more associations of work on different materials on the same tool.
Moreover, we have been able to verify the existence of a number of connections between the morphology and the use to which the implements were put. Thus the front of the end-scrapers was employed for scraping. Furthermore, these objects were preferentially used for working hide, although the degree of functional specificity of the end-scrapers varies from one assemblage to another.

The burins are clearly associated with the working of bone material, principally the transformation of antler. The dihedral and the trihedrals of the bevel of the burin were used to incise, while the facets were used for scraping.

The backed points and probably some of the backed bladelets were hafted as projectiles for hunting. A significant proportion of the backed bladelets, some of them pointed, were hafted as knives to cut meat or hide.

Certain long blades are truncated on one end, possibly hafted, and used for butchering or to work hide. Other truncated blades have active zones with an edge angle of nearly 90°, used for scraping hard materials.

Together with these morphological groups which have a relatively specific function, we encounter other types of pieces which are not intended for a given task. This is true of the denticulated pieces, the notched pieces and the retouched flakes and blades. Unretouched blades, although they were principally related to butchering, were also used for other activities. Unretouched flakes were used for different activities, except in the case of Berniollo, where the unretouched tools were primarily used to work wood.

The relation between form and function is not only linked to the general morphology of the tool, but also to the features of the active zone of the implement. Among these features the angle of the edge stands out in importance. We have been able to prove that this variable relates as much to the type of activity as to the material worked.

Longitudinal motions tend to be carried out with more acute edge angles than transverse motions. Furthermore, the softer materials, hide and meat, are worked with more acute edge angles than the hard materials, wood, antler and bone.

This correlation between the edge angle and the activity which is carried out with the active zone has been demonstrated for other archaeological assemblages, such as that of the site of Cassegros (Vaughan, 1985) or Verberie (Symens, 1986).

At Cassegros certain form/function associations were found, such as the use of end-scrapers for working hide and to a lesser extent, wood, but in general each typological group was used on different kinds of material (Vaughan, 1985, 89). This type of result is common to all functional studies of the assemblages of flint implements from the sites which date to the periods with which we are dealing (Moss, 1983; Vaughan, 1985; Plisson, 1985, etc.)

According to the data from the assemblages which we have analysed, despite the fact that there certainly exists a relation between form and function, such a relation is not strict. It is manifested in the fact that the same type of tool can be employed for different tasks and, at the same time, one type of work may be carried out with implements having different features (Kamminga, 1982).

The fact that several solutions may apply to the same problem or requirement implies that the set of factors which determine the choice of a tool for use is diverse and complex (González, 1933).

The features which define a tool are the lithic material of which it is made, the blank worked, whether flake, bladelet or blade, whether this particular piece is modified through retouch and, in the case of secondary shaping, the morphotype which may be created. These are the elements which the prehistoric maker is going to combine at the time of choosing a specific tool for a determined task, that is to say, in establishing the strategy for the exploitation of the tool.

At the site of Berniollo it has been possible to distinguish three main categories of lithic raw material. What are the factors which determine the differential use of the tools made from the three types of flint?

The choice of blank offers another field for investigation. The knapping of flint in the chronological period which we are studying is predominantly directed towards the acquisition of blade blanks. Nevertheless, during the knapping process plenty of flakes are produced, many of them being used later. Are there differences in the use of flakes and blades? If this is so, what criteria are used to choose one or the other?

We know that retouch can be used to condition an active zone, to facilitate hafting or the holding of the piece or it can be the result of resharpening a natural edge already worn out by use. Nevertheless, why, up to a point, are given activities carried out with retouched pieces or with unretouched pieces?

In the type of factors which we have chosen to define the use strategies of lithic tools, we have taken into account only economic ones. Thus, we will explain these procedures in purely economic terms. Certainly the economic factor cannot be disassociated from the rest of the cultural factors. The actions channeled into group subsistence are going to be conditioned by the type of social relations and the beliefs of the community studied. Thus variations have been demonstrated in the morphology of tools which concern ideological aspects. (Hodder, 1982). At the same time there has evolved a series of studies directed towards recognizing the style, the "particular way of making" which characterizes individuals or communities (Shenan, 1989; Ploux, 1990; Apellániz, 1991).

However, access to this information by means of the tools poses obvious difficulties of subjectivity (Clarke, 1968, 356). Moreover, the symbolic component of objects intended to satisfy daily needs, such as lithic tools, is always limited in comparison to other objects of the material culture.

This brings us, in the present state of theoretical-methodological development, to turn to an argument based on economic thinking. In studies such as the one we present, only the significant repetition of behaviour which cannot be explained in a purely economic manner can provide the possibility of applying explanatory hypotheses on levels beyond that of physical subsistence.

Therefore, the factors with which we will approach the economic mechanisms will concern the cost of manufacture and the benefit obtained. Both factors are found to be closely related, since in a rational economic mechanism, a balance would be created between them, so that the degree of effort invested has its correlate in the degree of benefit obtained.

The cost of manufacture is defined by the time and the effort invested in the procurement of lithic raw material, in knapping, retouch and, if such is the case, in hafting and resharpening of the tool. Furthermore, to these factors are added those concerning the overall activity context where the tool is made.

This last factor emphasizes that the value of time and the effort invested in the production ot the tool is not an absolute parameter. In fact, in contexts where the availability of time is greater than in others, the cost of manufacture is less, as long as the rest of the parameters are constant. In contexts where the need arises to achieve a determined activity, the main effort of the individual or the group will be directed towards satisfying that need, the carrying out of work which at that moment is considered secondary being put aside. Therefore, the cost of the completion of a task depends partly upon the importance which is attributed to such a task in relation to the rest of the work which is carried out in the same context.

In this sense it has been shown how among the Nuniamut tools are transported to the hunting posts already made, so that they could be used when the animal appeared. Therefore, the primary work in the manufacture of the implements does not appear in the contexts of hunting (Binford, 1980).

The benefit in the use is reflected in the efficiency of the tool, in the useful duration of the implement and in its adaptation to a given context of use.

Regarding the efficiency of the tool, Torrence (1989) states that the type of technology which is applied to a particular task is related to the extent to which the group depends on the success of a particular task and to the degree of difficulty which the execution of the work entails. The more necessary a task is for the subsistence of a group and the more risk of failure in its achievement exists, the more complex and specialized will be the technology needed for its implementation. In this way it is attempted to reduce the risk of failure, by increasing the efficiency of the tools. The search for efficiency in the tool can be seen in the degree of association between form and function, as the close relation between these two variables reflects a greater specialization of the tools (Torrence, 1986, 43).

The useful duration of use of the tool depends upon its resistence to wear and the possibilities of reconditioning the tool. We suppose that when the need arises to carry out prolonged and intense work the tendency will be to choose tools which last longer.

The context of use pertains to the place and the point in time in which the implement is used. An assemblage of tools may be intended to accomplish a task within a definite space and time, or to carry out work intermittently through time and in different places. In the latter case the technology developed will be more complex and specialized than in the former (Torrence, 1989).

When the work is done intermittently in different places, the need arises for portable and versatile tools, constantly ready for use. For this recourse is taken to standard forms, mainly blades, in which the relationship weight/length of the active edge is very favourable. Moreover standardized pieces allow easy transformation of the tools, so that they may be adapted into new forms when the need arises (Parry and Kelly, 1987).

A measure which can facilitate the acquisition of standard forms is the choice of high-quality raw materials, which allows for more precision and control in the production of blanks with ample possibilities of maintenance and hafting (Goodyear, 1989, 29; Edmonds, 1987, 163-165). Furthermore these tools can be resharpened frequently and thus be constantly available for use.

Having these portable and versatile tools available allows the prehistoric individual to carry out a task when necessary, independently of whether there is a supply of lithic raw material in the area to make a tool.

In this sense the tendency has already been demonstrated that highly mobile groups of hunter-gatherers have more elaborate tools, while sedentary communities usually have more expedient ones (Parry and Kelly, 1987).

L.R. Binford (1979) has grouped use-strategies for lithic tools into two categories, expedient tools and curated tools. The former are tools of limited cost of production, which are made and used in the same context; they are abandoned when the work is finished. Curated tools are more costly to make. They are not made for immediate use, but are intended to be used in different contexts over a long period of time, being transported from one settlement to another. This type of implement is usually produced from a high-quality raw material and tends to be hafted more often than expedient tools (Keeley, 1982; Hayden, 1989).

Although this classification is of unquestionable interest, we consider that these two categories do not exhaust the number of possible use-strategies of lithic tools found in archaeological contexts (Odell, 1996; Nash, 1996).

Moreover, it is probable that the archaeological assemblages with which we are confronted combine different use-strategies (Ebert, 1979; Nelson, 1991, 65). This is due to the fact that each tool, depending upon the characteristics of the task to be carried out, has a cost/benefit relationship adapted to each particular case (Binford, 1986).

In the ethnographic record this complexity of operation can be found in the example presented by L.R. Binford (1986) on the making of knives by the Alyawara in Australia. The knives used by the men are intended for numerous activities and contexts of use, and are carefully made at the lithic outcrop itself with a careful technique. On the other hand, for the manufacture of the knives used by the women, blocks of lithic material are brought to the settlement and are knapped when the necessity arises, the pieces obtained being discarded once the task is finished. In this case the same group applies curated and expedient use strategies for each type of tool.

Following the argument which we have presented in attempting to understand the characteristics of use of lithic tools, we think that the cost of making the tools in high-quality raw materials, whether scarce or distant, is greater than that of making them from materials which are abundant or nearby. The same can be applied to blades as compared to flakes, to retouched tools as compared to unretouched tools, and to those hafted and more elaborated as compared to those which are made simply.

Applying a rational economic mechanism, we deduce that the tools with the greatest effort invested in their production are expected to provide the greatest benefit when used: this concerns the efficiency of the work, the possibilities of extending the use, or the context in which the tools are going to be used.

The degree to which the factor of efficiency was important at the moment of choosing the type of tool will be reflected in the degree of association between the characteristics of the active zone of the overall tool and of its use. The greater the relation between form and function the more intention there was to adapt an active zone to a given task, aiming for efficiency in the work.

Making tools for prolonged use will be reflected in the intensity of use and in the degree of resharpening.

However, it must be taken into account that there does not have to exist a direct relation between the intensity of use and the degree of resharpening. Both the resharpening as well as the degree of wear of the working edge of the tool influence factors such as the effectiveness which one wants to give to the tool or whether it is going to be used several times again. Given an equal degree of wear of the working edge, an implement of which high effectiveness is demanded will be resharpened more intensively than one which is used for work which is less precise or important. Furthermore, if a tool is going to be stored for future use it is probable that it will be resharpened more frequently in the last stages of work than if it is going to be discarded, so that it is left in working condition for the next time of use.

The third type of utilisation benefit sought, designing tools for definite contexts of use, is more difficult to detect. Such a characteristic will be deduced from assemblages with a high cost of production which cannot be explained by criteria of efficiency for use or resistence to wear. These tools will have to present characteristics of portability and versatility.

4.1. DIFFERENTIAL USE OF THE RAW MATERIALS AT BERNIOLLO

In the two levels of the site of Santa Catalina a single variety of flint was used, black and fine-grained, so that these assemblages cannot be studied from the point of view of differential use of raw materials.

As we have previously explained, the lithic tools of Berniollo were mostly made from three different types of flint. Type 1 is black, type 2 is brown and type 3, also brown, presents a series of concentric streaks, called Lesengang rings. Type 3 flint is the least frequent on the site, while the other two types are relatively abundant.

Without a detailed study, we cannot define the exact origin of these lithic raw materials. We have located a flint outcrop near the site, in one of the cliffs which forms the river Bayas canyon, at the end of which is situated the site of Berniollo (Elorza, pers. com.). This is probably the place of origin of type 1 flint. Elsewhere, in the outcrops of Albardón, found in the Rojo river basin, 25 kilometers in a straight line from Berniollo, there are flint nodules which have characteristics definitely similar to types 2 and 3 (Lobo, 1986; Ortiz *et al.*, 1990).

The study of the knapping techniques (González, 1993), demonstrates the existence of different reduction models for flint. A clear association has been detected between the knapping models and the different types of flint.

Thus, flint 1 cores are unidirectional, with one platform and without rejuvenation elements, while for the knapping of types 2 and 3 complementary elements are used which facilitate the rejuvenation of the cores, such as opposing platforms and making of dorsal crests and platforms. Furthermore, flint 3 nodules have flat sides, which favours blade knapping, resulting in cores with narrow flaking surfaces and parallel sides. .Both blades and bladelets are made from flints 1 and 2, while flint 3 was used exclusively for making blades.

These systems of flint reduction have produced some blade blanks of different qualities. The flint 1 blades, produced by the least complex systems are wider and shorter than the flint 2 pieces. However, the flint 3 blades are those of the best quality, the most elongated and regular (González, 1993).

If we analyse the types of tools made from each group of blanks we observe that the retouched tools produced in flint 3 were made mainly on blades, while the retouched pieces in other types of flint were principally made on flakes .This can be seen as much in the end-scrapers as in the burins. The blade blanks from flint 3 are those most often retouched. Moreover, a marked high quality is observed in the flint 3 retouched pieces. The end-scrapers are very standardized and regular and the burins are large compared to the rest of the pieces.

According to what we have seen as far as the end-scrapers are concerned, the pieces in flint 3 show a higher incidence of fracture in the distal zone, a lower proportion of use traces in distal and lateral areas, standardization in size and are usually made on blades. All these factors indicate, probably, that the end-scrapers made from this material were more frequently hafted than the others.

At the site of Verberie there is also a tendency for the tools made from certain types of flint to be more frequently hafted. 62% of the tools in exotic flint were hafted, as opposed to 20% of the tools in local flint (Keeley, 1987).

All this implies that the cost of manufacture of the tools varied according to the type of raw material. For type 1, concepts of less costly manufacture were applied and blade products of lower quality were obtained. At the opposite extreme, a costly manufacturing process and more intensive and careful retouch were used for type 3, and they were preferentially hafted. This greater effort in the manufacture resulted in tools of quality, on blade blanks and of standardized morphology. The tools in flint 2 are to be found in an intermediate position, in the effort invested in the manufacture, as well as in the quality of the implements obtained.

We should probably associate the higher cost of manufacture for flint 3 with the greater cost of acquisition of the raw material. This consideration is based upon two factors: that it was considered to be a material of quality and that it is rare at the site.

If the principles of interpretation which we are using are valid, we expect that greater costs of manufacture are associated with greater benefits of use: greater use-efficiency, greater intensity of use or use in a particular context. This must be valid for flint 2 in relation to flint 1 and to flint 3 in relation to the two others.

Table 4.1. Activities and materials worked with flint 1 tools.

	Cutting	Scraping	Others	Total
Wood	0	12 *(20.7%)*	3 *(5.2%)*	15 *(25.9%)*
Osseous material	2 *(3.8%)*	15 *(25.9%)*	1 *(1.7%)*	18 *(31%)*
Butchering	7 *(12%)*	0	0	7 *(12%)*
Hide	0	12 *(20.7%)*	0	12 *(20.7%)*
Plants	1 *(1.7%)*	2 *(3.8%)*	0	3 *(5.2%)*
Projectile point	0	0	3 *(5.2%)*	3 *(5.2%)*
Total	10 *(17.2%)*	41 *(70.7%)*	7 *(12%)*	58 *(100%)*

Table 4.2. Activities and materials worked with flint 2 tools

	Cutting	Scraping	Others	Total
Wood	7 *(6.2%)*	24 *(21.2%)*	1 *(0.9%)*	32 *(28.3%)*
Osseous material	2 *(1.8%)*	23 *(20.3%)*	13 *(11.5%)*	38 *(33.6%)*
Butchering	11 *(9.7%)*	0	0	11 *(9.7%)*
Hide	4 *(3.5%)*	24 *(21.2%)*	0	28 *(24.7%)*
Projectile point	0	0	4 *(3.5%)*	4 *(3.5%)*
Total	24 *(21.2%)*	71 *(62.8%)*	18 *(15.9%)*	113 *(100%)*

Table 4.3. Activities and material worked with flint 3 tools

	Cutting	Scraping	Others	Total
Wood	2 *(8.7%)*	5 *(21.7%)*	0	7 *(30.4%)*
Osseous material	0	7 *(30.4%)*	1 *(4.3%)*	8 *(34.8%)*
Hide	0	7 *(30.4%)*	0	7 *(30.4%)*
Projectile point	0	0	1 *(4.3%)*	1 *(4.3%)*
Total	2 *(8.7%)*	19 *(82.6%)*	2 *(8.7%)*	23 *(100%)*

In comparing the type of work carried out with the tools made from flint 1 (table 4.1) and flint 2 (table 4.2) we observe that the tasks carried out with both assemblages of tools are similar, as much in the activities as in the materials worked. There are no substantial differences in the intensity of use or the degree of resharpening of the tools in the two types of flint. Therefore, the manufacture of higher-cost tools cannot be justified by the intent to make implements which could be intensively reconditioned and reused. Functional study alone cannot bring more data to bear on this question. With the information we have, we do not think that there exists a differential usage of the two types of flint.

In the analysis of the flint 3 tools it becomes evident that the tasks carried out with these pieces (table 4.3) differ little from those carried out with the other types of flint.

If efficiency in transforming a material does not seem a clear factor in the choice of these tools, neither does the possibility of carrying out intensive work with them. In fact, the tools in flint 3 show less intensity of use than the others, which can be demonstrated in the case of the end-scrapers, the burins and the unretouched blades, the only assemblages which can be compared.

In this case the more elaborated tools were used less frequently than the rest. In other archaeological assemblages cases similar to this have been observed, as in the Mesolithic site of Meer (Van Noten *et al.*, 1978).

We suppose that if the context of use of these tools had been the settlement itself the tools of quality would have been used more frequently and intensively than those made from flint 1 or 2. As it is to the contrary, we deduce that the quality tools tended to be curated without being used, and were probably intended for a context of use different from that of the settlement itself.

On the other hand the study of the end-scrapers indicates that the pieces in flint 3 were resharpened more often than the others. Taking this into account it could be proposed that the low intensity of use found for tools made from this material could be more an appearance than a reality, caused by intensive resharpening of the active edges, which eliminated the traces of successive use.

However, we do not think this is the case. It should be taken into account that this lesser intensity of use has been observed as often in the retouched active zones, which can be resharpened, as in the unretouched zones, which reflect all the uses of the tools. Thus, lower intensity of use is also seen in the unretouched active zones of the end-scrapers and in the unretouched blades.

The association between low intensity of use and frequent resharpening could appear contradictory in the first analysis. Nevertheless, it should be taken into account, as we have previously pointed out, that among the factors which condition the degree of resharpening there is not only intensity of use, but also the desire to keep the tools in good working order, an aspect into which subjective factors enter concerning the value which the maker places on his tool.

Consequently, the greater resharpening of flint 3 would reflect the intent to have an effective tool, ready to be used at any moment.

It has been demonstrated repeatedly that one of the principal advantages which blade blanks offer over the less standardized pieces is their portability, that is to say, their high use-potential in relation to their weight. The length and the regularity of their working edges and the versatility of the blanks allows the making of any specific tool when the need arises (Hayden *et al.*, 1996, 37). Blade blanks make it possible to carry out urgent work without the constraint of having to look for lithic raw material, or of making a suitable tool.

Therefore, in situations in which it is necessary to have a portable assemblage of tools with which to carry out a series of planned activities, an assemblage of high-quality blanks would be made in a type of raw material which is easy to knap, retouch and resharpen (Goodyear, 1979, 29; Edmonds, 1987, 163-165; Jeske, 1989, 36; Myers, 1987).

In different ethnographic contexts it has been proven that curated tools tend to be made in high quality materials, to be hafted and frequently resharpened (Keeley, 1982; Shott, 1986; Hayden, 1989).

Different data support the hypothesis that the objects of flint 3 were subjected to more intensive storage and transport than the rest of the objects. In general, the number of blades per core is low on the site, especially for flint 3 (González, 1993, 221). Taking into account that the most careful knapping occurs in this material, we reject the possibility that this fact is due to a low production of blades per core. We are inclined to believe, on the contrary, that a part of the blade production in flint 3 was transported outside the site, in the form of blanks as well as in the form of retouched pieces (González and Ibáñez, 1991).

Moreover, a high proportion of unretouched blades in flint 3 were kept unused. In the sample of unretouched blades which we have analysed, those of flint 1 and 2 show traces of use in half the cases, while those of flint 3 in only 10% of the cases. Furthermore the blades of flint 3 are the least fractured, which can be related to the curation of unused blanks, as we have already proved that blades were systematically fractured before use (Bergman *et al.*, 1987).

Finally, it is noteworthy that in the analysis of the spatial distribution of the lithic remains in Berniollo we have encountered groups of unfractured flint 3 blades, which are found outside the knapping areas. This phenomenon is also true of the retouched blade tools in flint 3. These associations have to be interpreted as a result of storage of these blanks and tools (González, 1993, 395-397).

According to the interpretations which we have made in this section, at Berniollo, different manufacture and use strategies of tools in relation to the type of lithic raw material were used. In the case of flints 1 and 2 we do not encounter differential use of tools, therefore we assume that the duality of sources of supply had no effect on the planning of use-strategies.

Curated tools were made with flint 3, and were kept for contexts of use where transportable, versatile, and ready tools were needed. For this, a high-quality raw material was chosen, which was carefully knapped to obtain stylized, regular blades. These blanks were concentrated in specific areas, ready for use. Before use, they were shaped by retouch and some were hafted. These tools were transported preferentially, to carry out work which was not especially intensive, in different places and at different points in time. The curated tools were subject to more intensive maintenance by resharpening and some of the retouched tools were stored.

4.2. DIFFERENTIAL USE OF TOOLS ACCORDING TO THE TYPE OF BLANK

On the sites we are studying most of the cores were blade-cores. This does not imply that the functional possibilities of flakes were not exploited. Flakes were used both retouched and unretouched. Blades are more costly to produce than flakes, because they are more specialized products, whose manufacture requires greater technical investment. The difference between the manufacture cost of blades and flakes will vary from one context to another. In contexts where there is a shortage of raw material or it is of low quality for knapping, the difference in cost will be great. The same occurs in contexts where the commitment in time and effort to knapping implies the neglect of other work fundamental to the group. In these cases, we expect that distinct differences exist in the concepts of use of lithic tools with respect to the kind of blank from which they are made.

In the differential use of flakes and blades not only are the production costs involved, but also the characteristics of the tasks which will be carried out with the tools. Thus the need to carry out work for which blade tools are especially efficient, or the need to have curated tools available, are going to influence blades being used in a different way from flakes.

During the study we have seen that blade tools are usually found associated with certain tasks. The sides of blades, whether retouched or not, were used in larger proportion for cutting than flakes, retouched and unretouched. This corresponds to a clear adaptation of the form of the active zone to the function, in that the sides of the blade usually have straight elongated edges, with an acute edge angle, which are very suitable for longitudinal actions.

It may be seen that in the three assemblages studied one of the tasks which the unretouched blades usually carry out is butchering, as they have edges suitable for cutting, and are large and strong enough to be easily held to exert the necessary pressure to cut through cartilage and tendons.

Aside from this, blades are also used in transverse actions, usually using an intentional fracture. Another use of the blade blanks is for backed points and backed bladelets.

Given their advantages, it is easy to imagine that among the unretouched blanks of Berniollo blades are more often used than flakes. However, most of the blades with signs of use are fractured, while complete blades are rarely used. This indicates that when a blade was wanted either a fractured one was chosen or the piece was broken intentionally, while the complete blades were kept unused. It is reasonable to suppose that the complete blades, which do not present traces of use, were stored and used now and again when it was necessary to make various tools.

In Berniollo, among retouched tools, those made from flakes were used more intensively than those made from blades, as may be seen for end-scrapers and burins. However, the degree of resharpening of blade tools is greater.

The spatial distribution of the remains at Berniollo offers interesting complementary information. The flakes are found associated more with knapping areas, defined by the presence of cores, core rejuvenation flakes, etc.. The unretouched blades are most often found away from the knapping places. These pieces are concentrated in areas where the blanks are used and in small deposits of unused blades. Furthermore, there are concentrations of retouched pieces, which are mostly blade tools (González, 1993, 389).

Thus, in Berniollo there is a differential use of tools according to the blanks from which they are made. The blades, products of a higher manufacture cost, are used for tasks for which their straight, sharp edges are especially suitable. Moreover, these pieces involve a more complex planning strategy insuring that a tool is available at the time and place necessary. Therefore, according to the concept discussed previously, the first factor above concerns the efficiency of the tool and the second the context of its use.

In the Azilian level the proportion of used unretouched blades and flakes is very similar, but when the tool is shaped by retouch those made from blade blanks are used more intensively. In this case the tools with a higher cost of manufacture were used more intensively.

In the Magdalenian level, concerning the proportion of used unretouched blanks, the flakes and blades did not differ significantly. For the retouched tools neither the intensity of use, nor the range of materials worked nor the degree of resharpening varied appreciably according to the type of blank from which the tool was made.

These slight differences in the use of flakes and blades in the Magdalenian level are perhaps due to the fact that the cost of manufacture of the latter was reduced, so that the value of both types of blanks in this context was similar. This low manufacture cost of the blades is explained, by the abundance of raw material and the availability of time for producing them. The large quantity of cores and knapping waste present in the Magdalenian level supports such an argument.

4.3 THE USE OF RETOUCHED AND UNRETOUCHED TOOLS.

Studies of use traces on lithic tools, including this work, have demonstrated the importance of the use of unretouched lithic implements. They have confirmed the hypothesis that, based on different ethnographic observations, a large range of functional possibilities is offered by unretouched blanks (White and Thomas, 1972; Gould, 1977; 1980; Hayden, 1977).

Secondary modification by retouch can have different objectives. It can serve to shape an active zone by adapting it to the work to be done, as in the case of the end-scrapers and burins. Also, retouch may be carried out to facilitate hafting, as occurs for backed bladelets or for some types of truncated blades. Finally, retouch can be the result of the resharpening of a worn edge which was originally used without any retouch.

In any case, the secondary modification of the piece requires a technical effort which must be compensated for in the use of the tool. The benefit that is expected from the retouch of the tool concerns the efficiency of the work, as much in the case of the retouch intended to shape active zones as that which facilitates hafting, and that used to extend the period of use of the tool, since more resistant active zones are obtained, which can be easily resharpened.

From this it is inferred that retouched tools will be preferentially used for tasks which need particular forms and edge angles or when it is expected that the task is going to be intensive, accompanied by heavy wear on the tool. Thus, the use of the unretouched pieces is related to the execution of tasks which are not very intensive, specialized or urgent (Celerier and Moss, 1983).

Furthermore, longitudinal actions tend to be carried out with unretouched active zones, while the retouched zones are associated with transverse actions, as well as engraving and perforation. This is due, according to what we have proven in our experiments, to the fact that natural edges are more effective for cutting tasks than retouched ones.

According to our data, woodworking and butchering tasks are carried out primarily with unretouched tools. The association between unretouched tools and wood is indicated by the data from Berniollo, where this relationship is very marked. However the tendency for butchering tasks to be carried out with unretouched tools is marked in all the assemblages, which reflects the intention to use the most efficient tools for a task.

Antler and hide are preferentially worked with retouched tools. As we shall see later, these associations are probably due to the complexity of the work procedures which characterize the processes necessary to working these materials.

In spite of the evident association between the type of task and the fact that the tool is retouched or unretouched, we realize that there is a certain overlap. Thus, butchering tasks were carried out as much with unretouched blades as with truncated blades which were probably hafted. Hide is worked preferentially with end-scrapers but in certain cases unretouched pieces are also used. In fact, all the materials (bone, hide, etc...) were modified using retouched and unretouched tools.

The importance of the unretouched tools in the tasks carried out at Berniollo was significantly less than that of the unretouched tools in the Azilian level at Santa Catalina. The Magdalenian level occupies an intermediate position in this respect.

In the three assemblages the proportion of unretouched pieces to retouched pieces is similar (12 to 1 in Berniollo, 10 to 1 in the Azilian level, and 11.5 to 1 in the Magdalenian level), so the proportion of unretouched pieces with use traces reflects the importance of unretouched tools at the sites. This proportion indicates that the assemblage in which the unretouched tools were most significant was the Azilian (see chapter 3).

This indicates that in the Azilian level less specialized tools, having a lower manufacture cost, were used more frequently than in the other two assemblages.

The data referring to the degree of selection of the used unretouched blanks intended for use will confirm this interpretation, the tendency in the Azilian level being for less elaborate tools to be used. The differences in size of the used blanks and the unused blanks are very slight, for blades as well as for flakes, which indicates absence of a careful choice of blanks most suitable for a given task. Nor are criteria clearly defined concerning the choice of particular edge angles.

On the other hand, in Berniollo the used unretouched blanks are clearly larger than the unused blanks and more precise edge angles were chosen, for each type of task. In the Magdalenian level, the degree of selection of the tools was intermediate between the two other assemblages.

According to these data, in Berniollo, unretouched pieces tended to be chosen to be used according to better defined criteria than in the Azilian level, and tools intended for use were more often retouched.

The factors which influenced the use characteristics of lithic tools in each of the assemblages must be considered in the context of manufacture and use, as we shall see later.

4.4. THE CHOICE OF MORPHOTYPES FOR USE

Association is observed between form and function in the more well-defined morphotypes, like burins, end-scrapers or backed bladelets and backed points, as a logical result of adaptation of form to function.

However, there are many cases in which such an association does not occur. Burins are predominantly used in the working of bone material, despite the fact that some pieces are used to work hide. For end-scrapers the association between form and function is more subtle than for burins, at least in Berniollo and in the Magdalenian level of Santa Catalina. In these two assemblages a significant proportion of end-scrapers were employed on different materials, such as hide, antler, wood and bone.

A few of the backed bladelets, apart from those which do not have any trace or those used to cut soft animal materials, were used to cut bone or other hard materials.

It is striking that, for example, bone materials were scraped as much with end-scrapers as with lateral facets of burins; the same occurs in certain cases for hide.

We believe that factors related to the tasks to be carried out or the context of use can explain part of the variability in the degree of morphotype/function association.

The burin is a tool specially adapted to the working of bone because the bevel zone may be used for engraving, the lateral facets for scraping and some lateral edges for cutting.. Therefore, they are suitable tools with which to carry out tasks involving a series of different connected actions on bone material.

Therefore, when the task to be undertaken with bone material is complex, burins were chosen, as tools with which a wide range of activities may be undertaken, and which can be used intensively, since they have a high number of active zones and are easily resharpened.

However, when the work to be carried out on bone material does not involve engraving, the task may be accomplished with an end-scraper or a truncated tool, which are as effective for this work as the facet of a burin.

Moreover, for simple work on hide, which does not involve extended use, the facet of a burin can be sufficient to accomplish this. If when confronted with the need to carry out an urgent and not very intensive task on hide, the prehistoric individual had only a burin on hand rather than an end-scraper, it is probable that he used the burin, rather than make the effort of producing an end-scraper.

Thus, we consider that close associations between form and function will be more frequent in contexts where high-costly tasks have been carried out, while in places where sporadic and not very intensive tasks are undertaken the association between form and function will be less.

4.5. CHARACTERISTICS OF TOOL USE IN SANTA CATALINA AND BERNIOLLO.

The analysis of the characteristics of lithic tools in relation to their function shows a series of attributes which are common to each of the assemblages which we have analysed.

The implements of Berniollo were used with greater intensity than those from the other two assemblages, as shown by data from end-scrapers, burins, and unretouched flakes and blades. The tools from the Azilian level show the opposite tendency, as it is the assemblage with the lowest intensity of use.

The intensity of use is conditioned by the availability of the tools and by the nature of the work to be undertaken. When the stock of tools is low the tendency will be to use to the maximum the implements available. At the same time, if the work to be done is costly the tools will show a greater degree of use than if the tasks are sporadic and light.

We consider that in Berniollo the scarcity of raw material was not significant in determining the extent to which the tools would be used: knapping was frequent on the site and cores are not knapped to exhaustion, with some of them still capable of producing flakes and blades.

Moreover, a significant proportion of flakes and blades analysed did not have use traces. Also we have confirmed that the complete blades were scarcely used, forming a deposit of blanks which were not exploited.

Therefore we suppose that the greater intensity of use of tools at Berniollo is related to the fact that work on this site demanded greater time and effort than in the Azilian level of Santa Catalina. The degree of intensity of use in the Magdalenian level is greater than that in the Azilian level and less than that at Berniollo.

In general terms we observe for the site of Berniollo a high measure of planning in the process of manufacture and use of the tools. In this site the tendency was to employ tools with a greater cost of manufacture than in the other two assemblages, especially the Azilian level of Santa Catalina. Thus, work was carried out more often using retouched tools at Berniollo, whereas unretouched tools were used more often in the Azilian level.

Moreover, in Berniollo the unretouched tools which were intended for use were selected more carefully, as much for size as for the angles and the morphology of the active edges. The complete blades were kept without being used, probably forming an assemblage of blanks which could be used when necessary. In this context we have detected the existence of expedient tools as opposed to another assemblage of curated implements. Curated tools tend to be made in flint 3 rather than in the two other types of flint, and made from blade blanks rather than flake blanks, and more often retouched rather than unretouched.

At Berniollo, where knapping is plentiful, less specialized blanks were chosen for daily activities. Meanwhile, the tools of better quality tended to be kept for more specific contexts of use.

In the Azilian level the work undertaken with lithic tools was, generally, less intensive than in Berniollo, the tendency being to use the tools which cost less to make. The blade tools were considered to be quality tools as compared with the flake tools, because they were used more intensively and for more specialized work..

Data from the Magdalenian level of Santa Catalina are intermediate in regard to the intensity of use, the degree of selection of the blanks and the frequency with which unretouched tools were used. Otherwise, there are no appreciable differences between the use-strategies of the blade tools and those of the flake tools, which is probably related to the abundance of lithic raw material and blade knapping which took place on this site.

CHAPTER 5

THE TECHNICAL PROCESSES

5.1. THE TECHNICAL PROCESS AS AN INSTRUMENT OF INTERPRETATION

The first objective of our analysis is to understand the economic organisation of prehistoric human groups. This implies an understanding of the purpose and the methods of manufacture which were practiced, that is, what they used and how they made it.

Thus far this analysis of use traces on lithic tools has provided information on the materials worked, the way in which the tools transformed these materials and the relationship of these two factors to the characteristics of the tools.

Each one of these technical actions was channeled into the transformation of a specific material. This transformation may have been oriented towards obtaining food, making tools with which to undertake other work, or providing clothing and shelter. The data referring to the type of material worked allows us to establish a primary division between subsistence or consumption activities and those of manufacture. Thus, butchering activities fall into the first group, while the working of antler, bone, wood and hide would pertain to the manufacture of objects. The case of plants is more ambiguous, as the transformation of this material can correspond to both types of activity.

In general, the transformation processes of each material entail the completion of a complex series of inter-connected actions (Leroi-Gourhan, 1964, 164). Because of this, the actions inferred from functional analysis do not acquire their full significance if they are not related to the actions which precede them and follow them.

We define a technical process (*processus technique*) as a set of inter-related actions to make an object or to prepare a material for consumption (Lemonnier, 1976 and 1983). A technical process is composed of an ordered succession of technical sequences (*châines opératoires*) which bring the form and features of the raw material closer to those of the final product (Pelegrin *et al.*, 1988; Karlin *et al.*, 1991).

The characteristics of a technical process, in regard to its complexity and the particular practices which are carried out, is fundamentally dependent upon the nature and initial condition of the raw material, the final product desired, and the technical knowledge and the material means of transformation possessed by the group (Geneste, 1991).

However, in spite of this variability, we can establish a general diagram of a technical process, from the concepts of (1) acquisition (2) transformation and (3) use/consumption of the materials. This comprises the following phases or technical sequences.

1. The acquisition of the raw material to be transformed. The most obvious activity is the procurement of animal resources through hunting.

2. The transformation of the material can include different sub-phases.

2.1. Roughing-out is a first treatment, in which the parts of the raw material which will not be useful in the transformation process are eliminated. The existence and intensity of this phase depend upon the nature of the material. The objective of this technical sequence can be to eliminate surplus or cumbersome material before continuing manufacture, to eliminate excess of weight to more easily transport the material to another work place, to check the quality of the raw material before continuing work on it, or, as in the case of hide, to prepare the material in order to insure its preservation.

2.2 The initial shaping involves making a rough form. This is the work phase in which there usually appears a large number of technical alternatives. The two principal alternatives are the reduction of the initial block of raw material, or the extraction of a fragment from a larger block. In this case we may speak of a phase of acquisition of a blank.

2.3. The shaping involves the passage between the initial outline and the pre-form. It includes more careful work, with more gradual modifications and with less elimination of material.

2.4 The finishing covers a series of final operations, concerning modifications of the surface of the object. Among these we include the smoothing of the surface of the object, treatment with mineral, plant or animal products and "decoration", which may have esthetic, functional or symbolic significance.

2.5 The repair and maintenance of the objects are intended to recover their original condition when any alteration has occurred through use, storage or transport. In some cases the damage to the object is so severe that repair means in practice a new phase of manufacture for the tool. This type of work will be called recycling, reserving the term repair for modification which is not substantial.

3. In the phase of use or consumption the expected benefit is obtained, which justifies the effort which has been invested in the technical process undertaken. If the technical process has culminated in the making of a tool, the use phase of this implement will coincide with the procurement phase or transformation phase of the material for which the tool has been conceived.

Each one of these phases or technical sequences is composed of different technical actions (*gestes techniques*). These technical actions correspond to simple movements or individual actions which the worker makes.

The archaeological record allows detailed reconstruction of the technical processes of flint knapping, as this material is perfectly preserved. Thus even detailed descriptions of knapping process have been made on sites such as Pincevent, Etiolles or Verberie (Cahen *et al.* 1980; Pigeot, 1987, Karlin, 1992). In most cases this reconstruction is also possible for bone material, whereas technical processes for perishable materials, such as wood, plants or hide cannot be known from archaeological remains. These are understood by means of the more specialized field of analysis of use traces (Karlin *et al.*, 1991, 115).

In the technical process, the data offered by use-wear analysis, concerning the movement of the tool on the material worked, are identified with the technical actions within the technical process. If we wish to give economic significance to the results of the analysis of use traces we must be capable of integrating the known actions, technical actions, into the technical sequences and the latter into the technical processes of transformation of each material.

It is imperative, first of all, to establish the different technical processes which have been undertaken in the chronological-cultural context studied and the different technical sequences by which objects could be made. This phase must be accomplished through ethnographic observation of technical processes in communities of hunter-gatherers, through experiments, and through deduction based on archaeological evidence. Secondly it is necessary to define the technical actions and technical sequences which must be practiced to carry out each technical process.

This logical outline, to establish the possible operations and define the type of material remains which each operation produces, is valid for any study on the reconstruction of human behaviour from archaeological data. However, such an outline must adapt to the nature of the information provided by the studies of use traces.

The nature of the data from use-wear analysis imposes serious limitations on the reconstruction of the technical processes undertaken with lithic tools. In the first place, limitations are posed concerning the aim of the technical processes. Use-wear analysis provides little evidence on the final product which is expected to be obtained with a particular technical action. It only gives information on the motion of the tool, the material worked and the intensity of use. Thus, a scraping action on antler could have been directed as much towards making a harpoon as a needle. In the case of some materials, such as bone, the preservation of the objects allows a knowledge of the range of final products. Technological analysis of these objects and of manufacture waste leads to an understanding of the technical process. However, other materials, such as wood and hide, are not preserved in most of these contexts.

Secondly, some of the stages of the technical processes can be carried out with tools made in materials other than flint, such as wood, bone material, or non-siliceous stone. Moreover, for some activities no tool is needed.

These limitations are going to condition the type of reconstruction of the technical processes, which have to be generic and comparative. By generic we mean that the reconstructions which we will attempt will not concern the making of particular objects, information which we do not have, but rather the work on each material. The data from use traces on lithic tools will allow us to understand the degree of technical complexity of the work carried out on each material in the three assemblages. This technical complexity cannot be measured in absolute terms, but has to be expressed in terms of comparison between the different assemblages which are analysed. Because of this, the reconstruction of the technical processes must be comparative.

Technical processes, as for any human behaviour, present a double space-time dimension (Geneste, 1991). The work takes place in a given place during a time which can be more or less extended. However, the processes do not have to be continuous, but rather breaks in the work can occur, which can be temporal, when the work is postponed, or spatial, when the half-made object is transported to another area where the work is continued, or both at the same time, when the activity is postponed to be taken up later in another place. As we shall see later, the degree of continuity or discontinuity of technical processes has an important impact at the time of defining the economic strategy of the group, since it reflects the degree of planning of the tasks of procurement, transformation and use/consumption.

The space-time dimension of technical processes also has an effect on the possibilities for its reconstruction from the archaeological record. Our study deals with technical actions carried out at a given geographic point, within the territory occupied by the group, and at a given moment within the annual cycle. Therefore, the activities carried out on the sites which we are studying do not have to show all the elements of the technical system of the group.

We should be able to distinguish whether the degree of technical complexity of a site is due to the characteristics of the technical processes undertaken by the group or to the development alone, in this particular context, of certain phases of the technical processes and not of others. In other words, it will be necessary to discern whether a context showing little technical complexity may be explained by the making of simple tools or only by the carrying out of phases of maintenance of complex tools.

After considering the limitations of use-wear studies of lithic tools for the reconstruction of the technical processes in which the tools were used, let us consider the possibilities. The reconstruction of the technical processes is tested from a set of criteria which use-wear analysis provides.

Knowledge on **the type of material worked** allows us to place each of the technical actions into the process of transformation of each material.

The state of the material worked makes it possible, in the case of hide, to distinguish different phases of the technical processes on this material, as we shall see later.

The type of technical action can relate to some of the phases of the technical processes. For example, chopping wood can take place in the first phases of procurement or in the initial shaping of this type of material, while boring has to relate to the last phases of shaping the object.

The variety of technical actions carried out by an assemblage of lithic tools is as much linked to the achievement of a larger or smaller part of the technical process on the site, as to the degree of complexity of the technical processes which took place.

The association of different technical actions in the same tool will be more frequent when the tool was used in technical processes in which it was necessary to carry out a series of connected actions. Therefore, in contexts where all the phases of manufacture of complex objects are carried out, more implements with more than one type of technical action (ie. cutting and scraping) are found than where simple tools are made or only repair work took place.

The length of active zones and the size of tools used in a task provide data about the volume of the material worked. This volume will be reduced progressively as the stages of manufacture advance and will be minimal in the last phases. At the same time, the length of the active zones and the mass of tools will also be less. It is necessary to point out that comparisons of the length of the active zone can only be achieved by taking into account the same type of actions. In detail work, belonging to the finishing and maintenance phases, the tendency will be to use implements of smaller size, more suitable for this work. For comparing two assemblages it is necessary to take into account the general size of the tools since the variability can be due, for example, to the availability of lithic raw material more than to the type of work for which the tools are intended.

The use of specific implements and specific active zones used in an activity is often related to the greater difficulty or importance of the technical sequence in which they are used. The retouched tools are considered more specific than the unretouched tools. Retouched active zones are more specific than those created by fracture or which are unretouched. Among the morphotypes, those which are more commonly associated with a particular task are considered more specific for the task (ie. end-scrapers and hide working).

The intensity of use of the tools tends to be greater for those tasks which require more time and effort, in general those encountered in the first phases of the technical process and especially in the more complex technical processes.

The above criteria are not only related to the type of task carried out with the tools. The cost of acquisition of lithic raw material, the cost of tool manufacture and the context of use are also relevant. Therefore, only the convergence of a high number of the above criteria, after consideration of other possible factors of variability, will lead us to come to conclusions concerning the technical processes which took place on the site.

5.2 THE TECHNICAL PROCESS ON THE DIFFERENT RAW MATERIALS.

Up to now we have established the general criteria which allow us to understand broadly the activities carried out with flint tools. As we have shown, a reconstruction must be achieved for each worked material, since we have no more specific information on the technical process. First, we shall establish the general characteristics of the technical processes for each material. Then, we will show the use characteristics of lithic tools which can be related to each of the technical processes and their technical sequences.

In the reconstruction of the technical processes, the intention is not to make an exhaustive description of all the systems of transformation which can apply to each material. The intention is, to set out the general work sequences, so that we can integrate the data from use-wear analysis into the technical process of each one of the materials. Therefore, the description of the technical process is adapted according to the nature of the data obtained from use-wear analysis. In this sense, we shall consider mainly those activities which can be understood through use-wear analysis of lithic tools.

Later we shall analyse comparatively the three lithic assemblages which we are studying, in order to characterize broadly the type of technical processes which took place in each one of them.

5.3. HIDE

5.3.1. **The technical processes on hide.**

Hide is a material intensively used among communities of hunter-gatherers. Its use as clothing is especially important in the cold and temperate areas of the world. The usefulness of hide is not restricted to personal clothing, but is used in all latitudes to make sacks for carrying, ropes to make nets and traps, rugs to cover the floors of tents, etc. (Delaporte and Roué, 1978, 220; Lee, 1979, 124; Spiess, 1979, 30; Kamminga, 1982, 37; Mansur-Franchomme, 1986, 147-149). The Nunamiut needed between 20 and 30 caribou hides to cover a tent (Gubser, 1965), which gives an idea of the importance of this material in this type of community.

Hide is a material which decays easily; also, when it dries it hardens, losing its flexibility. The main work concerning the transformation of hide is, therefore, directed towards its preservation and the prevention of loss of flexibility once the hide is dry. However, when neither flexibility nor long durability is look for, hide can be used without any treatment (Mason, 1889; Victor and Robert-Lamblin, 1989). The treatments which make possible the durability and flexibility of hide are called tanning. Hide must be treated with different agents, which can be animal, plant or mineral in origin (Pecheux, 1922, 12-17; Forbes, 1966). These agents can be applied by rubbing on the hide directly or by placing the hide in water in which the tanning agent or tannin is present. The tanning agent is fixed into the structure of the hide, stabilizing it (Chahine, 1991, 72).

There is some argument as to whether the level of technical development in hunter-gatherer communities allowed them to carry out proper tanning or only a pseudo-tanning (Delaporte and Roué, 1978, 220-221; Plisson, 1985, 211-213). In the second case the treatment is not definitive, so the characteristics of durability and flexibility can be lost once the hide is soaked (Halasz-Csiba, 1991). Whether authentic tanning or pseudo-tanning, a certain technical process must be carried out in order to favour the preservation of this material.

Beyond this necessary step the treatments will differ according to the properties desired in the final product. The qualities normally sought are flexibility/softness and impermeability, depending upon the use to which the hide will be put. At the level of technological development of hunter-gatherers, the accomplishment of these two qualities is to a large extent exclusive since treatments which achieve greater flexibility in the hide are detrimental to impermeability and vice versa (Delaporte and Roué, 1978, 221). Among the Baxtyari of Iran the

Photo 5.1. Defleshing a goat hide by scraping with a hafted end-scraper

hides which are used to make recipients for liquids, which require impermeability, are not scraped, while the hides intended for clothing or to carry objects, which require flexibility, are scraped, soaked and rubbed intensively (Digard, 1981, 114).

The bibliography in which the processes of preparation of hides are described for hunter-gatherer groups is lengthy (Mason, 1889; Gansser, 1951; Hatt, 1969; Reed, 1972; Le Mouël, 1973; Schaefer, 1973; Gallagher, 1977; Delaporte and Roué, 1978; Clark and Kurashina, 1981, *inter alia*). In these works the great diversity of techniques employed may be observed. However, it is possible to establish a series of steps necessary for the working of this material (Bienenfeld, 1988, 223; Plisson, 1991).

First, the type of hide to be worked must be chosen, taking into account the object to be made. The thickness, resistence, and size of a hide vary from one species to another, and also according to age, these being factors of utmost importance which condition the selection of the raw material (Gusinde, 1930/1982; Reed, 1972, 38-44; Delaporte and Roué, 1978).

The work begins with the acquisition of the raw material by **skinning**. Flint implements take part already in these first tasks as much to cut open the animal as to cut the connective tissue between the hide and the flesh (Gusinde, 1930/1982, 255). It is difficult to remove the hide from animals of medium or large size, such as goat, by simply pulling it. This task can also be carried out with bone knives, such as those used by the Ingalik Eskimos (Osgood, 1940, 92).

The traces which will be found on flint tools used to carry out this activity are indistinguishable from those produced from cutting fresh hide.

At this point begins a series of tasks aimed at preserving the material. The first is **defleshing**, the removal of fat and pieces of meat on the interior face of the hide, since these elements decay very rapidly (Reed, 1972, 48-49; Plisson, 1988, 157; Hayden, 1990, 92-93).

The defleshing stage is documented ethnographically for numerous groups, such as the Eskimo (Mason, 1889), the

Photo 5.2. Finishing a goat hide with a knife of backed bladelets.

Lapps (Delaporte and Roué, 1978), indigenous peoples of North America (Mason, 1889; Steinberg, 1966; McGrath, 1970; Belitz, 1973), Ethiopians (Clark, 1981) and Australian aborigenes (Kamminga, 1982, 39). This work is carried out by scraping, cutting or striking, for which a wide range of implements can be used. Among native North Americans it was common to employ a bone scraper. This occurred among the Ojibwa (Steinberg, 1966), the Crow, the Sioux, etc. (Mason, 1889; Belitz, 1973).

However, the use of bone implements is not exclusive. A study by K. McGrath (1970, 21), which covers twelve tribes of the southwest United States, demonstrates that for defleshing, implements of bone, antler, stone or simply the workers' hands were used. For the Australian aborigenes, J. Kamminga (1982, 39) affirms that though it was common to use implements of bone, shell or tooth, in some communities stone tools were also important (Kamminga, 1978, 361-362).

Other examples of the variety of tools used for this work may be seen among the Eskimos, who for the first scraping of bird hide used tools of stone, shell, metal, antler or bone (Mason, 1889, 560). Among the Iroquois, fresh hide was cleaned with a wooden blade or a stone end-scraper (Mason, 1889, 583), while the Navajos used a hunting knife (Mason, 1889, 575). The Shoshone of the north carried out this work with denticulated end-scrapers (McGrath, 1970).

Stone end-scrapers were also used for defleshing by the Paiute of Surprise valley, the southwest Yavapai, and the Walapai, who used them on the hides of deer, bear, badger, rabbit, coyote, etc. (Kelly, 1933, 119; Gifford, 1932, 222-223; Kroeber, 1935, 272-273; cited in McGrath, 1970, 97-98), the Selk'Nam and the Yamana of Tierra del Fuego (Gusinde, 1930/1982; Mansur-Franchomme, 1986, 147) and the Eskimos (Murdoch, 1892, 294, cited in Nissen and Dittemore, 1974). These end-scrapers were normally hafted. B. Hayden (1990, 92-93) points out that to deflesh, bone or stone end-scrapers were usually used, although abrasive stones were used for rubbing, observed among Eskimos, Bushmen, Navajos and plains Indians.

Defleshing is accomplished relatively easily when the hide is fresh (Belitz, 1973, 4; Moss, 1983, 54). In our experiments on goat hide, we proved the effectiveness of stone end-scrapers for defleshing fresh hides (photo 5.1.), also demonstrated by Kamminga (1982, 40) for other types of medium-sized animals. Defleshing hides of large animals, such as horse, demands other types of activities which complement scraping. For cleaning some areas of the hide, with thick layers of fat and subcutaneous tissue, it is necessary to use cutting actions, with sharp edges. Striking is also useful (Plisson, 1991,9), as demonstrated by O.T. Mason (1889) for various North American tribes. In our experiments, this striking action was carried out with a hafted end-scraper. Cleaning the hide of small animals, on the other hand, can be accomplished with the hands only (Murdoch, 1892, 294).

Defleshing fresh hide has been documented for Australian aborigenes (Kamminga, 1982, 39), North American Indians (Mason, 1889; Steinberg, 1966, 576-578; McGrath, 1970) and for the Selk'Nam and Yamana of Tierra del Fuego (Gusinde, 1930/1982, 233 and 482; Mansur-Franchomme, 1983a, 422). However, defleshing can also take place after the hide has dried, as occurs among Australian aborigenes (Kamminga, 1982, 39), Bushmen (Lee, 1979, 124), Tehuelches and, at times, the Selk'Nam and the Yamana of Tierra del Fuego (Gusinde, 1930/1982, 193 and 398-401; Mansur-Franchomme, 1983a, 421). In this case hide is commonly soaked, so that it becomes soft before being worked (Clark and Kurashina, 1981, 306; Plisson; 1985, 211-213). Seneca Indians endeavored to work hide in a fresh state, but, otherwise dry hide was soaked before being worked (Mason, 1889, 573). In contexts of pre-industrial craftsmanship dry hides are placed in water for 10 or 12 hours, being scraped afterward with a semicircular knife to eliminate fat and subcutaneous tissue (Pecheux, 1922, 19).

In the ethnographic record we observe a tendency for the hide to be defleshed in the fresh state. Systematic defleshing of hide in a dry state is carried out in dry regions, as the Kalahari, Patagonia or the arid deserts of Australia, where the risk of decay of the hide is low. M.E. Mansur-Franchomme (1986, 148) affirms that: "the hides which are already dry do not run the risk of decay in the arid and windy steppe of Patagonia". In the case of pre-industrial craftsmanship, the cleaning of hide in a dry state is justified by the long time lapse between the skinning of the animal and the cleaning of the hide.

When leather is desired, the next step is the **depilation** of hide (Mason, 1889; McGrath, 1970; Belitz, 1973). The hair removal can be accomplished by scraping while the hide is fresh, as among the Ojibwa (Steinberg, 1966, 580), the Nisqually, the Indians of the Columbia River (Mason, 1889, 569), the Iroquois (Mason, 1889, 583) and the Yamana and the Selk'Nam of Tierra del Fuego (Gusinde, 1930/1982, 194 and 398-401).

It is also common to remove hair after some decay has taken place as this facilitates its elimination (Mason, 1889, 562; Pecheux, 1922, 22; Delaporte and Roué, 1978, 233; Plisson, 1985, 212-213). The Naskopi, the Shoshone and the Crow favoured depilation after placing the hide in water for several days (Mason, 1889, 567, 569 and 572), while the Yamana sometimes substituted water with urine (Gusinde, 1930/1982, 398-401).

Depilation is generally carried out by scraping, although again at this stage of the work the variety of tools which may be used is large. K. McGrath (1970) shows for various tribes of Nort America the use of end-scrapers made of horn, bone and stone (see also Steinbring, 1966). The Naskopi carried out depilation with bone tools (Mason, 1889, 567), while among the Shoshone the hair was removed with wooden knives or with a rib bone (Mason, 1889, 572). The use of non-manufactured tools for depilation is also observed among the Navajos, who depilated hide with the sharpest part of an ulna , with a rib bone, or with blunt hunting knives (Mason, 1889, 575). A hide may also be depilated by pulling out the hairs by hand, when hide is slightly fermented, as among the Bactiari of Iran, who do this after having left the hide rolled up for several days (Digard, 1981), and among the Yamana (Gusinde, 1930/1982, 398-401).

In our experiments of hair removal we have obtained good results, after some decay of the hide, by scraping with end-scrapers as well as with the sharp edges of unretouched blades, although with the latter care must be taken not to damage the hide. The hide could also be depilated by pulling out the hair by hand, although this does not eliminate the epidermis, which is detrimental to the preservation as well as to the esthetic appearance of hide. H. Pecheux (1922, 22) indicates in a handbook on the craftsmanship of leather that depilation can be facilitated by the addition of sand or ashes; however we have not been able to verify this point, since the addition of these materials, in our experiments, brought no advantage to the efficiency of the work.

The removal of flesh and hair from the hide, if carried out when it is still fresh, will result in the appearance of tools with traces of scraping fresh hide. In some cases, these two phases can also occur when the hide is dry or soaked, so that scraping of hide in these states can also be associated with these work phases. If the hide has been cut in the first phases of work traces of cutting fresh hide will be produced.

Afterward the hide is left to dry for a variable amount of time, being kept stretched on the ground, on a frame, different stages, between which the hide is usually washed with water (Mason, 1889; Steinberg, 1966; McGrath, 1970; Belitz, 1973; Delaporte and Roué, 1976; Lee, 1979; Bienenfeld, 1988).

In some cases during these phases **tanning** is carried out, by the addition of tanning agents which can be of animal origin (brain, fats, entrails, urine), plant origin (bark and leaves of different species) or mineral origin (Mason, 1889; Pecheux, 1922, 12-17; Forbes, 1966, 5; Moss, 1983, 53; Chahine, 1991, 72).

The **softening** of hide can be achieved in various ways. The simplest is continuous rubbing between the hands, as may be observed among the Kung (Lee, 1979, 124), the Selk'nam (Gusinde, 1930/1982, 194; Mansur-Franchomme, 1986, 147; Hayden, 1990, 92-93), the Zulus (Forbes, 1966, 10), the Eskimos (Mason, 1889, 560) and the Naskopi of North America (Mason, 1889, 567). This procedure was principally used for softenning hides from small animals (i.e. fawns or young doe) (Tepper, 1994, 54). Another way to render the hide flexible without tools is to chew it continuously, known among the Eskimos (Mason, 1889; Forbes, 1966, 6; Balicki, 1970; Hayden, 1990, 93) and the Selk'nam (Gusinde, 1930/1982, 194). Stretching hide on a frame, with the legs and the hands of the worker, twisting or rubbing it around a stick are also useful (Digard, 1981, 114).

When tools are used, making the hide flexible can be accomplished by breaking the fibres with sharp sticks, beating with sticks, abrading with stones or scraping with various tools (Hayden, 1990, 93), which normally have blunt edges (Gassin, 1993, 248) . Rubbing dry hide with pebbles of sandstone, quartzite or delerite to soften and even it out before making various objects, such as sacks, cords, etc., has been observed (Osgood, 1940, 92; Roux, 1985, 51; de Beaune, 1989, 63). Among various tribes of North America, softening is accomplished with end-scrapers made of stone, bone or antler (McGrath, 1970). Among the Ingalik Eskimos a bone scraper is used for this task, although large hides are scraped first with stone tools and finished with bone tools. In preindustrial craftsmanship the hide is beaten with wooden hammers to soften it (Pecheux, 1922, 24). For our part we have carried out sucessful softening of badger, goat and horse hides with flint end-scrapers and with pebbles, as well as kneading with our hands.

Hide scraping with lithic tools results in softening and thinning, while tools with blunt active zones are useful only for softening. For thinning hide, shell scrapers can also be used (Mansur-Franchomme, 1986, 147-149). Rendering the hide flexible and thinning it are usually carried out when the hide is dry or in the process of drying (Mason, 1889, 569; Belitz, 1973, 6; Gusinde, 1930/1982, 194 and 398-401). Furthermore, such phases can be carried out after greases have been applied to the hide for impermeability and preservation.

The application of mineral materials to hide is a current practice which favours preservation. The Ingalik Eskimos apply an altered mineral, a compound of feldspar and clays, leaving the hide dry afterward (Osgood, 1940, 81). More common is the use of ochre, a mineral with a high percentage of iron oxide. F. Audouin and H. Plisson (1982) have pointed out the antiseptic properties of ochre, which stop or slow down the decay of organic materials, as well as favouring impermeability. These authors contribute various ethnographic references concerning the use of ochre in the treatment of hides, as does Steinman (1906), who asserts the use of ochre and fat in the treatment of guanaco hides among the Tehuelche, with the aim of insuring their preservation. Sollas (1924) makes the same observation in referring to the working of hides among the Tasmanians, while Peabody (1928) specifies that such treatment is to make the hide waterproof.

The Yamana and the Selk'nam of Tierra del Fuego greased their hide objects with fat and red coloured earth. When questioned as to the reason for this treatment, they offered a purely esthetic justification. However, M. Gusinde believes that there also exists a practical reason, the preservation of the object and the maintenance of its qualities of impermeability and flexibiity (Gusinde, 1930/1982, 181-182 and 232).

Although ochre is not a tanning agent (Plisson, 1991, 15; Philibert, 1993), it facilitates the drying of hides, which indirectly stops the decay processes, which begins again when the hide is soaked. In this way, ochre helps the dessication of hide, which, even if it does not produce an authentic tanning, would contribute to the preservation of this material.

The experiments which we have carried out with hides seem to point in this direction. We applied ochre to one area of a horse hide, leaving it to dry for a month. We observed that the areas with ochre dried much more quickly. Ochre absorbed the humidity and natural fat of the hide, which is crucial for the preservation of this material (Halasz-Csiba, 1991, 35). The application of ochre to the hide helps to dry it, but also protects it against humidity, as once introduced into the hide it will cause permanent dryness.

In some cases the ochred hide is rubbed with pebbles, as has been observed among the Tuareg (de Beaune, 1989), in order to introduce the mineral into the hide and to rend it flexible. In some late Magdalenian sites in the Basque country, pebbles with hide rubbing traces have been observed, some of them having residues of ochre (Ibáñez and González, 1994; González and Ibáñez, in press). The scraping of ochred hides with flint tools has been archaeologically confirmed at the Perigordian site of Le Flageolet I (Kimball, 1989, 99), the Epipalaeolithic site of Balma Margineda (Philibert, 1993) and the late Magdalenian site of Laminak II (González and Ibáñez, 1994b).

In some groups the process of hide preparation ended with smoking, which favours the impermeability and preservation of the material (Mason, 1889; Gansser, 1951, 1167; Albright, 1984, 57; Gassin, 1993, 247). Among the Australian aborigenes engraving on hide is documented, the purpose being not only decorative but also favouring flexibility. For this activity tools of shell, tooth or stone were used (Kamminga, 1982, 43).

The manufacture of hide objects is carried out once the preservation of the hide is assured and the desired properties have been obtained (McGrath, 1970, 21). During the making of the objects cutting and sewing take place (Kamminga, 1982, 44 and 45). For this reason, the traces of cutting and boring of dry hide are related to this phase of the technical process. K. McGrath (1970, 21) points out the use of a stone knife, a bone awl, a cactus thorn, and tendon for thread as tools for making hide objects among various tribes of the south-west United States. The use of needles, found on late Upper Palaeolithic sites, as tools for sewing hides is highly probable, as they are used by the Australian aborigenes (Bird and Beek, 1980).

The Selk'nam made a long leather strap by cutting a hide in a spiral from the centre outward. The hide was wetted as it dried, since completely dry hide is difficult to work. Then the strap was scraped lightly to even out the parts which needed it (Gusinde, 1930/1982, 233). Scraping actions which are not very intensive and are associated with longitudinal movements may be related to the phase of manufacture of objects.

It is often necessary to carry out maintenance work on the hides. In many cases, the treatments on hides only achieve a pseudo-tanning so that the objects suffer deterioration, especially from humidity or parasites (Chahine, 1991b). When this occurs, part of the preparation work must be repeated (greasing, smoking, scraping, application of ochre...) until it has returned to its initial condition.

Use-wear traces on flint tools from softening, thinning and mantaining hide can be associated with traces of scraping dry hide, to which may be added fat, ashes, ochre or other minerals.

The various phases of hide working do not always have to be carried out in the same place. Among the Ojibwa there are examples of hunters who go away for a time, returning to camp with the hides half-prepared (Steinberg, 1966, 880). K. McGrath (1970) states that various North American tribes usually carried out the work of conditioning hides in the residential camp. Once prepared, hide was transformed into objects right away or stored to await future necessity. Among the Naskopi Indians, hides were stored until the appropriate moment arrived to prepare them (Mason, 1889, 567). Among the Nlaka'pamux, if the hide was not to be tanned immediatly, it was defleshed, washed, rubbed with grease and oil and rolled up for storage (Tepper, 1994, 52)

Ethnographic information offers us a frame of reference for reconstructing prehistoric technical processes of hide working, but, at the same time, some clear limitations arise. We find cases in which the technical processes of hide preparation are extremely simple. When it is attempted to achieve waterproof hides without considering flexibility or when rigidity of the material is sought, as in some cases occurs for floor rugs, tent coverings or quivers for arrows, the technical processes may be simplified. In these cases the treatment is reduced to drying the hide (Gusinde, 1930/1982, 194 and 398-401; Delaporte and Roué, 1978, 228-229).

Secondly, we find technical actions which are not practiced with tools, such as soaking and greasing hides or rubbing by hand or chewing to achieve flexibility.

Thirdly, despite the fact that many of the tasks mentioned can be accomplished with flint tools, tools made in other materials can also be used. In this sense, J. Kamminga (1982, 39) points out some cases of Australian aborigene groups who carry out the entire process of transformation of hide without any stone tool, although in contexts of great lack of lithic raw material. The relation of bone needles and probably awls to the making of hide objects is evident. Furthermore, numerous ethnographic references document the use of tools in bone, antler, horn, wood, shell or abrasive stones for the cleaning and depilation of hides.

With these difficulties it may be assumed that the reconstruction of the technical processes in the working of hide can only be partial, limited to the work undertaken with tools made in unperishable materials, the remainder of the technical operations being hypothetical.

5.3.2 The working of hide in Berniollo and Santa Catalina

The importance of the non-flint tools in the working of hide in the archaeological context which we are going to study offers a field for discussion. Bone tools intended for scraping hide present a very particular morphology, having a non-cutting edge for transverse rubbing (Mason, 1889; Steinbring, 1966; McGrath, 1970). This function could be covered by bone spatulae which are found on late

Upper Palaeolithic sites. However, these pieces are not very frequent (they are not present in Santa Catalina), so that, although this was their function, the distortion that they would produce in the image of hide-working offered by lithic tools would be very small.

The use of shell scrapers or knives does not seem very probable, since it is difficult to undertake this work with "patellae" or "monodontae", the type of shells which appear almost exclusively on the coastal site of Santa Catalina. The use of pebbles for rubbing hide can be found by the analysis of use-wear traces present on these objects. As we have observed in our experiments, the rubbing of dry hide, with or without the addition of ochre, produces facets with striations or lustered surfaces, according to the characteristics of the pebble. These traces has been observed on some pebbles from late Magdalenian sites in Vizcaya, including Santa Catalina (Ibáñez and González, 1994; González and Ibáñez, in press).

A major problem involves the possibility that non-manufactured bone tools were used for this work, or objects of wood with rounded edges, aspects which cannot be addressed or which have been scarcely so to date (cf. Peltier and Plisson, 1986).

Bearing in mind these limitations, the study of lithic tool function offers considerable information concerning the working of hide. In the three assemblages we observe different proportions in the working of hide according to its state. At Berniollo working dry hide was substantially more frequent than working fresh hide, since it accounts for 90% of hide-working.

In the Magdalenian level of Santa Catalina, the predominance of dry hide is not so evident, representing somewhat more than 60% of the activities in which the state of the hide could be identified. On the contrary, in the Azilian level of Santa Catalina working fresh hide is predominant over dry hide in a proportion of more than two to one.

This variability can have different meanings. The different proportions in the working of hide according to its state can (1) represent variations in the technical processes between one context and another or, (2) reflect that in each context some phases of the technical process occurred more frequently than in others, the technical processes for the three assemblages being similar.

In the first case, it may be that a less frequent working of fresh hide was the consequence of (1a) the existence of a drying phase occurring before the removal of flesh and hair, as we have seen in various ethnographic examples, or (1b) the removal of flesh and hair from the fresh hide with tools other than those of flint.

We consider the first alternative (1a) as improbable, since systematic drying before defleshing occurs in arid zones, which differ from those which we are studying. Moreover, if this were so, the drying before defleshing would have been carried out in some of the processes, but not in others, as demonstrated by the fact that in all the assemblages there is fresh hideworking to a greater or lesser degree. Concerning the second possibility (1b), our present evidence suggests that neither pebbles nor non-modified bones were used for defleshing, although, lacking systematic research, this possibility cannot be totally rejected.

Although we cannot rule out that the differing proportions in the working of hide according to its condition are due to changes in the technical processes, various indications lead us to think that the differences in the state of the worked hide correspond to the accomplishment of different phases of the technical process in each context. This hypothesis would fit in better with the data on use intensity, degree of specificity of tools, length of the active zones and size of the tools employed in this work.

Table 5.1. Average number of active zones for each piece in tools which worked on these materials.

	Berniollo	S.C. Azilian	S.C. Magdalenian
Fresh hide	---	2.3	2
Dry hide	2.9	2.7	2.7

Table 5.2. Development of the polish on tools used on dry hide.

	Berniollo	S.C. Azilian	S.C. Magdalenian
Very developed	7 *(26.9%)*	2 *(18.2%)*	10 *(23.3%)*
Quite developed	16 *(61.5%)*	8 *(72.7%)*	33 *(76.7%)*
Little developed	3 *(11.5%)*	1 *(9.1%)*	0

Table 5.3. Development of the polish on tools used on fresh hide.

	Berniollo	S.C. Azilian	S.C. Magdalenian
Very developed	---	0	0
Quite developed	---	24 *(85.7%)*	22 *(73.3%)*
Little developed	---	4 *(14.3%)*	8 *(26.7%)*

Tabl4 5.4. Percentage of activities on fresh hide and dry hide carried out with end-scrapers

	Berniollo	S.C. Azilian	S.C. Magdalenian
Fresh hide	---	60.7%	66.6%
Dry hide	96%	45%	79%

Table 5.5. Length of the active zones for end-scraper activities on hide.

	Berniollo	S.C. Azilian	S.C. Magdalenian
Fresh hide	---	18mm.	19mm.
Dry hide	28mm.	14mm.	22mm.

Table 5.6. Length x width of tools for working hide.

	Berniollo	S.C. Azilian	S.C. Magdalenian
Fresh hide	---	27x21mm.	32x20mm.
Dry hide	36x31mm.	26x21mm.	36x22mm.

The analysis of use intensity for tools used to work dry hide indicates that this material was worked with greater intensity in Berniollo than in the other two assemblages, demonstrated by the greater average number of active zones per tool (table 5.1) and the existence of more developed use-polish (table 5.2).

In comparing use intensity for fresh hide, we set aside the data from Berniollo because of the low number of active zones found. In Santa Catalina, tools which worked fresh hide in the Azilian level showed a higher average of active zones per tool and a more intense development of use-polish than in the Magdalenian level. These data for fresh hide are opposite to those for the rest of the activities for both levels, as a greater use intensity is observed in the Magdalenian level (tables 5.1. and 5.3.).

The degree of specificity of the tools used also offers interesting results. Again the data for work on fresh hide in Berniollo are not considered. In the Azilian level, fresh hide was worked with more specific tools than dry hide, since end-scrapers were used more frequently (table 5.4). In the Magdalenian level the specificity of the tools is reversed, since dry hide is transformed with end-scrapers more frequently than fresh hide. In Berniollo, dry hide was worked almost exclusively with end-scrapers.

In analysing the length of the active zones for fresh vs. dry hide working, we observed an inversion of values in the Azilian and Magdalenian levels. While in the Azilian the active zones for working fresh hide are larger than those for dry hide, in the Magdalenian level the results are the other way around. The active zones which scraped dry hide in the Magdalenian level are up to 33% greater than those of the Azilian level. The active zones which worked dry hide in Berniollo are greater in length (table 5.5).

The measurements of the tools which worked hide offer similar results. In the Azilian level of Santa Catalina those which worked fresh hide are slightly larger, as opposed to that which occurs in the Magdalenian level. Again, the tools which worked dry hide in Berniollo have the largest dimensions (table 5.6.).

According to these data, we find clear associations between the importance of working hide in a particular state, the intensity with which the tools were used to transform this material, the degree of specificity of the tools used, the length of the active zones and the size of tool employed. These data confirm the importance of working fresh hide in the Azilian level of Santa Catalina. On the contrary in Berniollo, the transformation of dry hide was more important. In the Magdalenian level the preparation of dry hide is more important than in the Azilian level, but less important than at Berniollo.

We have found associations between the higher percentage of hide worked in one state or the other, greater intensity of use, higher degree of specificity of the tools, the length of the active zone and the size of the tools. We believe that there would have been no reason to find these associations if the percentage in which the hide was worked in different states was a result of differences in the technical processes on hide. With these data, it seems more probable to us that the importance of hide-working in one or the other state is in relation to the phases of the technical process which was dominant in each of the archaeological contexts.

Thus, in Berniollo it was principally the final phases of hide preparation (softening and thinning) which took place, after the drying. A similar interpretation has been suggested by A.L. van Gijn (1989, 77) for the Neolithic site of Beek-Molensteek. The systematic and intensive work on this material caused the use of relatively large end-scrapers, resulting in intensive use, with broad contact zones between the tool and the worked material.

In the Azilian level of Santa Catalina, however, the preparation of fresh hide (defleshing and perhaps dehairing) was more relevant, in the moments immediately after the skinning of the animal. In the Magdalenian level more of a balance is observed in the proportions, which would indicate a similar importance for the initial and final phases of the process.

The analysis of the type of activity practised on these materials can give us an idea of the objective of the work carried out.

In Berniollo, scraping is the main action on dry hide (table 5.7.). Scraping dry hide was for softening and thinning the material before the manufacture of objects, as well as for maintaining hide objects in good condition. The tasks involved in the manufacture of hide objects had less importance, as there is a scarcity of evidence for cutting, engraving or boring dry hide. This is confirmed by the fact that only 20% of the tools used to work hide show evidence of more than one activity associated in the same tool. Eight of the ten pieces which worked dry hide only show scraping actions, for softening the hide. The two end-scrapers on which cutting and scraping are associated can be related to more complex actions in the manufacture of objects. The two end-scrapers of Berniollo with traces of scraping fresh hide would indicate that the first phases of hide preparation were carried out.

In the Azilian level of Santa Catalina, scraping fresh hide and a part of the cutting (table 5.8.) represent de-fleshing and removal of superflous parts in the first stages of the preparation of the hide. It is possible that some of the "fresh-hide cutting" activities correspond in reality to skinning activities. This is more probable in the case of two unretouched flakes which only show one active zone with traces of cutting. Most of the fresh-hide working tools, end-scrapers in the majority, show only scraping activities (7 out of 10), which relate to defleshing and depilation of the hide in a fresh state. The two

Table 5.7. Activities carried out on dry hide.

	Berniollo	S.C. Azilian	S.C. Magdalenian
Cutting	2 *(7.7%)*	4 *(36.4%)*	16 *(37.2%)*
Scraping	24 *(92.3%)*	6 *(54.5%)*	24 *(55.8%)*
Boring	0	0	1 *(2.3%)*
Engraving	0	1 *(9.1%)*	0
Indeterminate	0	0	2 *(4.7%)*

Table 5.8. Activities carried out on fresh hide.

	Berniollo	S.C. Azilian	S.C. Magdalenian
Cutting	0	6 *(21.4%)*	8 *(27%)*
Scraping	8 *(100%)*	22 *(78.6%)*	22 *(73%)*

end-scrapers and the retouched blade in which cutting and scraping are associated can be explained as tools that were used for skinning (cutting) and afterwards for defleshing (scraping) or as tools used in a defleshing task that included cutting and scraping motions. Similar interpretations have been proposed for several endscrapers from the Perigordian level of Le Flageolet I (Kimball, 1989, 99)..

In the working of dry hide in the Azilian level (table 5.7.), the proportion of cutting activities is high, indicating the manufacture of objects in hide. The proportion of pieces which worked dry hide and show more than one associated action is 50%, which indicates that the tasks related to manufacture were relatively complex. The bone needle and the two awls found in the level may be associated to this stage of hide working.

In the Magdalenian level, although the final phases of hide-working predominate, the whole process appears to be represented, since there is a certain balance between the quantity of activities on fresh hide and dry hide. Among the eight end-scrapers used on fresh hide, four show associations of longitudinal and transverse activities. The significant presence of cutting and boring of dry hide (table 5.7.) indicate the manufacture of objects. In addition, the percentage of pieces with associated actions is 42%. Seven out of thirteen end-scrapers show associations of cutting and scraping, while four tools were used only for cutting, which leads us to believe that aside from the softening of hide, objects were made. The presence of a bone needle and three awls in the level support this.

We propose in our line of argument an explanation of the differences in hide-working activities in the three assemblages based on differences in the phases of the technical process predominant on each of the sites. However, we do not reject the possibility that changes in the technical processes also occured. This point will be checked in future investigations, since, if the cause is due to changes in the technical process more than to the phases of the technical process carried out on each site, the differences between the Magdalenian and Azilian levels will be repeatedly of the same nature.

5.4 THE BONE MATERIALS

5.4.1. **The technical processes on bone materials**.

The manufacture of bone tools in the three assemblages led to the making of a wide range of tools, awls, needles, harpoons and spearheads.

5.4.1.1. Raw material

Bone and antler are the materials selected to make these tools. Worked antlers in the late Magdalenian and Azilian of Vizcaya and Guipuzcoa come from deer (Múgica, 1983). This is the animal predominantly hunted in these contexts, providing raw material. The deer antlers are renewed each year, the old ones falling off at the end of the winter. Thus, antler was acquired by removal from the hunted animals or by gathering the fallen antlers. In the neighboring and contemporary archaeological sites both activities are observed. If the use of fallen antlers is more common (Múgica, 1983), those from slaughter are also frequent. In fact on the site of Santa Catalina, in the late Magdalenian level, two worked basal fragments of antler were found, with part of the top of the skull attached, which indicates that the antler was extracted from the forehead of a hunted deer. The gathering of fallen antlers probably implied the existence of search parties at the end of winter, which probably coincided with hunting (Billamboz, 1977, 100; Sidera, 1993, 111). Antlers from slaughter were extracted from the head of the animal by fracturing the top of the skull around the antlers, through percussion, demonstrated by the two examples from Santa Catalina.

Antler and bone have different mechanical features, which dictate the methods of manufacture and the type of tools made. Antler is more flexible than bone, and better resists impact without breaking. In fact to break antler a force three times that necessary for bone is required (McGregor, 1985). However, bone is more mineral and thus harder, and can be made into tools with sharper points and edges (Guthrie, 1983, 278) which better resist wear by abrasion.

The different characteristics of both materials were appreciated by prehistoric communities, which tended to make hunting tools, which had to sustain strong impact, out of antler, and tools used in pressure action principally out of bone. This duality has been demonstrated for the bone tools of Ksar Akil (Newcomer, 1974) or Natufian tools (Campana, 1989). If we look at more nearby contexts, M. Julien (1982, 127) points out that the majority of Magdalenian harpoons were made of antler, while bone was chosen for making needles (Stordeur, 1979).

Deer antler is composed of a central stem which spreads out in branches. The thickness of the cortical zone varies from one zone to another. Most of the tools in antler from the contexts we are studying were made from the central stem, to obtain wide dense shafts. Often, shafts obtained from the stem are slightly curved. The flexibility of antler allows them to be straightened, the shaft being wetted and left to dry in the correct position, or using heat (Julien, 1982, 129).

Such flexibility is not a feature of bone, and the blank cannot be straightened once obtained. Thus it is necessary to carefully choose the bone which is to be made into a tool. In general, long bones were chosen, especially the metapodals.

While antler can be worked directly, bones must be cleaned of flesh, cartilage and tendons. This may be carried out easily after cooking, soaking in water, or putrefaction, although in this last case the bone material undergoes drying, which renders the working of it afterward difficult. Bone may also be scraped with flint tools in cleaning. In some archaeological contexts, striations on bones have been interpreted as caused by scraping for cleaning (Olsen, 1979, 344; Sidera, 1993, 111).

Bone materials offer a high resistence when worked, especially when found dry, thus they are usually treated in different ways. One of the simplest is soaking the material, by placing it in water for a time, or simply wetting it constantly as the work goes on. This technique is recorded among the Salihs (Clive *et al.*, 1938, cited in Campana, 1989, 117). Moreover its efficiency has been proven by a large number of experimentors, including ourselves (Billamboz, 1977, 100; Newcomer, 1976; Campana, 1989, 117; inter alia). The difference in resistence found in the dry material and the wet material is considerably greater in the case of antler, which has a greater capacity for absorbing water (Bonnichsen and Will, 1980; Guthrie, 1983, 278).

In other cases treatments using heat have been recorded. The Maori moistened whale bones and submit them to the action of water vapor, as a method of softening them before work. (Campana, 1987, 117). In Poland, in the High Middle Ages, antlers were placed in hot water to soften them (Cnotliwy, 1970, cited in Billamboz, 1977).

Given the simplicity of the technique of soaking and the evident advantages obtained, it is very probable that this technique was used in prehistory. The material could be kept constantly in water while being worked or simply moistened with saliva, a technique known to numerous work processes.

5.4.1.2. The blank

For making some tools in bone material it is necessary to obtain a blank, usually elongated, from bone or antler. Pounding or crushing can be carried out to obtain bone slivers for modification afterward. With these fragments quite acceptable awls can be made through scraping of the active point of the tool. Among the Micmac and the Salihs bone was left to heat up until it cracked, the fragments thus obtained being worked later (Campana, 1989, 25).

The manufacture of more sophisticated tools implies the development of more complex techniques for obtaining the blank. In the Abri Fritsch, a technique for obtaining blanks from antler was identified, consisting of cracking the branch through indirect percussion with flint wedges (Allain *et al.*, 1974).

However, this technique of obtaining a blank is not current, the acquisition of elongated blanks by making longitudinal grooves in the bone material being more common. The technique consists in making two deep parallel grooves, which define an elongated zone which is extracted later. This technique was used particularly in working antler, to obtain shafts which were made into harpoons, spearheads, etc. (Clark and Thompson, 1953; Julien, 1982, 217).

The parallel grooves define the shaft longitudinally, but it is also necessary to define it transversely, at both ends. This can be done by sawing the body of the antler at the point at which it is desired to define the shaft at one of its ends. (Delporte and Mons, 1992, 12).Another system of defining the end of the shaft consists of making the parallel grooves converge in the same point (Delporte and Mons, 1992, 12). Finally, the longitudinal grooves can be complemented by other transverse ones at the ends, by sawing (Rueda, 1985, 12). These techniques of defining blanks longitudinally and transversely can also be applied to bone, as has been done experimentally (Newcomer, 1976). The technique of engraving grooves and sawing transversely has been observed in the making of bone spatulae at the Grotte du Placard (Mons and Stordeur, 1977, 15) and in a reindeer rib from the same site (Mons, 1972).

Once defined, the removal of the shaft or blank from the antler is accomplished by extracting an end and pulling it towards the exterior, since the flexibility of the material allows the operation to be undertaken without the shaft fracturing. To facilitate the total or partial removal of the shaft, wedges can be used, pushed into the grooves by striking with a hammer. These wedges can be of antler, demonstrated by the example of an antler with a fragment of a wedge in this material introduced into a groove, found in the site of Santimamiñe (Múgica, 1983). However, a flint wedge can also be used as we have demonstrated in our experiments. A fragment of flint rather than antler was observed in a similar find, from Valby, near Copenhagen (Clark and Thompson, 1953).

Unlike antler, the shaft in bone cannot be removed by pulling, as it is not a flexible material. Removal can be facilitated by striking the bone (Newcomer, 1976), or by using wedges, whose traces may be seen in the Placard rib, worked by parallel grooves (Mons, 1972).

Obtaining blanks for the making of needles presents some peculiarities. Although the parallel grooves technique may be used, the small size of the blank means that less costly techniques can be used. In using a bone sliver, one side of the blank may correspond to one side of the sliver. The other side is defined by lined-up grooves on the internal and external face of the bone, thus making it possible to separate the blank desired from the rest of the sliver. The use of small bones, of birds or rodents, was also common and enabled the manufacture of very fine needles with little effort (Stordeur, 1979, 109; 1984).

5.4.1.3. Shaping

For the manufacture of awls from the metapodals of herbivores, mainly deer or goat, a preliminary phase for obtaining a blank is not necessary, as the tool is made from the complete metapodal (Murray, 1979). In this case the preliminary shaping consists of sectioning the bone longitudinally, making pointed later one of the ends from which the epiphesis has been removed. The end opposite to the point, the grasping zone, may retain the complete epiphysis, only half of it, or can be removed (Stordeur, 1976a). The longitudinal sectioning can be carried out through indirect percussion (i.e. wedges) or grooving (Camps-Fabrer and D'Anna, 1976; Stordeur, 1984). In percussion, the bone is placed upright, the epiphysis of one end pressed against an anvil, while a wedge, which can be flint, is introduced between the two apophyses of the epiphysis at the other end. The wedge is then struck so that the metapodal is split, eliminating one of its lateral halves on part of its length.

Longitudinal grooving allows a more controlled primary shaping. Two grooves are made on both faces of the metapodal which run all its length from one end to the other. The deepening of these grooves leads to the sectioning of the bone in two halves. However, the use of grooving and indirect percussion are not at all exclusive, but at times complementary. In some South African sites, indirect percussion was used to begin the longitudinal groove (Smith and Poggenpoel, 1988, 106-107). In some Natufian sites, the longitudinal grooves were finally removed thanks to the use of wedges for indirect percussion (Campana, 1987, 117). In general, primary shaping by direct percussion is associated with less elaborated tools, while grooving is related to technically more complex tools (Stordeur, 1984).

Some awls made from metapodals show the removal of the epiphysis opposite to the point. This task to eliminate one of the ends of the metapodal can be undertaken by sawing or indirect percussion (Mons and Stordeur, 1977; Smith and Poggenpoel, 1988, 106-107; Maarleveld, 1985, cited in van Gijn, 1989, 109).

The use of abrasion for the making of awls from metapodals has been recognized, but this technique belongs to the Neolithic, being unknown in Upper Palaeolithic contexts (Poplin, 1974). In fact, the use of abrasion for shaping all types of bone objects was a Neolithic technique, which progressively substituted scraping with lithic tool edges (Stordeur, 1983; Caspar, 1985, 73; Cahen *et al.*, 1986, 77; Campana, 1989, 130; Senepart, 1992, 330; see, however, Rueda, 1985).

After the primary phase, the metapodal had to be sharpened and if more elaborated forms of awl were sought, the shaft had to be scraped with a flint tool.

The shaping phase for needles consists of smoothing the shaft, sharpening the point, and boring. The first two tasks are carried out by scraping, while the hole or eye can be made by different methods: by boring, through precise concentric pressure or by grooving followed by boring (Stordeur-Yedid, 1979, 109).

Antler shafts may be made into harpoons or spearheads. In both types of tool a previous phase of rough shaping by scraping takes place, which eliminates the spongy material and the rest of the lateral grooves which are usually present in the antler shaft.

In harpoon-making the sides of the blank are reduced by scraping, creating a beveled section where the teeth are to be made, an area which may be separated from the central shaft by a longitudinal groove. Then the teeth are made, marked out by engraved grooves on both surfaces of the object. If the teeth are close to each other the same groove defines the interior area of one tooth and the exterior of the adjacent one. If the teeth are more distant from each other it is necessary to remove the material existing between each of them, for which a blade edge may be used as well as a pointed tool (Duvois, 1974, 80-83; Berke, 1977, 86; Julien, 1982, 131).

The process of making a spearhead is more simple. Scraping allows the shaping and sharpening of the blank. The base can be made in different ways: by scraping in the case of beveled bases (Delporte and Mons, 1992), by engraving coinciding grooves for those with a forked base (Delporte and Mons, 1992) or by chopping in the case of the foreshortened base (*à base raccourci*), as we have been able to prove experimentally.

Photo 5.3. Finishing a harpoon on antler with the facet of a burin.

5.4.1.4. Finishing and maintenance

The last step in the work process on bone material may be the decoration of the object. The grooves which decorate bone tools could have had a functional purpose (to facilitate grip, or hafting or the introduction of flint elements) (Alain and Rigaud, 1986), or a symbolic one. Such grooves are made by engraving with a pointed tool or with a blade through a cutting or sawing movement.

There is some ethnographic information about the treatment of bone tools after manufacture, although such operations have not been observed in the archaeological record. Among the Atapascan bone points are boiled in bear grease, so that they withstand better the agression of natural agents and the impact of use (Guthrie, 1983, 282).

During use, friction or impact of the tool against the materials which are worked alters the morphology of the object, so that it loses efficiency. The active zones which have been blunted or somewhat broken by use are repaired by resharpening. These actions are not intensive and are on limited areas of the object, so tasks involved in bone tool maintenance are similar to those related to the finishing phase in the process of manufacture. If the break in the tool is so important that it is necessary to reshape the tool, this repair means the reintroduction of the piece into the manufacture process. We may establish in this way the difference between manufacture and maintenance as much by the nature of the activity (creation of an *ex novo* tool as opposed to the repair of one already made) as by the scope of the work undertaken (retouching definite areas as opposed to the modification of the general form of the tool).

It has been proven experimentally that the areas of hunting tools most frequently damaged are above all the tip and the base, which are the parts most easily repairable (Guthrie, 1983, 291). This point is confirmed by M. Julien (1982, 134) for Magdalenian harpoons, many of which show resharpening of the tip, and in some cases also on the base. Some of the harpoons present a conical tip, while others are beveled. The latter are shorter than the former, which indicates that the beveling of the tip is the result of resharpening harpoons with conical tips. Harpoons in which the tip coincides with the first line of teeth are also interpreted as a result of resharpening. Conical tips present a very regular finish, whereas it is coarse in the beveled tips, which indicates that while the making of the harpoon is careful, its repair is rapid and expeditive (Julien, 1982, 64).

Concerning the spearheads, it is probable that those with a foreshortened base were the result of rapid repair by chopping. This could explain the fact that in Placard and Laugerie-Basse they are not as long as other types of spearheads, although in Isturitz the length is similar (Mons, 1992).

Among needles there are also cases of repair of the tip, with resharpening, or of the base, with a remaking of the perforation (Stordeur, 1976b; Stordeur-Yedid, 1979, 109). Awls are always repaired by scraping (Mons and Stordeur, 1977, 15).

As we have observed in the course of the description of the technical processes on bone material, there is a certain relationship between work phase and type of activity. This is clearly indicated by the study of the bone industry on the site of Telarmachay, where percussion on an anvil, chopping and grooving are associated with roughing-out phases, whereas the finishing of the objects was carried out by retouch, scraping and boring (Julien, 1978-80).

Indirect percussion takes place in the first phases of work, in the extraction of shafts or in the primary shaping of metapodals. Engraving can be used to obtain blanks (groove and splinter technique), for secondary shaping (making harpoon teeth or spearhead bases) and to decorate objects. Boring takes place in the phase of secondary shaping of the tools, for making holes in needles and in some harpoons. Sawing can be used to remove antler branches, to saw the main stem of the antler or define the ends of the shaft, in the initial shaping of awls or in the decoration of tools. Scraping is used mainly in the secondary shaping of objects and in their repair, although in some cases it can serve to clean bone during the initial phases of preparing the material to be worked.

In addition, we have been able to establish that the working of antler was primarily to produce hunting weapons, while bone was worked to produce domestic tools.

5.4.2. **Bone tools in Santa Catalina**.

The study of remains of bone tools and of manufacture waste from the sites analysed can provide a first approach to the question. This is possible only in the two levels of Santa Catalina, since at the open-air site of Berniollo bone tools were not found, because of lack of preservation.

Table 5.9. Bone tools in the Azilian level of Santa Catalina.

	Harpoon	Awl	Needle	Spearhead	Other
Bone	2	2	1	0	2
Antler	1	0	0	0	1
Indeterminate	0	0	0	1	0

Table 5.10. Bone tools in the Magdalenian level of Santa Catalina.

	Harpoon	Awl	Needle	Spearhead	Other
Bone	0	3	1	1	0
Antler	1	2	0	5	1
Indeterminate	0	2	0	2	1

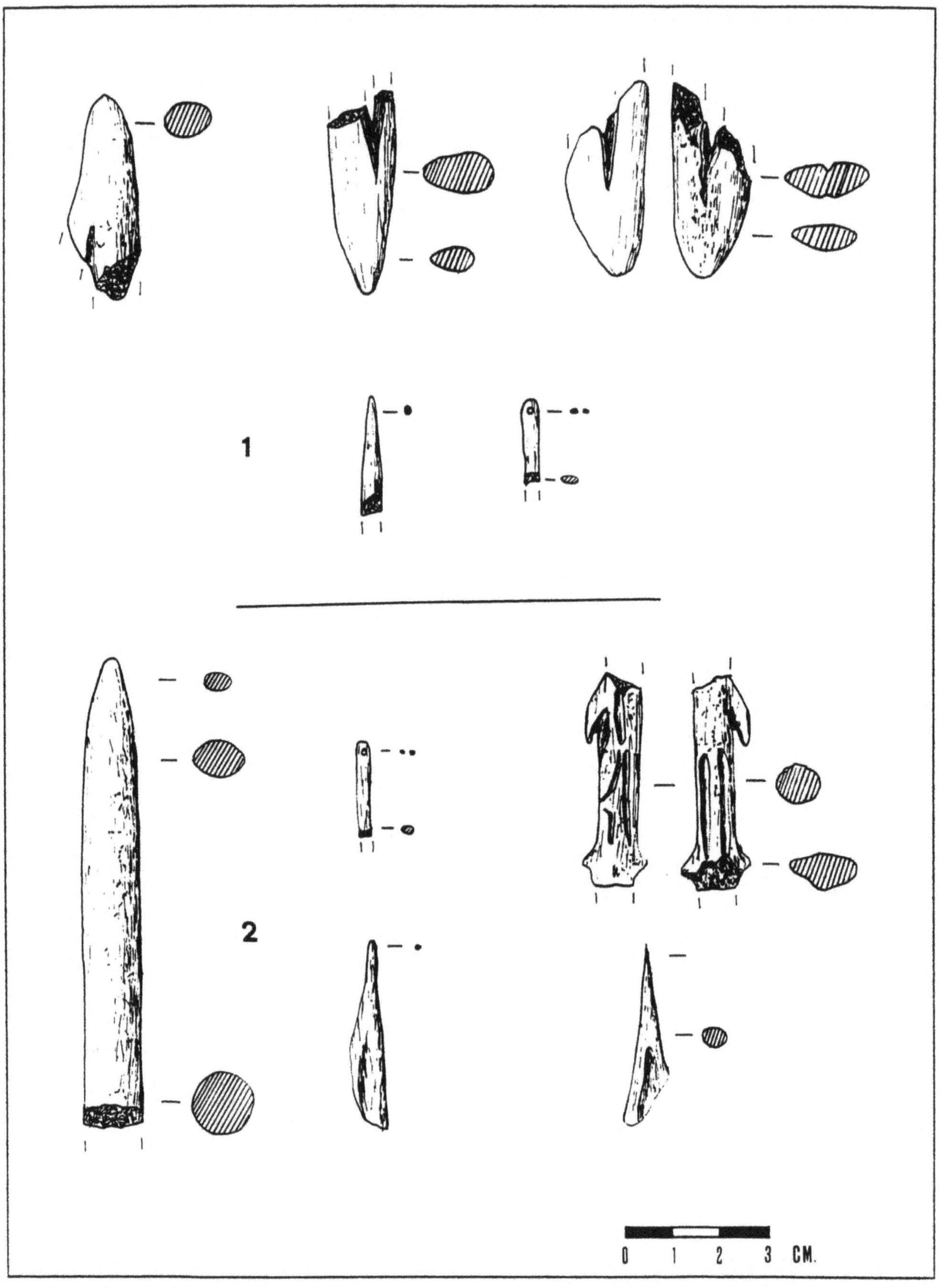

Figure 5.1. Bone and antler tools in the Azilian (1) and Magdalenian (2) levels of Santa Catalina.

We can observe the difference existing between both levels concerning bone material used to make tools (tables 5.8 and 5.9). In the Magdalenian level antler is predominant, while in the Azilian level bone is more abundant.

This change in raw materials agrees with the data from use-wear analysis of lithic tools. Thus, in the Magdalenian level the bone material which was most worked was antler; while in the Azilian it was bone. J.A. Múgica (1983) has pointed out the existence of a substitution process of antler in favour of bone as a preferred raw material in the manufacture of tools, for the chronological and geographic context with which we are concerned.

In Santa Catalina there is a clear association between the raw material and the function of the tools in bone material. Hunting weapons, spearheads and harpoons were made principally in antler, while bone was used for other sorts of tools such as needles and awls. This same difference in function was found by Newcomer (1974) in his study of bone tools from Ksar Akil, by Campana (1989) for Natufian tools and by M. Julien (1982) for Magdalenian harpoons.

The reason for this may be found in the characteristics of the two materials, since bone is more fragile and harder, while the toughness and flexibility of antler is more suitable for work involving impact or blows, the purpose which it fulfills in nature (Newcomer, 1974; Julien, 1982; McComb, 1989).

5.4.3. The working of bone and antler in Berniollo and Santa Catalina.

The criteria we employ to reconstruct the technical process on bone materials from use-wear analysis of lithic tools produced the following data: (tables 5.11 and 5.12.):

The analysis of the results concerning work on bone reveals clear differences between the two assemblages. In the Azilian of Santa Catalina, the only activity found was scraping, which we believe did not cover the work of making complete tools. The lack of association of activities in the same tool indicates that the technical actions tended to be isolated, without the channeling of actions which characterize the manufacture of a complete object.

Furthermore, the work was not very intense and was carried out with tools which were hardly specific, as may be observed in the high proportion of natural edges used and the slight importance held by scrapers and burins. The actions took place on limited fragments of material, as indicated by the small length of the active zones. With these activities it is improbable that the more complex types of tools which appear on the site (harpoons, needles...) were made. All these data indicate that these tools were brought already made, the actions carried out being intended mainly for repair of bone tools.

The results from the Magdalenian level are opposite to these. The activities carried out were quite varied with a significant amount of engraving and cutting actions. Moreover, in contrast to the Azilian level, there is a high degree of association of activities in the same tool. The specificity of the active zones reinforces this image, as much in the small proportion of natural edges as in the importance which the scrapers and burins had. We interpret these data as the work of manufacture of complex bone tools, with the execution of all the phases of the technical process, from the procurement of the blank to the finished tool.

In Berniollo, the scarcity of information prevents us from reconstructing the technical processes carried out on bone.

In general terms, the characteristics of the activities carried out on antler with lithic tools in the three levels are similar, without the clear differences observed in the work on bone. In all of them the actions are varied, though with a predominance of scraping. A high percentage of tools show traces of more than one activity, which seems to reflect the existence of a complex technology in the three assemblages. The work was intensive and was carried out with specific active zones, using mostly scrapers and burins.

The conclusions for the three assemblages are similar, and reveal quite complex technical sequences of tool-making, going through all the steps of the technical process.

In Berniollo, it is probable that the higher proportion of engraving, the higher intensity of use and the use of more specific tools implies a higher presence of the phases of obtaining blanks in relation to the two other assemblages.

The importance of scraping in the Magdalenian level would imply that in this assemblage the steps of finishing or repair of tools in antler had more relevance. Among the five unretouched tools which worked antler, four show exclusively scraping actions. Probably, these unretouched tools were employed for the reconditioning of tools made of antler. On the other hand burins and scrapers show a higher proportion of associated activities, which would relate these tools to the manufacture of antler objects.

		BERNIOLLO	SC AZILIIAN	SC MAGDAL.
TYPE AND VARIETY OF ACTIVITIES	Engraving	0	0	6 *(23.1%)*
	Sawing	0	0	5 *(19.2%)*
	Boring	0	0	2 *(7.7%)*
	Scraping	3 *(100%)*	9 *(100%)*	13 *(50%)*
ASSOCIATION OF ACTIVITIES		0	0	50%
ACTIVE ZONE AVERAGE		1	2.2	3.2
DEGREE OF POLISH DEVELOPMENT	High	33.3%	0	19.2%
	Medium	0	88.9%	65.4%
	Low	66.6%	11.1%	15.4%
LENGTH OF THE ACTIVE ZONE		24,3 mm	11.1mm.	20mm.
NATURE OF THE EDGE	Natural	66,6%	77.8%	46.2%
	Fractured	0	22.2%	15.4%
	Retouched	33,3%	0	38.5%
END-SCRAPERS AND BURINS		33%	22%	87%

Table 5.11. Characteristics of work on bone

To summarize, we propose that in the Azilian level repair work of bone tools and the making of tools in antler in all its phases was practised, perhaps with particular attention to obtaining blanks.

In the Magdalenian level the work on bone material entailed the making of complex tools, in bone as well as antler. This information is confirmed by the presence in this level of two basal fragments of deer antler. On these, we may observe clear traces of the removal of shafts, carried out to the exhaustion of the raw material. In antler, the tasks of finishing and repair of tools had special relevance.

The conclusions on the working of bone material at Berniollo are based exclusively on the activities found in the flint tools (no bone tools were found). The data seem to indicate the making of simple tools in bone, and a more complex technical sequence in antler, in which the first phases of obtaining blanks were especially important.

		BERNIOLLO	SC AZILIAN	SC MAGDAL.
TYPE AND VARIETY OF ACTIVITIES	Engraving	9 *(24.3%)*	4 *(18.2%)*	8 *(7.5%)*
	Sawing	4 *(10.8%)*	1 *(4.5%)*	13 *(12.3%)*
	Boring	1 *(2.7%)*	1 *(4.5%)*	1 *(4.5%)*
	Scraping	23 *(62.2%)*	16 *(72.8%)*	82 *(77.4%)*
ASSOCIATION OF ACTIVITIES		50%	50%	40.5%
ACTIVE ZONE AVERAGE		4.1	2.3	2.9
DEGREE OF POLISH DEVELOPMENT	High	18.9%	13.6%	6.6%
	Medium	78.4%	59.1%	75.5%
	Low	2.7%	27.3%	17.9%
LENGTH OF THE ACTIVE ZONE	Cutting	24mm.	13.1mm.	20mm.
	Scraping	16mm.	15mm.	24mm.
NATURE OF THE EDGE	Natural	36.3%	13.7%	41.5%
	Fractured	7.3%	13.7%	12.3%
	Retouched	56.1%	63.6%	46.2%
END-SCRAPERS AND BURINS		100%	79%	82%

Table 5.12. Characteristics of work on antler.

5.5. WOOD

If we look at the importance of working wood in each of the assemblages studied, we observe that in the two levels of Santa Catalina it is somewhat marginal, while in the Epiplaeolithic site of Berniollo it is the material most worked with lithic tools.

Traces of working wood are very scant in the late Upper Palaeolithic sites, in the Magdalenian as well as the Azilian, demonstrated by, among others, the evidence from Pont d'Ambon, Pincevent (Moss, 1983; Plisson, 1985), Verberie (Audouze *et al.*, 1981; Symens, 1986) Andernach, Niederbieber (Plisson, 1985), Paglicci (Donahue, 1985) and Wojnovo (Winiarska-Kabacinska, 1988).

However, it begins to be more frequent in Epipalaeolithic settlements, as G.H. Odell (1977) has shown for Bergumermeer, J. Dumont (1988) for Mount Sandel and Star Carr and H. Juel-Jensen for Vaenget Nord and Ringkloster (Juel Jensen and Brinch Petersen, 1985; Juel-Jensen, 1982; 1986).

This more frequent use of wood in the post-Palaeolithic periods would have been favoured by the abundance of wood brought about by reforestation in the Holocene. However, from our point of view this explanation is insufficient to explain the phenomenon. It will be necessary to provide a larger vision of the problem to clarify to what point the working of wood satisfied new needs or was rather the result of a substitution process for tools in antler and bone.

5.5.1.. **The technical processes for wood.**

It is problematic to establish general criteria for a technical process on wood on account of the great variety of objects which can be made, from habitation structures to miniscule elements which are part of composite tools.

However, in analysing the ways of working this material it is useful to employ the schema of procurement, transformation, and use-maintenance which we have applied for other materials.

With the technical means available in the final Upper Palaeolithic it would have been difficult to bring down large hardwood trees and later to cut them up for transportation. If we take into account the type of tools which we find in late Upper Palaeolithic levels, where tools for chopping are not usual, we can hypothesize the absence of tree-felling as a common activity in this period.

The appearance of the polished axe must have implied a substantial change in this regard. It has been demonstrated experimentally that with this tool it is possible to fell in about fifteen minutes trees having trunks of 20 cm diameter (Coles, 1979, 102). However, the increase of the diameter of the trunk multiplies the working time exponentially and, thus, the felling of a tree of one meter in diameter can require several days of work (Carneiro, 1979, 47, figure 1.15; Coles, 1979, 104). Trees can also be brought down by burning the base

The Australian aborigenes cut down trees by chopping with large retouched flakes, although these are trees of small diameter (Gould, 1970, 10-11; Thompson, 1964, 409). This is achieved by making two opposite cuts and then applying pressure (Hayden, 1979). The appearance of post holes in Azilian levels (Velter and Lamothe, 1976, 102) could be explained by this technique of felling small trees.

Ethnographic observations have shown that many groups of hunter-gatherers use trees and branches which have fallen through natural causes. Branches were collected and cut by hand or with a stone knife (McGrath, 1970, 16).

There are many references to the cutting of young trees or branches to obtain blanks for making lance shafts (Osgood, 1940, 101). The Yamana of Tierra del Fuego chose a suitable branch, made an incision in the proximal zone, put a leather strap around the branch and held on to it, until the branch gave way and broke at the place of the incision (Gusinde, 1930/1982, 454).

The Australian aborigines use branches, stalks, and roots of the acacia for lance shafts (Gould, 1970, 10-11). This strategy not only reduces the cost of obtaining the wood, but also substantially limits the work necessary to make the blank, as the natural form is very similar to the desired form. The Lacandon Indians use canes as arrow shafts (Nations and Clark, 1983).

As a solution to the technological limitations of felling and cutting up trees there is an ingenious system of obtaining blanks for making wood objects. A fragment of wood is obtained by breaking it off from the living tree. First grooves are made by striking with a lithic tool, snapping off the part delimited by the grooves in a second phase, using wedges (Kamminga, 1988, 27; Stewart, 1973, 116). There exist different variations of this technique. Thus, among some groups of Australian aborigines two longitudinal and parallel grooves are made on the tree (Hayden, 1979, 29-37), while in other cases transverse cuttings are made on the trunk, one on the upper zone and another on the lower zone, delimiting the future blank, or only one on the lower (Gould, 1970, 27). In Nevada (U.S.A.) there are trees still remaining with clear extraction scars on the trunk. As in the Australian example two transverse cuttings were made through striking, which defined in length the blank which was sought. The incisions cut off the passage of the sap, so that the part which was wanted dried out after a few months and could be broken off easily with wedges and levers (Wilke, 1988).

When felling tools are used for obtaining wood, these may carry out three types of actions. First is direct violent striking with an axe or adze. These can be massive tools made on a block (i.e. choppers) or on large flakes, both retouched and unretouched. It is also possible to use lithic tools with relatively sharp edges using a longitudinal movement, for sawing branches or trunks of small diameter. Finally, the raw material can be obtained with the help of thick flakes used in indirect hammering, as wedges (Hayden, 1979, 11).

Photo 5.4. Primary shaping of a bow by adzing.

When tools with which wood has been obtained are not very elaborate it is common to abandon them at the foot of the tree once the task is finished (Hayden, 1978, 189). Thus wood-obtaining tools (hammers and wedges) have been observed next to trees with extraction scars, in Australia (Gould *et al.* 1971) and in the U.S.A. (Wilke, 1988, 15-16).

Once obtained, the wood begins to be transformed. The initial roughing-out, if it involves the elimination of a great quantity of superflous material, can be carried out in the same place of acquisition, to diminish the cost of transport (Coles et al, 1978; cf. the case of the Havasupi in Spier, 1928, 147, cited in McGrath, 1970). Chopping, using wedges or the controlled use of fire are effective actions in these cases. More complex phases of manufacture are usually carried out at the residential settlements on the already roughed-out material.

The initial shaping of wood produces the preform, an outline of the desired object obtained by a significant reduction of the volume of the original blank. This phase can be carried out modifying the volume of the blank or its length, or both at once.

The volume of the blank may be reduced by using the physical characteristics of wood, which breaks more easily along its fibres. The wood is more commonly split or cracked longitudinally. This is usually done with wedges which allows control of the point of fracture. The wedge used can be stone but wooden or antler wedges can also be effective, such as those used by the northern Shoshone to obtain blanks for arrows (Lowie, 1909, 192, cited in McGrath, 1970). The Ingalik Eskimo made a groove 1.5 cm deep on the wood to be split, where they placed wedges which were then struck alternately (Osgood, 1940, 101). The Indians of the American Northwest use this technique to obtain boards for construction; the larger wedges are of horn, while the smaller can be of bone or hard wood (Stewart, 1973, 116). The Lacandon Indians and the Yamana split sticks in half with wedges, to obtain blanks for making bows (Nations and Clark, 1983, 37; Gusinde, 1930/1982, 454).

Photo 5.5. Secondary shaping of a bow by scraping with a hafted blade.

R.A. Gould (1970, 10-11) observes that among the Australian aborigines, when the volume of a blank is to be reduced to make a lance, cuts are made in the wood at intervals of a foot, eliminating the material between the cuts with wedges. However, the same author suspects that these techniques came after contact with Europeans, whereas before, young trees, with trunks more suitable in size for making spear shafts, were more frequently employed, which considerably reduced the work of making the object.

Chopping is also used to obtain the preform. The ethnographic examples of this technical action are numerous in the initial shaping of objects: in the manufacture of canoes as well as in making smaller objects such as plates, spears or propellants (Miles, 1963/1986; McGrath, 1970, 111; Hayden, 1977, 185; Richardson, 1988, 58; Hayden, 1979; Knutsson, 1988, 15).

The suitable length of a blank can be obtained by sawing or chopping. Sawing is useful for limiting the size of the wood when working with fragments of small section, as is the case for foreshafts of arrows (McGrath, 1970, 23). Chopping is a more rapid method of cutting a blank. Using this method the Australian aborigines made two opposing cuts so that the blank could be broken in the desired place (Hayden, 1979).

Apart from these actions of primary shaping accomplished by stone tools, actions needing no tools can also be carried out. Much of this work is done by manipulating the wood with feet and hands, with the use of heating and soaking.

The secondary shaping defines the morphology of the object sought. Scraping is the most common action in this phase of manufacture since the initial form may be gradually turned into the desired one by this method. Among the Indians of the southwest United States wood is scraped in the manufacture of bows, arrows, digging or throwing sticks, pincers, splints for basketry, etc...(McGrath, 1970, 23 and 109). The Australian Aborigines scraped wood when making tools such as spears, digging sticks, handles or propellants (Hayden, 1977, 183; Knutsson, 1988, 15).

It is common to burn the wood to make its shaping by scraping easier. Among the Ingalik Eskimo this technique was much used, especially in the making of handles (Osgood, 1940, 76-77). This method is also chosen by the Paiute of Surprise Valley to shape arrows (Kelly, 1933, 142-144, cited in McGrath, 1970, 108) and for various groups of Australian aborigines, who sharpened the ends of their spears with successive burning and scraping (Gould, 1970, 10-11; Kamminga, 1988, 30).

The boring of wood occurs in the production of more complex tools. L.H. Keeley (1983, 252-254) and D.Cahen *et al.* (1986, 79) relate this activity to the manufacture of assembled pieces, in Neolithic contexts. J. Kamminga (1982, 65) records boring work in the manufacture of propellants and other tools in Australia. Among Northwest Coast people of Canada, wood was joined by sewing. Wood pieces were bored and sewn together by means of sprice root threads (Kew and Goddard, 1993, 26)

Sawing is necessary for example in the making of barbs on wooden spears, as occurs among the Pintupi, Australian aborigines, who incise the wood with oblique sawing accompanied by chopping (Hayden, 1977, 185; Kamminga, 1982, 62).

The finishing of the object is carried out in many cases without the help of flint tools. Rubbing with abrasive stones is common (Gusinde, 1930/1982, 454; Hayden, 1979, 11; Nations and Clark, 1983, 37), and in some cases the use of knapping "debris" for filing or rasping is documented (cited in Knutsson, 1988, 14).

Decoration is usually carried out in the final phases of making the object. Gould (1970, 10-11) records that among the Australian aborigines the final phase of manufacture consists of making the personal incision. For this work engraving, sawing or boring may be used (Miles, 1963/1986), although always with less intensity in this task than in the previous manufacture phases.

Methods are used which guarantee the preservation and effectiveness of the object, through the addition of an agent or the transformation of the wood surface. Thus, superficial burning hardens the wood and prevents its rotting, practised by the Benihoaritas of the island of La Palma, who used it to finish their spears, called "mocas" or "varas tostadas" (Abreu Galindo, 1602?/1977). The Selk'nam greased some of their wooden objects with fat and red coloured earth, to help preserve them (Gusinde, 1930/1982, 181-182 and 212-216).

The ethnographic record also provides information on the state of the material when worked. Wood is worked more easily when it is fresh (Digard, 1981, 110). Among the Australian aborigines, if they did not manage to finish the work in a short space of time and the wood dried out, they soaked the wood for two or three days before continuing working. They kept the wood in sand during the night to preserve the humidity (Kamminga, 1988, 28). The Indians of the southwest United States also soaked wood or placed it in steam during most of the work (McGrath, 1970, 23; although exceptions in McGrath, 1970, 109).

The work of maintaining wooden objects, when lithic tools are used, is accomplished by scraping the wood. This work takes place above all in the sharpening of arrows, spears and digging sticks. The cracks which are created in the object by deterioration of the wood can be repaired by applying resin. Among the Australian aborigines, when the crack was large, its two ends were perforated, to avoid its opening further, resin being intoduced into the crack afterward (Gould, 1970, 37).

As occurs in working other materials, many of the activities of obtaining and transforming wood are carried out without tools and others are achieved with tools of wood, antler or bone or with unknapped rocks. Moreover, the technical actions achieved with stone tools do not always belong to only one stage of the technical process; thus scraping is used during the shaping, the finishing or the maintenance of an object, while the use of wedges is possible in the procurement and initial shaping of wood, for example.

Finally, it must be pointed out that the reconstruction of technical processes on wood cannot be based on the study of the objects and the manufacture waste in this material, as it can for bone material. These circumstances limit the interpretive possibilities, as we shall see later.

5.5.2. **Wood working in Berniollo and Santa Catalina**

The characteristics of working wood in the three assemblages are shown in table 5.13.

In the first place, the scarcity of evidence on how the wood was obtained is notable. This characteristic has been observed in many archaeological contexts, including those where wood was plentifully worked (see Hayden 1987,174). In the Carriéres Thomas (middle-late Acheulien) it is the material more often worked, but it was worked only by scraping or brushing (Beyries and Roche, 1982, 273). In the Musterian levels analysed (Anderson-Gerfaud, 1981; Beyries, 1987), the activities which can be related to procurement stages are also little

		BERNIOLLO	SC AZILIAN	SC MAGDAL.
TYPE AND VARIETY OF ACTIVITIES	Engraving	2 *(3.7%)*	1 *(9.1%)*	3 *(10.8%)*
	Sawing	9 *(16.7%)*	1 *(9.1%)*	5 *(17.8%)*
	Boring	2 *(3.7%)*	0	2 *(7.1%)*
	Chopping	0	0	2 *(7.1%)*
	Scraping	41 *(76%)*	9 *(81.8%)*	16 *(57.2%)*
ASSOCIATION OF ACTIVITIES		27.2%	20%	29.4%
ACTIVE ZONE AVERAGE		3.3	1.7	1.8
DEGREE OF POLISH DEVELOPMENT	High	3.8%	0	2.2%
	Medium	74%	36.4%	65.5%
	Low	22.2%	63.6%	32.3%
LENGTH OF THE ACTIVE ZONE	Cutting	29mm.	12mm.	26mm.
	Scraping	19mm.	14mm.	19mm.
NATURE OF THE EDGE	Natural	53.7%	54.5%	35.7%
	Fractured	16.7%	0	17.9%
	Retouched	29.6%	45.5%	46.4%
END-SCRAPERS AND BURINS		38%	46%	49%

Table 5.13. Characteristics of work on wood.

represented. In Cassegros, where wood is the second most worked material, only one chopping action was observed (Vaughan, 1985, 58). In Star Carr, chopping and sawing do not reach even 0.5% of the work on wood according to the calculations of J. Dumont (1988, 129-130). The same occurs in the Neolithic sites studied (cf. Cahen et al, 1986, 76). Only at Mount Sandel (Dumont, 1988, 130) and in El Castell (Vila, 1987, 121) have a larger number of lithic tools been related to the procurement raw material.

It is not very probable that chopping and cutting activities on wood in the Magdalenian level (table 5.13) reflect procurement on a large scale. One of the pieces used for chopping was used as an adze, while the other, which was used as an axe, is relatively small in size, which leads us to think that they were intended for the primary shaping of wood. The tools with cutting traces show active zones of short length.

The absence of flint tools which reflect the procurement of wood is explained in part by the possibility of collecting without the need for flint tools. It is also possible that these tools would have been left behind in the place of activity, as with the Australian Aborigines (Kamminga, 1982, 60) or the Indians of Nevada (Wilke, 1988). This explanation has been suggested by P.C. Anderson (1981, 72) for the Musterian and D. Cahen *et al.* (1986, 78) for the Neolithic.

At Berniollo, working wood was the activity which most required lithic tools. The importance of the transformation of wood in the other two assemblages was considerably less.

In the Azilian level there is less variety of activities than in the two other assemblages, with a clear predominance of scraping. Moreover, there is a lesser degree in the association of activities in the same tool, in the intensity of use and in the length of the active zone. All this evidence seems to indicate that the working of wood which took place in this context was not very complex, whether in the type of tool made or in the phases of the technical process which were carried out.

Comparatively, working wood was more complex in the Magdalenian level and in Berniollo. In the latter assemblage the use of unretouched tools for working wood stands out, while in the two levels of Santa Catalina retouched tools are used more often.

We have considered the use of unretouched tools to be one more criterion which characterizes the carrying out of simple tasks. However, this assessment cannot be applied to the working of wood in Berniollo, since the rest of the criteria indicate the opposite. Therefore, in Berniollo, unretouched tools were used as specific tools for working wood. Unretouched flakes and blades are perfectly suitable tools for this task, as demonstrated by ethnographic observations (Gould *et al.*, 1971; Hayden, 1986).

In other archaeological contexts, a relation between unretouched tools and wood working can be established, as on the Mesolithic site of Vaenget Nord (Juel Jensen and Brinch Petersen, 1985), in relation to blades, and in the Neolithic of Darion (Cahen *et al.*, 1986), in relation to flakes.

In Berniollo, a high proportion of the tools used on wood presented several active zones, and different types of activities were carried out with the same tools. The five pieces on which traces of wood-cutting were found also present scraping traces. This association of different activities in the same tool indicate that interconnected technical sequences were carried out. The low specificity of the active zones, the scarcity of boring and engraving, and the length of the active edges indicate that these objects were relatively large and simple.

In comparative terms, during the Magdalenian occupations at Santa Catalina, working wood was carried out in order to produce more complex tools as shown by the greater variety of activities.

5..6 MEAT

5.6.1 **The technical processes related to meat**

Butchering is a fundamental part of the economies of hunter-gatherers. We have discovered traces produced by this activity in the three assemblages studied.

The behaviour of hunter-gatherer communities in regard to butchering offers a series of similar characteristics. If the animal killed is large, such as in the the case of deer, horses, etc., a primary quartering of the animal is carried out, so it can be more easily transported to the place of processing or consumption. Among other examples, this type of activity is practised by the Nunamiut Eskimos (Binford, 1978), the Kung bushmen (Lee, 1979, 219), the indigenous peoples of Tierra del Fuego (Mansur-Franchomme, 1983a, 412), the Indians of the American Southwest (McGrath, 1970, 12-13) and the Mitassini (Spiess, 1979, 23). Small animals are usually carried whole to the place of habitation.

The Selk'nam of Tierra del Fuego dissemboweled guanacos in the place of slaughter, then transport them whole to the encampment. However, they quartered the animal in situ if for reasons of distance, bad weather or fatigue they could not bring the animal whole to the settlement (Gusinde, 1930/1982, 255).

The principal action related to butchering is cutting, employed to open the abdomen of the animal, to skin it and to dismember it at the joints.

Besides cutting at the joints, some bones can be broken to make the butchering easier. This can be carried out without tools, simply with muscular force, using leverage on the bones. The Nunamiut break the proximal zone of the ribs where they connect to the spine, opening the thoracic chamber and forcing one side of the ribcage outward. They use a similar technique to cut up the spine, pushing vertically on the ground; pressure is exercised towards the exterior until it breaks (Binford, 1978, 46-80 and 94-97; 1981, 147). Bones can also be broken by blows with blunt objects. The Australian aborigines use wooden sticks to strike the proximal zone of the ribs and separate them from the vertebrae, while the Nunamiut use similar instruments to break the leg bones, especially when animals are frozen (Binford, 1981, 144). It has been pointed out that Australian aborigines butcher large animals with wooden wedges, use rocks to crush and small flakes to cut hide and tendons (Gould *et al.*, 1971, 157). In the Mesolithic level of the site of Poeymau (Pyrenées Atlantiques, France) traces were found which indicate the use of relatively blunt tools for butchering, in order to section the spinal column (D. Helmer, pers comm.).

An axe can also be used for butchering, as do the Hadza, in order to separate the ribs from the breastbone and the spinal column and to section the spine (O'Connell and Hawkes, 1988, 118-120), and the Navajos (Binford and Bertram, 1977, 91-93). Traces of butchering by axe are not rare in Neolithic levels (Helmer *et al.*, 1987); however, we have no references on the use of this technique in the Upper Palaeolithic (Rueda, 1993).

After primary butchering a secondary cutting up of the animal occurs, preparing the meat for consumption or processing it for preservation. Various authors make a distinction between the first phase, "heavy" butchering, and the second, "light" butchering (Mansur-Franchomme, 1983a, 404; Plisson, 1985, 209). The traces resulting from the first show signs of contact with bone and cartilage.

The final step in butchering is the breaking of the long bone to extract the marrow. This work is efficiently carried out by striking the bone against an anvil which has a blunt edge or striking the bone with a hammer held in the hand (Binford, 1981, 148-163). We do not reject the possibility that flint tools could have been used for this work, in direct or indirect hammering, but we consider this improbable, given the efficiency of techniques and tools which are simpler.

5.6.2 **Butchering in Berniollo and Santa Catalina.**

We have observed tools with traces of butchering in the three assemblages studied. In the majority of cases, unretouched edges were used, which are more effective than the retouched edges (Walker, 1978, 710). In regard to the morphological types of the tools, backed bladelets, untouched tools and truncated blades were chosen, showing a diversity which probably has a functional meaning, as we shall see later.

Most of the unretouched tools used for butchering are blades (80%). The use of unretouched blades for butchering occurs in many archaeological contexts, as in Habitation I at Pincevent (Plisson, 1985, 209), Niederbieber (Plisson, 1985, 288), Roc de Migdia (Rodríguez, 1993, 69) and in various sites of the LBK Neolithic (Cahen *et al.*, 1986, 52). This show a logical adaptation between form and function, since the blades present sharp, straight edges, suitable for this work. Some of the blades used in butchering are truncated. We interpret the truncation of these tools as preparation for hafting (see figure 3.7., p. 38).

Besides this type of tool, we have discovered traces of butchering on backed bladelets. In general, the traces of meat-cutting in the backed elements are fainter than on the larger pieces, with less microscarring and fewer polished areas from contact with hard materials, such as bone or cartilage.

Probably, these two morphologically different types of tool with a similar function is indicative of an adaptation of the lithic tools to different phases of butchering. For the primary butchering the tendency was to choose larger tools and strong edges, preferably unretouched blades. For a secondary butchering, for consumption or processing, composite knives made with backed bladelets were used.. Thus the lesser intensity of traces on the backed bladelets and the dual morphology of the butchering tools would be explained.

Observation of the edge angles of the tools reinforces this hypothesis. The average angle of the backed bladelet edges having traces of butchering is 24 degrees, while the average increases to 38 for the rest of the tools with traces of butchering. Tools for primary butchering present a more robust edge that those for secondary butchering.

Butchering was carried out more often in the Azilian level than in the Magdalenian. In Berniollo this activity is least represented, although it is necessary to take into account that probably the greater degree of alteration of the tools would make the identification of this task difficult, as it leaves traces of little intensity on the active edge.

5.7. PLANTS

5.7.1. **The technical processes on plants**

The non-woody plants consitute one of the principal raw materials manipulated by most groups of hunter-gatherers. Plants represent an important part of food resources and are also basic to the manufacture of many objects and structures (Keeley, 1992; Harlan, 1992). B. Bocek (1987, 110-112) shows that in the ethnographic record of California are found hundreds of plants collected for reasons other than consumption, fuel, clothing, ropemaking, containers, construction materials,...

Most of these plants are collected without the help of any kind of tool, simply by hand. Bushmen women, when they go out to collect, carry a large hide to carry the plants collected and, at the most, a stick to help them (Lee, 1979). Similar references are common in all sorts of contexts (McGrath, 1970; Hayden, 1977; Nelson, 1981; Kamminga, 1982; Kelly, 1985).

Wild cereals are gathered mainly by shaking the ears with a stick and gathering the grains that fall in a recipient, as has been documented in communities in North America, in the Sahel and in India (Sigaut, 1985; Hillman and Stuart Davies, 1992). This system of gathering is possible because the wild cereals have a fragile rachis, which easily releases the grain when the ear is ripe. The harvesting of wild cereals when ripe would thus be ineffective. However, when the ear is still green or half-ripe, it may be harvested efficiently with stone implements as has been proven experimentally (Anderson, 1992). There are some ethnographic examples of this. R.L. Kelly (1985, 68-69) records the use of stone knives for grain harvesting in the central United States as J. Kamminga (1982, 54) does for cutting still-green cereals by the Australian aborigines. For Natufian sites, various authors have recognized the harvesting of wild cereals (Anderson, 1992; Unger-Hamilton, 1988 and 1989; Fujimoto, 1988), which could be a key to the domestication of cereals (Hillman and Stuart Davies, 1992).

The transformation of plants for consumption does not usually require the use of stone tool edges. However, the Australian aborigines of Arnhem Land scrape and cut yams with certain types of shell, in a context of great lack of workable lithic material (Kamminga, 1982). Most plants are treated without tools, although in some cases they are ground for better digestion, for mixing with other foods or in order to make storage easy. Grinding is carried out with mortars and pestles of stone or wood. .

Plants are also abundantly used among communities of hunter-gatherers to make textiles, basketry and cord. In the Near East the existence of rope and basketry is plentifully documented for the 8th millennium (Stordeur, 1989; Barber, 1990). For earlier periods we have a fragment of cord found at Lascaux, dated between 15,000 and 14,000 B.C. (Delluc and Delluc, 1979)). Z. Castro Curiel (1990) interprets some engravings at Parpalló with serpentine lines, intertwinings and interweavings as a demonstration of rope, basketry and textiles at the end of the Pleistocene (11, 000-10,000 B.C.).

There is an enormous quantity of plants which can be used as textiles, basketry and cord, to which may be added the bark of different trees (Miles, 1963/1986; Kuoni, 1981; Rodriguez Santana, 1989; Schick, 1989; Stordeur, 1989; Bocquet, 1989; Rivera and Obón, 1991, etc.). The gathering of a good part of such materials implies the use of cutting tools. For the gathering of pita fiber and mescal a digging stick is needed as much as a knife to cut stems and leaves (McGrath, 1970, 14; Nelson, 1981, 15). In the North American Southwest there are plenty of references to the use of sickles in horn to gather grass in great quantities, for caning beds, making screens, basketry, etc. (Heitzer, 1951).

The preparation of these materials implies new possibilities for the use of stone tools. Scraping is common on different plant species such as pita fiber, esparto grass, sisal, broom, nettles..., to make fibers with which to fashion cord, textiles, etc. (McGrath, 1970, 24; Nelson, 1981, 16; Miles, 1963/1986, 109); these plants are usually worked fresh or soaked, if they are already dry.

Plants used for making baskets are treated in a similar way. Willows and brambles can be scraped to remove the outer skin (McGrath, 1970, 111; Kamminga, 1982, 54; Godeau, 1989; Blanc and Gury, 1989; Hardouin, 1989; Rivera and Obón, 1991), although the first two species can also be used whole (Digard, 1981, 116). During manufacture of basketry , other than an awl which eases the passing of the bands, a cutting instrument is necessary. The knife can be of stone (McGrath, 1970, 24) and it is used to cut the stalks which are too long or to section longitudinally. These technical actions continue to be practised by basketry craftsmen, as we have been able to confirm in various places in Spain.

In other cases basketry and braiding are carried out without any help from cutting or scraping tools. Among the Yamana and the Selk'nam reeds are pulled up by hand. Afterward roots are eliminated by burning. The only tool which is used in basketry with reeds is an awl which helps in the braiding (Gusinde, 1930/1982, 232 and 487).

Another group of plants, basically canes, are used for making arrows. They are chosen for their natural suitability in many regions of the World (Nations and Clark, 1983; Gassin, 1993, 191). In these cases, the transformation process is reduced to little modifications by scraping.

5.7.2 The working of plants in Berniollo and Santa Catalina.

In Berniollo, the two tools used in the working of plants are small. These are two denticulated bladelets, which were probably hafted. One of these was used to scrape plants, as much by the denticulated edge as by the opposite, unretouched edge. The functional inference for the other piece to the working of plants is not sure, as it could also have been used to cut soft wood when fresh.

The working of this material is of little importance in the Azilian level of Santa Catalina. We have found only a short blade with traces of scraping plants on one side.

In the Magdalenian level the working of plants is more in evidence. There are three flakes with a total of five active zones, of which four are for scraping and one for cutting. The work was carried out with unretouched flakes of medium size, about 30x30 mm. Also, we have a burin with traces of plant-scraping on an unretouched edge, while the opposite edge (the burin facet), show traces of scraping antler. Probably the two uses were separated in time, although we do not know the order in which they happened. The work was carried out with unretouched flakes of medium size, about 30x30 mm.

A. van Gijn (1989, 81) interprets transverse movements on plants to be for scraping the fibres used for different manufactures, and the longitudinal movements for collecting plants for technological purposes.

For scraping plants, tools with different types of edge angles were used, from 17 to 90 degrees, denticulated edges as well as crude. There was a tendency to use unretouched pieces, except at Berniollo. In general this was delicate work and not very intense. This scraping work would have to be interpreted as preparation of the plant fibers for braiding.

The two pieces with traces of cutting would probably also reflect technological activities, although we have no data to determine whether this was gathering or manufacture.

The results obtained indicate that the working of non-woody plants with lithic tools was not very important, being somewhat more significant in the Magdalenian level of Santa Catalina. Plant working with lithic tools is very scarce in late Upper Palaeolithic sites. This activity was not found at Verberie (Audouze *et al.*, 1981), at Pincevent (Cahen *et al.*, 1980; Moss, 1983; Plisson, 1985), in area 4 at Niederbieber nor in the "Upper Palaeolithic" level at Andernach (Plisson, 1985, 289). Furthermore, it is very scarce in the Magdalenian and Azilian levels of Pont d'Ambon (Celerier and Moss, 1983, 99) and in the Magdalenian levels of Andernach (Plisson, 1985, 267) and Cassegros (Vaughan, 1985, 58).

Previous use-wear analysis have shown that the use of plant resources was more intense during the Epipalaeolithic/Mesolithic than in the Upper Palaeolithic, this process leading to the Neolithic as has been demonstrated for the Near East (Anderson, 1988 and 1992; Unger-Hamilton, 1989; Fujimoto, 1988, 171) and northern Europe (Juel Jensen, 1988, 246; Juel Jensen and Brinch Petersen, 1985, 47; Juel Jensen, 1994).

CHAPTER 6

SETTLEMENT ECONOMY

For the process of reconstructing the economic activities which occurred in the three assemblages studied, we have first approached questions concerning the use of lithic tools, and then those concerning the materials worked by the tools. Following this approach we must then reconsider the significance of the above data so as to reconstruct the overall image of each assemblage.

According to their particular cultural tradition and the determining factors of the environment, prehistoric communities developed a system of decision-making which rationalized the acquisition and consumption of resources. Each type of behaviour is not an isolated event, but rather is linked to those preceding and following it. We refer to this system of decision-making as the strategy of exploitation of the environment.

R.B. Lee (1979, 117) affirms that among the common features of hunter-gatherer communities are those of mobility and the existence of certain group structures sufficiently flexible to adapt to changing opportunities.

In fact, mobility and the flexibility of group structures constitute mechanisms of response to an irregular distribution of resources, as much in time as in space. The technology of food conservation is very little developed, which limits the storage of food, a process making sedentarism possible. The high group mobility generates a large number of settlements, while the flexibility of social an economic structures will bring about a high variability in the characteristics of each of the settlements.

The economic behaviour carried out in any one settlement represents only a part of the overall picture of the exploitation of the environment or landscape in general. Thus, the functionality of a habitation cannot be understood in isolation, but rather as a coherent assemblage which acquires its functional sense in the strategy of exploitation of the environment carried out by prehistoric human groups.

Continuing from this premise, we consider that any attempt to understand settlement economy should begin by establishing the type of economic strategies which are possible in communities of hunter-gatherers. Once the possible outlines are established it is necessary to decide which type of settlements correspond to each model, and to define each category of settlement according to the activities which would occur there. In this way the significance of behaviour detected in an archaeological context can be situated within an overall system of reference.

6.1 THE STRATEGY FOR EXPLOITATION OF THE ENVIRONMENT.

Since the 1960's, with the development of the "New Archaeology", efforts have been made to follow the path of using ethnographic observation as a point of departure.

Within strategies of exploitation of the environment a first model was established in which three possible types of settlement were presented: "base camp", "work camp", and "transient camp" (Binford and Binford, 1966, 268-269). The base camp is a place of year-round continuous habitation, in which the tasks of manufacture and repair of the tools as well as the preparation and consumption of food are carried out. Groups go out from the base camp to acquire specific resources, such as meat and hides, lithic raw material, etc. The places where these groups stay are specialized camps. Along the way to places where resources are acquired, a series of transient camps where the group rests come into being.

This framework was based mainly upon observations made of Eskimo peoples. However, the study of the !Kung bushmen (Lee, 1979) demonstrated that in these hunter-gatherer communities there is no creation of settlements during activities of acquisition of resources. The change of habitat occurs only when the resources of the zone have become exhausted (Yellen, 1977; Lee, 1979).

These observations were used by L.R. Binford, who in a later work distinguished two systems of economic adaptation common to hunter-gatherer societies: the "forager" system and the "collector" system (Binford, 1980). The "collector" groups or groups of logistic mobility are characterized by the storage of food, at least for part of the year. A logistical organization, involving provisional camps, occurs for groups which go out to acquire provisions. The products are partially processed in provisional camps, then transported to the residential base. Therefore, this system is characterized by three types of settlement: "residential base", "field camp" and "station".

The field camp is the place where the provisionment group lives while carrying out a particular task. In the station the movements of prey are observed and the strategy to be followed is established. Thus, settlements appear for different reasons.

In "forager" or residential mobility communities there is no storage of food, so it must be collected daily. These groups exploit the territory directly from the site, which is abandoned when the resources run out. In this scenario there may be two types of areas used: the "residential camp" and the "location". In the latter defined tasks are carried out, but if these activities are not intensively repeated in the same place these activity areas are difficult to detect. The variability of settlements, if this occurs, will reflect a seasonal differentiation in the activities and/or variations in the duration of the habitation. This model is illustrated by the bushmen or the Australian aborigines of the Central Desert.

However, the duality established by Binford does not exhaust the possibilities of organization of hunter-gatherer groups. In fact the ethnographic record presents an extremely complex picture, with numerous intermediate possibilities between both strategies. The behaviour of some communities does not even fit into this scenario of mobility as there exist hunter-gatherers who are practically sedentary. Because of this, ethnographic references will not be used here as examples for a set typology of economic strategies. They are useful in that they at least present models for possible strategies, and above all indicate which factors define the variability of such strategies.

To study prehistoric communities we start with the elements which define the strategy of exploitation of the environment. In our view this is defined as the distribution of the human group and the activities of acquisition, transformation and consumption of resources in places of habitation and work in a definite time and area. Therefore, the economy of a settlement and the strategy of exploitation of the environment are understood using these two factors: the human group and the activities which occurred. These two factors evolve in spatial and temporal dimensions.

The concept of "human group" refers as much to the size as to the composition of the group which occupies the settlement. There are numerous ethnographic examples which indicate that these factors vary during the annual cycle (Damas, 1968, 111; Yellen, 1977, 38-39; Lee, 1979, 350; Balikci, 1968, 80). In general terms, the settlements could be made up of extended families, nuclear families or a collection of individuals to carry out a specific task (Sahlins, 1968; Watanabe, 1969, 73; Binford, 1978).

The second element within the strategy of exploitation of the environment is the technical system, the organization of the activities of the acquisition of resources, their transformation and consumption. The execution of these activities within the general economic system can occur in many different ways. The dimensions in which resource acquisition activities are carried out can be temporal or spatial. In the first case, procurement of food and raw materials varies during the different seasons of the year, in relation to the availability. This general behaviour is very common in hunter-gatherer economies, the way in which it occurs differing according to the relation between what resources they have and which ones they want. Ethnographic examples can illustrate this type of system. The Athapaskan Kutchin of Alaska collect forest fruits in the autumn, as well as fishing and hunting certain animals, which continues during the winter. In the spring, fishing and the hunting of ducks, geese, bears and other animals begins again, while summer is a period of little activity (Nelson, 1973, 27-30). Systems of this nature, although with different characteristics, occur among the !Kung San (Lee, 1979; Yellen, 1977), the Nunamiut (Binford, 1978), the Ojibwa and the Mistassini (Odell, 1980, 415), among others.

Acquisition activities have a spatial dimension in systems in which there are settlements specialized in the procurement of one or several resources. In these systems, a group of individuals leaves the established settlement in order to stockpile a raw material, which can be lithic, a particular type of animal, etc. Later the material collected is taken to the base camp. The distance between the acquisition area and the central habitat may be sufficiently far or the time of acquisition long enough to require a provisional settlement. In this case a simultaneous network of settlements is established (Binford and Binford, 1966).

The Ainu of Japan exploit the river valley and the terraces from the principal settlement. However, hunting deer and bear, which occurs in mountainous and hilly zones, is carried out from hunting cabins periodically occupied by groups of hunters (Watanabe, 1968, 71). The Nunamiut establish hunting camps during the caribou migration towards the north in spring complemented by places for slaughter, quartering and storing the meat (Binford, 1978, 171-178; Binford, 1983, 117-118). This type of behaviour is also found among the Athapascan Kutchin (Nelson, 1973).

The work of transforming the materials can be related in different ways to the work of acquisition. Among the !Kung the process of manufacture continues throughout the year. They work sporadically and for a short time, and thus transport the products half-made from one camp to another (Yellen, 1977, 76). In this case the manufacturing activities are fully integrated into procurement activities.

In other cases the work of manufacture takes place preferentially at moments when procurement activities are not so prevalent. Thus tools are made which will be used at a later time. G.H. Odell (1980, 415) points out that among the Canadian Round Lake Ojibwa and Mitassini tribes, the autumn is a period of great activity, during which, among other tasks, the repair and manufacture of equipment and clothing is carried out, in preparation for the winter season. According to this author, this type of seasonality is common to many peoples of the central and northern zone of North America, such as the Athapascan Kutchin (Nelson, 1973, 27-30).

In general, specialized settlements are usually intended for the procurement of a particular resource. The work of manufacture, on the other hand, takes place in the stable settlements. Carrying out costly tasks in the specialized habitat would interfere with the principal activity, the acquisition of the desired resource. Thus, the Nunamiut bring their tools to the hunting settlements, ready to use when an animal appears. Because of this, evidence of the first phases of manufacture of tools rarely appears in these places. However, during waiting periods, broken tools are repaired or half-made tools are finished (Binford 1983, 129-130).

As reflected in the examples which we have presented, the technical system of a group of hunter-gatherers is defined by the characteristics of the organization of the activities of acquisition, transformation and use/consumption over time (annual cycle) and in space (territory exploited). It is necessary to take into account as much the distribution of economic activities in time and space as the way in which these activities are interrelated. The technical systems can be more or less complex, according to the degree of planning of the economic activities carried out by the prehistoric group.

Planning of economic activities implies the existence of discontinuities in the technical processes. The development of each activity occurs in the place and at the moment most favourable to the success of the work. For the planning of economic activities to be profitable it has to be accompanied by transport and storage technologies. In effect, concentrating the acquisition of a resource in one particular time and place can increase the quantity of the resource obtained, but this does not make sense if the surplus cannot be stored. Also, the surplus has to be transported to the place where the resource will be transformed or consumed. In this sense, the fundamental factor which differentiates the more planned (collector) technical systems of hunter-gatherer communities from the less planned (forager) is that the former seek to bring the resources to the consumers, whereas the latter are the consumers, who move according to the resources (Binford, 1982). As we have seen, planning occurs in a temporal dimension, when the different economic activities are distributed throughout the annual cycle, or in a spatial one, when the activities are carried out in different zones of the territory exploited. Both types of planning are not exclusive, but rather complementary in many cases.

The archaeological consequence of technical systems with a high degree of planning will be the functional diversity of the settlements of one group. As we are studying non-sedentary groups, technical systems with a high degree of planning lead to settlements in which all parts of the system are not represented in a balanced way. Rather, a part of the work of acquisition, transformation and use/consumption of the different resources predominates over others, demonstrating that the carrying out of some economic activities is subordinated to that of others which are more important. The hierarchical organization of economic activities carried out in each settlement is reflected by the presence of discontinuities in technical processes of working a material.

In regard to lithic tools, the degree of planning of the technical system will have recognizable effects. In systems with a high degree of planning, discontinuities are observed in the knapping process, as well as in the technical processes for other materials. The fact that definite points in time are intended for knapping, in preparation for future necessities, will imply the appearance of stocks of blanks or finished tools. As we have mentioned above (see Chapter 4), the strategy of using lithic tools must insure that the suitable tool is available in the place and at the time necessary. In a strategy with a high degree of planning, tools must be available to be transported from one place to another.

The hierarchical organization of economic activities carried out with stone tools will imply a differential exploitation of these tools: the tools which were more costly to make being destined for the work which is considered most important in each given context. As we have seen above (Chapter 4), this differential exploitation can be based on the raw material used to make the tool, the blank, the fact that the tool is retouched or not, and its morphology. In systems with a high degree of planning a more careful selection of tools intended for principal tasks occurs, with more specialized tools appearing. In systems with a low degree of planning the type of selection of tools in relation to the different economic activities will be very similar. The use of specialized tools can have as an objective efficiency in the transformation of a material, or to obtain resistant edges for longer use or suitability for a definite context of use. An example of this last objective is curated tools, whose presence implies planning for the activities to be carried out with these tools.

6.2. ACCESS TO INFORMATION ON SETTLEMENT ECONOMY: POSSIBILITIES AND LIMITATIONS OF THIS STUDY.

Besides data from stone tools, the study of other archaeological material provides information on the economic activities of prehistoric groups. The analyses of technology and function of bone tools, of the strategy of exploitation of animals as seen from faunal remains, or of the plant remains preserved on the sites, offer important information.

For Santa Catalina and Berniollo, this information is not yet available, and therefore, we cannot use it in a detailed reconstruction of the economy of the settlement

However, we can evaluate the degree to which the analysis of the use of lithic tools can be useful for a precise reconstruction of the economic strategy of the settlement and of the general picture of the exploitation of the environment.

In the first place, the data obtained refer exclusively to the economic activities carried out. The size and composition of the group and the point in the year when the settlement was occupied are unknown factors. Our study can only produce information on the second factor which defines the strategy of exploitation of the environment and the economy of the site: the economic activities carried out. It is necessary to emphasize also that the nature of the analysis is such that it only provides understanding of behaviour involving lithic tools.

On the other hand, in each settlement a sequence of decisions is represented which are integrated in an overall economic picture, a picture for which our vision of the general questions is still limited, given that only three assemblages have been studied. It is necessary, therefore, to widen our perspective through the analysis of a larger number of archaeological assemblages.

The nature of the archaeological record imposes even more lack of definition. In the case of the Azilian and Magdalenian assemblages of Santa Catalina we are dealing with sediments of a certain thickness. It is logical to suppose that both levels are the result of a succession of settlements. This leads us to think that we could be considering together moments of occupation which, in fact, represent different economic strategies. If this was the case, the overall consideration of the attributes of settlements for different purpose would lead us into error.

This problem is not so acute in the case of Berniollo, an open-air site with a single level of occupation. Although we cannot confirm with absolute certainty that it represents a single episode of occupation, we consider this hypothesis to be the most probable, and in any case the number of settlement phases was very limited (González, 1993).

E. Camilli (1989, 18) points out ethnographic examples in which occupations of different function are superimposed. Among the Nunamiut, some sites are the result of different residential occupations in particular years, but served as settlements for specialized activity at other times. R.K. Hitchcock (1982) describes the same situation among the Kua groups, who on hunting trips established camps next to abandoned stable settlements. R.A. Gould (1980) shows similar patterns of reuse for the Ngatatjara of western Australia.

We are conscious of the fact that when we speak of the economy of a settlement for one of these levels we are considering the general tendency of the different episodes of occupation. It is possible, however, that the same settlement could have been used successively with very different purposes. If these superimpositions were produced in the archaeological record, the result could be the appearance of assemblages of ill-defined economy. Therefore, when an ill-defined economic pattern is found, it will be necessary to raise this possibility. If the economic pattern is well-defined, we will assume that there was a tendency for the site to be used in the same way over a long period of time.

Given the questions which we have raised, the development of studies on prehistoric economic strategies must involve the following lines of investigation:

- It is necessary to make progress in the construction of theoretical models of the strategies of exploitation of the environment by societies of hunter-gatherers. For this the behaviour observed in ethnographic contexts must be studied systematically. Then characteristics can be established for interpreting data in the archaeological record corresponding to each one of the models previously established.

- A large number of archaeological assemblages need to be analysed, so that patterns of types of settlements for a particular period can be found. Comparison of ethnographic models and the archaeological reality will allow for better interpretations.

- We need to integrate different fields of knowledge contributed by the archaeological record, such as information on sources of procurement of flint, and on lithic and bone technology, in order to reconstruct technical operations with lithic tools in greater detail.,

However, we believe our data already allow outlining some of the features of the economic patterns of the settlements within the period and the geographic context with which we are dealing.

The possibilities offered by these data for establishing general aspects of settlement economy are based on concepts of the technical process and management of the tools.

The study of the management of lithic tool (chapter 4) allows us to understand differences in their use according to whether the tool was retouched or not, whether it is a flake or blade, and what raw material was used to make it. Such differences in the use of the tools indicate the existence of a hierarchical organization of economic activities. In general, the tools with a higher cost of manufacture are used for the activities considered crucial to the group (Torrence, 1983).

The concept of technical process allows integration of technical actions within a dynamic process, that is, within space-time dimensions. The characterization of the technical processes carried out in each of the assemblages studied (chapter 5), allows us to understand the degree of planning in the activities involved in the economic system of which the settlements studied form a part.

Thus, in those cases where discontinuities are found in the technical processes carried out in an archaeological context, we can assume that the phases before or after these discontinuities occurred at another time and place. Therefore, activities carried out in a settlement can be connected with other ones which took place outside the context studied. The existence of these discontinuities indicates planning of economic activities, in which tasks are not carried out on a continuous basis but deferred.

If we do not find discontinuities in any of the technical processes, we can deduce that the degree of planning of economic activities is slight. We will assume that the settlement had centralized a sequence of tasks involving procurement and transformation of resources.

Concerning the tools, systems with a high degree of planning are characterized by the existence of discontinuities in the processes of making the tools, in the appearance of caches of blanks and of finished tools, and in the differential exploitation of the tools.

Taking these considerations into account, we observe, on the sites inhabited by groups with technical systems having planning, the following characteristics:

- imbalance in the proportion of activities intended for procurement, transformation and use/consumption of resources,
- discontinuities in the technical processes carried out on the different materials,
- storage of resources and
- differential exploitation of the tools.

The reverse of these indicates technical systems with little planning.

In systems with a high degree of planning, the planning can occur throughout the annual cycle, or throughout the expanse of the territory exploited. In the first case, the diversity in the function of the settlements will be reflected in the type of resources obtained, which will differ according to the resources available at the time of year the occupation occurred. The systematic association in the various settlements studied of occupations occurring at a given period of the year, with the procurement of seasonal resources and the occurrence of certain activities of transformation, will reflect planning of a technical system throughout the annual cycle.

Planning in regard to space will imply the existence of an organized network of settlements, with a series of central or base settlements and other specialized ones. Concerning the technical processes, on the specialized settlements procurement of one or several resources occurs and the procurement phases are better represented, than those of transformation and use/consumption. Also the technical processes or sequence of operations which are not directly related to the principal activity of the settlement tend to be avoided. In base settlements all activities of the technical system are carried out, although to a lesser degree for those which have to do with the procurement of resources. In specialized settlements tools more costly to make are used for the major activities of the settlement, and manufacture and repair activities of the tools are limited to those used for the principal tasks.

In the base settlement all types of tools are made, some of which will be stored as blanks or as finished tools. If there is no hierarchical organization of tasks it implies that there is no differential exploitation of the tools.

Finally, a relationship exists between the technical systems which entail planning in space and the presence of curated tools. The presence of this type of tool does not imply the existence of base and specialized settlements (Bamforth, 1986, 40), but on the contrary, systems of exploitation of the environment with planning in space need curated tools (Binford, 1977, 35). The base camps are areas of manufacture and storage of this type of tool, whereas the specialized camps are areas of use and discard. This implies that the characteristics of curated tools can provide information on the economy of the settlement: In specialized settlements, tool manufacture waste is rare, and curated tools appear to have been more intensively used than the other tools. In the base settlements, debitage and retouch products will be present, storage of blanks and finished tools will be found, and curated tools will be less intensively used than the rest of the tools used for daily activities.

At present, we do not have a series of objective criteria which allow us to define in absolute terms the degree of economic planning in a settlement, according to the type of specialization which it reflects. Because of this, planning is discussed in relative terms, compared to the assemblages we have studied. Therefore, an assemblage will be characterized as more or less specialized as compared to the other two.

Using this approach, the information on the use of lithic tools carried out in Berniollo and Santa Catalina allows us to begin to understand the economic patterns of the groups studied, to be verified by future investigations.

6.3. THE AZILIAN LEVEL OF SANTA CATALINA

The observation of the type of activities which are carried out in this assemblage, in comparison with the others, indicates that although data for hunting activities are scarce in absolute terms, these activities are far more frequently attested in this level than in the other two assemblages.

Furthermore, butchering and the working of fresh hide are more prevalent here than in the Magdalenian level and in Berniollo.

These data indicate that, relatively speaking, the settlements which occurred in Santa Catalina at the phase of the Azilian were oriented predominantly towards hunting and to the first stages of transformation of animal resources. Such factors indicate a greater specialization of this assemblage.

The characteristics of the series of operations involved in the manufacture of tools seem to confirm the specialized orientation of the economy of the settlements. It shows the limiting of a effort invested in activities which were not directly related to the main settlement economy.

Concerning flint knapping, blade production was principally oriented towards the production of bladelets, which represent about 75% of the blade blanks. These pieces tended to be used only with retouch, as backed bladelets and backed points. Therefore, knapping was intended for making tools destined for hunting or for butchering.

According to R. Donahue (1985; 1988) the Italian site of Paglicci shows a clear economy of hunting stands, since butchering is the predominant activity. Tool-making is very poorly represented. In this assemblage, as in our case, lithic production is oriented principally towards the manufacture of bladelets.

The technical processes for making tools and other objects in bone and wood seem to reflect the aim to simplify to the maximum certain technical processes which were carried out in the settlement.

In this level, complex tools in bone found at the site, such as awls and needles, could not have been made with the type of technical actions which we have found. Therefore,.the technical actions carried out with flint tools on bone were to repair the bone tools.

The case of wood is similar. The known activities using stone tools reflect simple technical processes, compared to those of the other two assemblages.

For working antler, however, the desire to simplify the technical process is not found, since all the phases of the technical process are represented. Perhaps it is due to the particular function of tools made in antler, as they were used for hunting. We have already discussed in the section on bone materials the factors involved in choosing antler over bone for tools which are to sustain impacts.

Therefore, the making of antler tools on the site was probably caused by the necessity to replace hunting tools when necessary. Such behaviour would be similar to that of flint knapping which was oriented principally towards the production of bladelets to make projectiles and composite knives.

The characteristics of the technical process for hide support the idea of the specialization of the settlement. The first phases of work, defleshing of fresh hide, are well represented, while the later phases, of working dry hide, are scarce.

This association between specialized sites and the first steps in the process of conditioning hide has been discussed by L.H. Keeley (1988) for Magdalenian sites. In their study on the function of lithic tools at Verberie, Rascaño and El Juyo discuss the greater proportion of working of fresh hide in the two first sites, which appear to be more seasonal occupations with a more specialized economy.

Simplification of technical processes which are not related to the predominant function of the settlement avoids the practice of costly work interfering the principal activity, and allows this activity to be increased.

The tools were brought already made, being repaired when necessary. Thus, by reducing the effort expended for certain manufacture activities, procurement of animal resources and the first stages of transformation of these resources could be increased to the maximum (Myers, 1989). The ethnographic correlate of this behaviour is the conduct cited for the Nunamiut, who transport the tools already made to the settlements specialized for hunting (Binford, 1983, 130).

Functional analysis of the French site of Cassegros offers similar data (Vaughan, 1985), presenting another situation in which a specialized site shows manufacture activities which are not very complex. In this settlement, which served basically as a place for working hide, there was hardly any work on wood and bone materials except scraping, this being carried out using tools which are not very specific.

The use characteristics of unretouched blanks can be explained within this general framework. We have seen that the tendency was to use unretouched tools in greater proportion than in the other two assemblages. Moreover, the criteria of selection of the unretouched blanks were not very stringent and the intensity of use was slight.

The performance of uncostly and unplanned tasks would lead to solutions of expeditive use. Little-specialized, unretouched tools were chosen with little attention to selection criteria, and were abandoned once the activity had been carried out.

However it is to be stressed that for working fresh hide, a fundamental activity in the site, tools of greater cost were used, such as end-scrapers, which is in contrast to the data from the Magdalenien level. The same can be shown for the tools used to work antler, mainly burins and end-scrapers. Therefore, even though we see in this assemblage a general tendency to use little-specialized tools, this is not applicable to the working of fresh hide and antler, activities which are strategically important to the function of the settlement.

According to the data we have, the human groups which occupied the cave came there with an assemblage of tools already made. During the time of the occupation they devoted themselves mainly to hunting and the gathering of marine resources. Once killed, the animals were butchered.

The first phases of working hides were carried out by scraping the hide in a fresh state. These products were stockpiled and transported later to another place, where work continued on the hide. The first treatment insured preservation until the moment at which the work process could continue.

We have no proof that this conduct of storage and transport also occurred for meat, although such a hypothesis should be verified from study of the bone remains found on the site.

As the hunting tools became unusable, knapping would be carried out to make backed points and backed bladelets with which to replace them. It is probable that this conduct of tool-making was also applied to the working of antler.

For the rest of the tools, simple technical activities would suffice to maintain them. More costly technical operations were avoided, to foster the main activities and because there was no urgent need to have tools not directly related to the predominant function of the settlement.

The specialization of the settlement and the existence of fragmentary technical processes indicate that the different moments of occupation of the cave were not prolonged.

This conduct implies a certain planning of the economic system. The existence of one or several settlements was necessary to carry out the activities which were finished at Santa Catalina. In this other settlement, bone tools would have been made which were only repaired in Santa Catalina. On the other hand, the final phases of working hide which were not carried out in Santa Catalina were completed in another settlement.

It is worth questioning whether the planning occurs in time or space. In fact, given the data we have, two explanations are possible. In the first, bone, wood, and probably stone tools which were brought already made to the site would have been manufactured in an earlier phase of the seasonal economic cycle. Hide, transported to another settlement at the moment when the cave was abandoned, would have been finished in the following phase of this cycle.

The second explanation is that Santa Catalina functioned as a specialized settlement, secondary to another base site, which was contemporaneous. The finished tools were brought from the base settlement to the specialized one, and part of the resources procured were transported back to the base site.

In any case, we probably have a system of exploitation of the environment in which a certain functional complementarity was established between the different settlements of the same groups, as shown in time and/or space.

If this interpretation is correct, we should find settlements of the Azilian period which show the type of specialization found in the Azilian level of Santa Catalina.

Moreover, there must exist other settlements in which economic activities are carried out which are complementary to those in the specialized occupation. At these base settlements the economic activities will become unbalanced, the manufacture activities of all types of tools taking precedence over those involved in the procurement of animal resources and those of the first phases of transformation of these resources. Also, the most costly phases of manufacture will be carried out, such as the working of dry hide and the making of tools in antler, bone and wood, from the first phases of production onward.

The function of these hypothetical types of occupation (settlements where activities are planned which are going to be completed on other more specialized ones) could imply use of curated tools, made in these settlements, but kept for other contexts of use.

6.4. THE EPIPALAEOLITHIC SITE OF BERNIOLLO

The analysis of the economic activities which have been discovered in Berniollo offers us an image opposite to that which we observed in the Azilian level of Santa Catalina. In Berniollo the work of hunting, butchering and working fresh hide are hardly represented. On the other hand, the making of tools in wood and bone materials is predominant as well as the working of dry hide.

In Berniollo, high cost technical processes were carried out. For flint knapping, the technical study indicates that all the phases of the technical process for knapping were carried out on the site, from the shaping of the nucleus to the retouch of the blank. Knapping was used mainly to obtain blades, and to a lesser degree, bladelets (González, 1993).

The working of wood and antler shows a certain complexity, which indicates that sequences of technical operations for making complex tools were carried out..

In the case of hide we see that the final phases of conditioning were practised, for making the hide flexible and softening it; this work requires much time and effort, as we have pointed out in an earlier chapter.

The use characteristics found for flint tools show that they were used for work requiring considerable effort. As much for the burins as for the end-scrapers or the unretouched flakes and blades, the data show a greater intensity of use on this site than in the other two assemblages. This greater intensity is accompanied by a careful selection of the type of blanks to be used as tools.

We assume that the characteristics of the tasks to be accomplished determined the criteria for selecting the lithic tools. The necessity of finishing intensive work, reflected in the intensity of the traces, led to using tools of a specific forms, with a greater cost of manufacture than those from the Azilian level of Santa Catalina. Thus, the work tended to be carried out more often using retouched tools, and there was also a more careful choice of the flakes and blades used.

On the other hand, we have data which suggest that the economic behavoiurs shown by the activities carried out at Berniollo was highly planned.

In the technical process of hide-working we observe two clear discontinuities. On one hand, almost no initial conditioning work of the hide (defleshing) is represented. Furthermore, no activities associated with the making of objects are to be found, which would have taken place after the conditioning. This indicates that the work on hide involved its systematic softening, in a specific phase of the conditioning, without appreciable making of hide objects taking place.

The first phases of conditioning the hides must have occurred in another place, contemporary to Berniollo or occur in a phase preceding the annual cycle. The work of making objects from the hides was probably carried out in a phase following the occupation of Berniollo.

All the phases of flint knapping are present on the site. However, a part of the blade blanks were not found at the settlement, since the number of blades found is low compared to that of the cores, suggesting that some were taken to another place. Therefore, we consider that not all the blade blanks were made to be used on the settlement, but rather that a part were intended for work elsewhere.

Thus, we are looking at a lithic production with anticipation, for necessities which were not immediate, for later use. In any case, though we see all the production phases of lithic tools, the use phase is under-represented.

This planned behaviour can also be found in the presence of specialized tools. According to what we have seen, an assemblage of lithic tools was kept for use contexts out of the settlement, involving the factor of mobility and, thus, that of portability of the tools. The probability that a tool was used in a curated way rises when the tool was made from quality flint, from a blade blank, and when it was retouched. Here we see another example of the tools which were made at Berniollo, but which were probably used in another context of use.

In general terms, the importance of manufacture activities in relation to those of subsistence (hunting and butchering) indicates that all the phases of the economic system of the group were not evenly represented, suggesting that the activities carried out at Berniollo were complemented by others carried out in other settlements.

According to our data, the human group which inhabited the settlement at Berniollo carried out a wide range of tasks of manufacture. For the tasks carried out at the site, they used preferentially unretouched tools, flakes and blades, and also retouched tools made from flakes.

A part of tool production was used for activities which took place outside the settlement. For this, it was necessary to have transportable and effective tools, which could be transformed and sharpened easily. Therefore, we find on the site an assemblage of tools of expedient use and another of curated use.

The presence of curated tools indicates that in Berniollo a series of economic activities were planned which took place in contexts of use at a distance from the site.

Furthermore, an assemblage of hides that had been prevoiusly defleshed was transported to the settlement. Intensive hide conditioning was then carried out, scraping the dry hides. It is probable that, at the moment in which the settlement was abandoned, the hides were carried already conditioned to another settlement where the making of objects took place.

We may conclude that the variety of activities and the intensity of the work indicates that the occupation of Berniollo was relatively stable.

On the other hand, with respect to the degree of planning of economic activities, difficulty arises in establishing whether such planning took place in time or in space. That is to say, whether such planning implies the existence of an organized network of settlements or not. With the level of information which we have, both hypotheses are possible, as we have seen for the Azilian assemblage of Santa Catalina.

Our functional interpretation implies the existence of contemporary settlements with very particular characteristics. Settlements with characteristics similar to those of Berniollo will be found, as well as others with a complementary function. These will show a spectrum of activities in which hunting, butchering and working of fresh hide predominate, and there would have been an effort to simplify the technical processes carried out on bone, wood and stone. Work on dry hide would have been only slight. This framework fits well with the results obtained in the Azilian level of Santa Catalina.

In this specialized settlement the lithic curated tools would predominate. The tools would be mainly blades, retouched and made in flint type 3, also showing a clear standardization. This type of tool would show a high degree of use.

6.5. THE MAGDALENIAN LEVEL OF SANTA CATALINA

The features of the economic activities represented in this level present an image which is intermediate between the two previous assemblages.

Hunting and the preliminary activities of transforming animal materials, which predominated in the Azilian level, are less represented in this level. However, their importance is greater than at Berniollo.

In addition, activities of tool manufacture are more intensive than in the Azilian level, but the quantity is less than at Berniollo.

Comparatively, there is no clear predominance of one particular phase of the economic system of the group, but rather that they are all represented.

The analysis of each one of the work operations, according to the materials, does not indicate the presence of any break in the process of transformation.

Thus, flint knapping may be seen from the first phases of roughing-out of the nodules. Furthermore, production is directed as much towards obtaining blades as bladelets.

The use traces on the flint tools which worked wood, antler and bone show that the activities carried out were complex and intense, which leads us to suppose that all the phases of the technical processes were carried out, from the first phases of transformation.

Also continuity is observed in the working of hide. On this material the initial and final phases of conditioning were carried out: scraping of fresh and dry hide, softening the hide by rubbing with pebbles (Ibáñez and González, in press), and the making of objects from hide.

The use characteristics of the lithic tools also present results intermediate between the other two assemblages, as much in regard to intensity of use as to the degree of specificity and choice of the pieces.

The variety and intensity of the work carried out in the Magdalenian level leads us to assume that the occupations that formed this level tended to be relatively long in duration, in any case more so than for the Azilian level.

There is not a clear specilization in economic activities carried out in the Magdalenian level of Santa Catalina.

At this point two lines of explanation are possible. The first is that the limited functional definition of the settlements which formed the level is more apparent than real. In this case, the superimposition of different settlements of clear function would generate an archaeological record confusing to read.

The second explanation would assume that the general function of the different moments of occupation reflects homogeneous and continuous behaviour. In this case, 1- the absence of discontinuities in the technical processes, 2- the development in the site of all the activities in the technical system and 3- the lack of differential use of the tools with respect to the blank (blade or flake) would indicate, in comparison with the other two assemblages, less planning of economic activities.

The more recent sites (the Azilian level of Santa Catalina and the Epipalaeolithic site of Berniollo) show more complex economic behaviour. Further analysis will show if the hypothesis of an increase in the economic planning between the final Magdalenian and the Epipalaeolithic is real. If this is the case, we should elucidate whether the higher planning is developed principally along an annual cycle, with specilized economic activities carried out in each period of the year, or across the landscape, with the existence of base and specialized settlements.

It will be possible to confirm this hypothesis if other contemporaneous assemblages show characteristics similar to those observed at Santa Catalina and Berniollo.

CONCLUSION

This work is oriented towards the reconstruction of economic strategies of hunter-gatherer groups at the end of the Palaeolithic in the Basque country. The study is based on materials from two sites, Berniollo and the cave of Santa Catalina, whose human occupation is situated between approximately 12,000 and 10,000 B.P..

The information used is provided by the study of the use of chipped stone tools: what these tools were and how they were used, what materials they worked and in what way. This information is relevant to the question, but not sufficient.

Two types of limitations are the most evident. On the one hand we lack knowledge of the technical and economic behaviour and choices which can be deduced from other material remains. On the other hand, the activities of hunter-gatherer groups are variable in time and space. To understand the economic organization of these groups it is necessary to observe this variability of activities in a large number of sites. The study of more sites is the only way to correct this limitation.

Due to these two limitations we decided to center this study on the question: How can functional information from knapped stone tools help us to understand the economic organization of hunter-gathers? Basically we are proposing a strategy for "reading" the results of the functional analyses from this perspective.

For this we establish three levels of interpretation. In the first we relate the characteristics of the tools to the function which they fulfill. This consists of understanding the choices concerning the characteristics of the tool (the lithic raw material used, the blank chosen and the type of retouch) in relation to the intended use (the material worked, the motion of the tool, the intensity and the context of use). This level corresponds, at least in part, to the approaches concerned with questions such as raw material economy or blank and tool management systems.

On the second level we attempt to reconstruct the activities carried out with the stone tools: what materials were worked, what type of work took place, and with what objective. To complement this reconstruction we have turned to archaeological and ethnographic references on the working techniques of hunter-gatherer groups. Concepts such as technical sequence (*chaîne opératoire*) and technical process (*processus technique)* have been very useful in organizing this information. This level of interpretation is limited by the fact that many activities take place without tools or with tools which are not knapped stone. This level of inference should be reinforced and complemented by the functional and technological analysis of other types of tools, on bone, antler, pebbles, etc.

Finally, on a third level, we attempt a reconstruction of the economic strategy of the groups studied. We focus primarily upon the degree and the type of planning of the activities of procurement and transformation of resources. This is based upon the interpretations established in the two previous levels. In the first, we observe how management of the tools is organized and according to what criteria. In the second, we reveal the presence or absence and the intensity of the various phases of transformation of each material worked, that is, which activities were carried out and which were omitted. These choices have a logic which is established thanks to ethnographic knowledge of the economic organization of hunter-gatherer groups. On this level other types of information are still necessary concerning seasonal mobility of these groups. The territory of procurement of lithic raw materials and other resources, the animals which were hunted, the inter and intra-site spatial organization, along with many other aspects, provide useful information. It is indeed in the integration (not in the juxtaposition) of the various data concerning the techno-economic choices of these groups that a more complete interpretation should be based. This work attemps to offer a framework to integrate information on the use of lithic tools with the rest of the data on the economic strategies of hunter-gatherers.

APPENDIX 1

CRITERIA FOR FUNCTIONAL INFERENCE

0. All active zones

Determination: Location and morphology of the active zone: linear, pointed, bevelled.
Criteria:
a. extension and concurrence of the different types of trace: microscars, striations, rounding and polish.
b. to separate contiguous active zones, existence of discontinuity in the extension or intensity of the traces.

1. Linear active zones.

Determination: way in which force is applied: pressure or percussion (see diagrams 5, 12 and 18 *vs.* 1, 6, 10 and 15).
Criteria:
a. More than 20% of the microscars have more than 2 mm. in length and more than 1.2 mm. in width: = percussion. The contrary: = pressure.
b. (If edge angle less than 30 degrees) Less than 10% of the microscars with snap terminations: = percussion. The contrary: = pressure.
c. (If edge angle between 30 and 45 degrees) More than 45% of microscars with hinge/step terminations: = percussion. The contrary: = pressure.
d. (If edge angle more than 60 degrees) Less than 30% of microscars with hinge/step terminations: = percussion. The contrary: = pressure.
e. Rounding in the active zone: = pressure.

1.1 Linear active zones, percussion actions.

Determination: type of percussion, direct (chopping, adzing) or indirect (wedge).
Criteria:
a. Small tools with traces of striking in the part opposite to the active zone: = indirect. The contrary: = direct.
b. Unequal distribution of traces (number of microscars and extension of polish) on the two faces of the active edge: = adzing. The contrary: = chopping.

1.1.1., 1.1.2. and 1.1.3. Chopping, adzing and wedges.

Determination: material worked.
a. These actions are used for working medium-hard or hard materials.
b. Absence of isolated microscars: = hard materials (ie. bone materials). The contrary: = medium-hard materials (i.e. wood).
c. Generally, it is difficult to distinguish the worked materials more specifically because of the faintness of the polish. When it is more developed the criteria established in section 1.2.2 are used.

1.2. Linear active zones, pressure actions.

Determination: type of movement of the tool, longitudinal (cutting) or transverse (scraping) (see diagrams 1, 10, 14, 19 and 22 *vs.* 2, 6, 7, 13, 15 and 20).
Criteria:
a. Edge angles higher than 60 degrees cannot be used for cutting soft materials.
b. Striations or linear components parallel to the edge: = longitudinal. The contrary: = transverse.
c. Balanced disposition on both faces of the microscars (6/4 or more balanced) and polish extension: = longitudinal. The contrary: = transverse. These criteria do not refer to edge angles between 30 and 40 degrees, in which transverse actions also produce balanced distributions of traces.
d. (If edge angle less than 55 degrees) Balanced distribution of rounding: = longitudinal. The contrary: = transverse.
e. (If edge angle less than 30 degrees) More than 25% of microscars with snap termination: = transverse. The contrary: = longitudinal.
f. Random distribution of polish: = longitudinal.

1.2.1. Linear active zones, pressure actions, longitudinal movement.

A. Determination: worked material by categories: hard and soft (see diagrams 1, 10, 14 19 and 22).

Criteria:

a. Random distribution of polish, in separate bands on the edge: = hard materials (bone, antler, stone and hard wood).

b. More than 50% of microscars without contact with polish: = hard materials.

c. More than 10 microscars per centimetre of active edge: = hard materials (bone, antler, stone and hard wood) . The contrary: = soft (meat, hide and plants).

d. More than 15% of microscars with length more than 1 mm.: = hard materials. The contrary: = soft.

e. Less than 10% isolated microscars and more than 10% superimposed microscars: = hard materials (bone, antler, stone and "heavy" butchering). Between 10% and 20% of isolated microscars and between 5% and 15% of superimposed microscars: = materials of medium hardness (wood and tough plants, such as *phragmites)*. More than 20% of isolated microscars or less than 5% of superimposed microscars: = soft material (skin, meat and soft plants, such as *graminae*).

f. (If edge angle between 30 and 45 degrees) Less than 10% snap terminations: = hard materials.

g. Scarcity or absence of striations: = wood, non-woody plants (except harvesting), fresh hide and meat.

B. Determination: specific material worked, antler, wood, non-woody plants, hide, meat, butchering, shell.

Criteria:

a. Polish with flat topography; compact pattern; random distribution of polish; abundant striations: = stone or shell.

b. Polish with smooth topography; association of compact and closed patterns; narrow network; medium rounding: = bone (photo 3 in Appendix 3).

c. Polish with smooth and smooth-undulating topography; compact, closed and semi-closed pattern; moderate network: = antler.

d. Polish with undulating topography, association of all the range of patterns; wide network: = wood (photos 1, 2, 5, 10 and 11 in Appendix 3). With topography to smooth-undulating and with dominance of semi-closed and open patterns: non-woody plants.

e. Polish with irregular topography; patterns from closed to open, narrow network; many micro holes; high or medium rounding: = dry hide.

f. Isolated spots of polish with smooth topography; association of compact and closed pattern, product of contact with bone and cartilege, next to more extended areas of polish with irregular topography and semi-closed or open pattern: = butchering.

g. Polish with irregular topography, patterns from closed to open, moderate network; weak rounding: = fresh hide and meat.

1.2.2. Linear active zones, pressure actions, transverse movement.

A. Determination: hardness of material worked, hard and soft materials (see diagrams 2, 6, 7, 13, 15 and 20).

Criteria:

a. More than 15% microscars more than 0.7 mm. in length: = hard materials (stone, bone, antler and wood). The contrary: = soft materials (hide, meat, plants).

b. More than 10 microscars per centimetre of active edge: = hard materials. The contrary, soft materials.

c. (If edge angle between 30 and 45 degrees) More than 10% of microscars with snap termination: = soft materials. The contrary: = hard materials

d. (If edge angle higher than 60 degrees) More than 60% of microscars with hinge/step terminations: = hard materials. The contrary: = soft materials.

e. More than 20 microscars per centimetre of edge: = hard materials (stone, bone or antler).

f. (If edge angles between 60 and 90 degrees) More than 70% of the microscars without contact with polish: = hard materials (stone,bone or antler). The contrary: = softer materials.

B. Determination: specific material worked, bone, antler, wood, non-woody plants, hide, meat, butchering, shell.

Criteria:

a. Polish with smooth topography; compact/closed pattern; narrow network, extension up to 0.015 mm.: = bone (photos 12 and 13 in Appendix 3).

b. Polish with smooth and smooth-undulating topography, compact/closed/semi-closed pattern, network moderate, with extensions between 0.01 mm. and 0.03 mm.: = antler (photos 14 and 15 in Appendix 3).

c. Polished zones with smooth-undulating and undulating topography; association of patterns from open to compact; extension of polish between 0.02 mm. and 0.06 mm.: =wood.

d. Polished zones with smooth-undulating and undulating topography, patterns from compact to open, with greatest extension of patterns semi-closed and open; network wide: = non-woody plants.

f. Polish with topography irregular or irregular-undulating, patterns closed and semi-closed, extension from 0.04 to several millimetres, network narrow, presence of micro-holes; moderate or pronounced rounding; more than 5 striations per centimetre of used edge: = dry hide (photos 6 and 16 in Appendix 3).

g. Polish with topography irregular, pattern closed and semi-closed, extension more than 0.04 mm, and network moderate; low rounding: = fresh hide (photo 17 in Appendix 3).

2. Pointed active zones.

Determination: way in which force is applied, pressure or percussion.

Criteria:

a. Microscars of more than 3 mm. parallel to the axis of the point: = percussion. The contrary: = pressure.

b. Striations or polish with linear disposition parallel to the axis of the point: = percussion. The contrary: = pressure.

c. Presence of developed polish in the ridges which form the apex: = pressure.

2.1. Pointed active zones, percussion actions.

Determination: type of percussion: direct, indirect or thrown.

a. Traces of striking in the zone opposite to the active edge: = indirect. The contrary: = direct or thrown.

b. Impact traces, i.e. striations (photo 9 in Appendix 3) and fractures (figures 24 to 27 in Appendix 4), in pointed tools used for direct percussion (dagger, pike) or thrown (javelin, arrow) are similar. Information about the morphology and size of the tool is crucial for determining the method of use.

2.1.1., 2.1.2. and 2.1.3. Pointed active zones , percussion actions, direct, indirect and thrown.

Determination: material worked.

Criteria: Generally, these actions do not produce enough signs to recognize specific worked materials, because of the slight development of the polished areas. When these present sufficiently extended areas, the material worked is determined with the criteria used in section 1.2.2.. Indirect percussion is only valid for working hard or medium hard materials. Thrown percussion is explained only by actions of hunting or war.

2.2. Pointed active zones, pressure actions.

Determination: type of work, engraving or boring (see diagrams 3, 8, 11, 17 and 21 in Appendix 4).

Criteria:

a. Presence of striations perpendicular and parallel to the axis of the point: = boring.

b. Disposition of microscars:

- Boring in one direction: = microscars more numerous on the passive surfaces, polish more developed on the active (leading) surfaces.
- Boring in two directions (figure 2.1., n. 2, p.10): balanced distribution of the number of microscars and the development of polish.
- Engraving with one surface leading (figure 2.1., n. 5, p.10): polish more developed and extensive on the facet leading; microscars more numerous on the passive surface.
- Engraving with a ridge leading (figure 2.1., n. 6, p.10): greater number of microscars and greater development of the polish on the vertex of the passive surface.

2.2.1. and 2.2.2. Pointed active zones, pressure actions, engraving and boring.

Determination: material worked (see diagrams 3, 8, 11, 17 and 21 in Appendix 4).

Criteria:

a. More than 10 microscars per centimetre of active edge: = hard materials (stone, bone and antler); between 5 and 10 microscars per centimetre: = medium-hard materials (hard wood, shell, dry hide); less than 5 microscars per centimetre: = soft materials (soft wood and fresh hide).

b. More than 15% of microscars more than 2 mm. in length: = hard and medium-hard materials (stone, bone, antler, shell, wood and dry hide). The contrary: = soft materials (fresh hide)

c. Less than 5% of isolated microscars and more than 25% of superimposed microscars: = hard materials (shell, bone and antler). The contrary: = softer materials.

d. For criteria based on use-polish see section 1.2.2. except the extension of the polish.

3. Bevelled active zones.

Determination: type of action, boring or engraving (see diagrams 4, 9, 16 and 23).

Criteria:

a. Disposition of use-traces:

- Boring in one direction: greater number of microscars on passive surfaces and greater development of polish on leading surfaces.
- Boring in two directions: balanced distribution of the number of microscars and of the development of the polish.
- Engraving with the burin moving longitudinally (engraving in V; figure 2.1., n.3, p.10): balanced distribution of microscars and of the development of the polish on the two surfaces which form the bevel of the burin; striations parallel to the bevel.
- Engraving with the bevel of the burin moving transversely (engraving in U, figure 2.1., n. 4, p.10): greater number of microscars on one of the surfaces that forms the bevel; perpendicular striations on the bevel. The distribution of the polish depends on the working angle.

3.1. and 3.2. Bevelled active zones: Boring and engraving.

Determination: material worked.

Criteria: The same criteria as in section 2.2.1. and 2.2.2. are used.

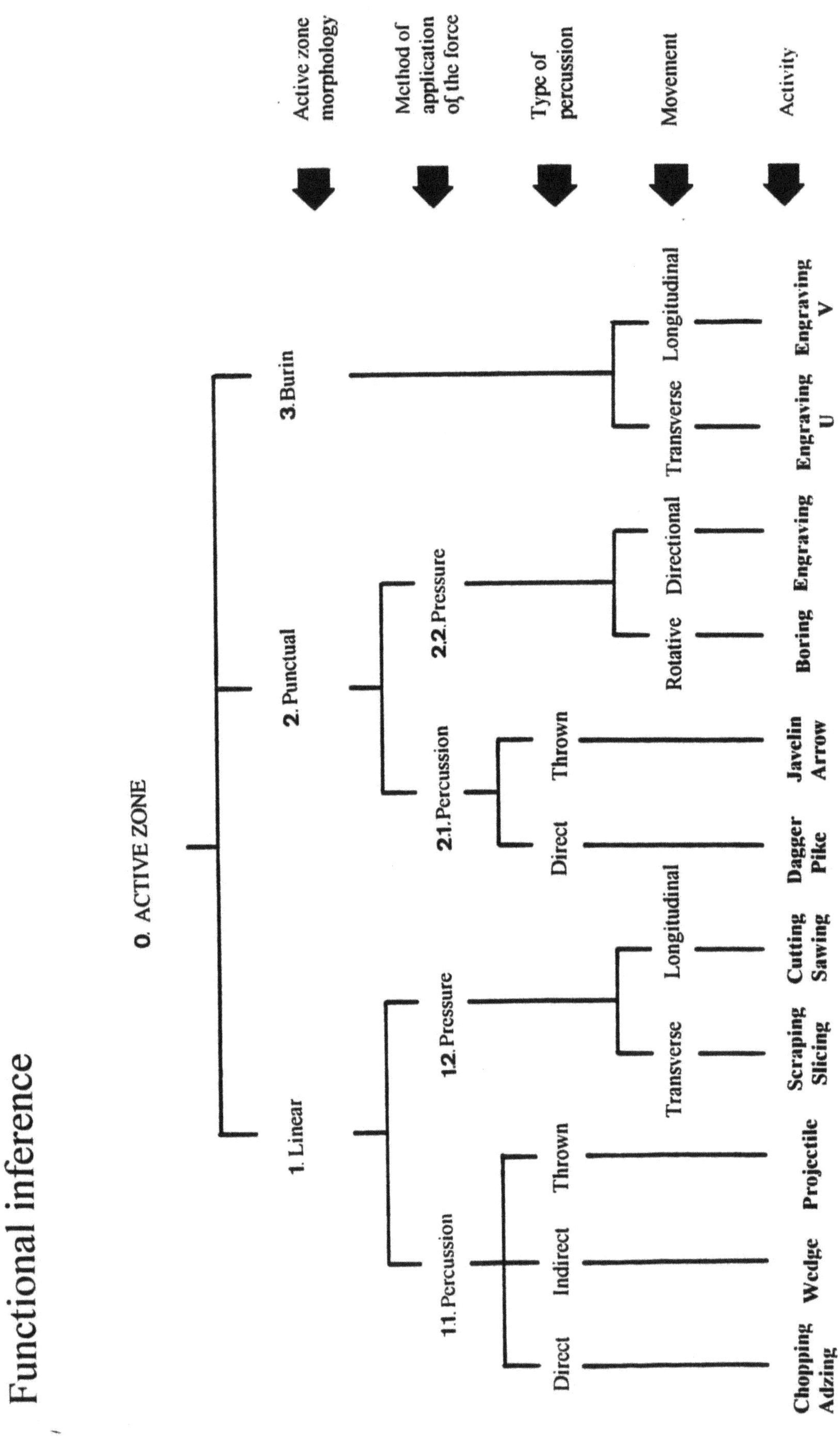
Functional inference
0. ACTIVE ZONE
1. Linear
2. Punctual
3. Burin
1.1. Percussion
1.2. Pressure
2.1. Percussion
2.2. Pressure
Direct
Indirect
Thrown
Transverse
Longitudinal
Direct
Thrown
Rotative
Directional
Transverse
Longitudinal
Chopping Adzing
Wedge
Projectile
Scraping Slicing
Cutting Sawing
Dagger Pike
Javelin Arrow
Boring
Engraving
Engraving U
Engraving V
Active zone morphology
Method of application of the force
Type of percussion
Movement
Activity

APPENDIX 2

THE EXPERIMENTS

CUTTING

	Worked material	State	Flint.	Time of use	Edge Angle	
1.	fish	fresh	Kur.	60	22	
2.	goat	fresh	Alb.	20	43	Butchery
3.	goat	fresh	Kur.	10	55	Butchery
4.	goat	fresh	Kur.	30	35	Butchery
5.	goat	fresh	Kur.	15	33	Butchery
6.	goat	cooked	Kur.	20	51	Butchery
7.	goat	cooked	Kur.	15	59	Butchery
8.	goat	fresh	Kur.	30	60	Butchery
9.	goat	fresh	Kur.	20	66	Butchery
10.	hair		Kur.	60	37	
11.	hide	dry	Kur.	18	25	
12.	hide	dry	Kur.	2	33	
13.	hide	dry	Kur.	3	57	
14.	hide	dry	Kur.	5	18	Hafted
15.	hide	dry	Kur.	60	33	
16.	hide	dry moist.	Alb.	25	31	
17.	hide	dry moist.	Kur.	60	40	
18.	hide	dry moist.	Kur.	60	38	
19.	hide	dry &ocher	Alb.	50	28	on wood
20.	hide	dry &ocher	Kur.	12	23	
21.	hide	dry &ocher	Kur.	15	25	
22.	hide	dry &ocher	Motr.	10	32	
23.	hide	dry &ocher	Ber.	12	47	on stone
24.	hide	dry &ocher	Alb.	15	21	on wood
25.	hide	dry &ocher	Alb.	18	27	
26.	hide	dry &ocher	Kur.	34	18	on stone
27.	hide	dry, hair side	Kur.	X	70	
28.	hide	fresh	Alb.	25	16	on stone
29.	hide	fresh	Kur.	60	32	
30.	hide	fresh	Kur.	X	39	
31.	hide	fresh	Kur.	5	28	
32.	hide	fresh &ocher	Alb.	40	15	on wood
33.	hide	partly dry	Alb.	25	28	on wood
34.	hide	partly dry	Kur.	15	16	on stone
35.	hide	partly dry	Ber.	30	24	on stone
36.	hide	p.dry &ocher	Alb.	25	22	on wood
37.	hide	p.dry &ocher	Kur.	20	20	on stone
38.	hide	tanned	Kur.	60	35	
39.	hide	tanned moist.	Ber.	10	16	
40.	hide	very dry	Ber.	5	2	
41.	hide (fat)	fresh	Alb.	45	25	
42.	meat	dry	Kur.	60	34	
43.	meat	fresh	Kur.	10	15	Butchery
44.	meat	fresh	Kur.	60	24	
45.	meat	fresh	Kur.	60	28	
46.	meat	roasted	Kur.	60	36	
47.	vegetal	fresh	Vau.	75	25	Bramble
48.	vegetal	fresh	Vau.	40	21	Reeds
49.	vegetal	fresh	Vau.	40	16	Esparto
50.	vegetal	fresh	Vau.	50	30	Esparto
51.	vegetal	fresh	Vau.	50	35	Esparto
52.	vegetal	fresh	Kur.	60	36	Ferns
53.	vegetal	fresh	Kur.	90	22	Grape
54.	vegetal	fresh	Kur.	60	35	Grass
55.	vegetal	fresh	Kur.	60	33	Willow
56.	vegetal	fresh	Syr.	75	17	Willow
57.	vegetal	fresh	Vau.	60	18	Rush

Worked material	State	Flint.	Time of use	Edge Angle	
SAWING					
58. antler	dry	Alb.	15		
59. antler	dry moist.	Alb.	10	34	
60. antler	dry moist.	Alb.	48	28	Hafted
61. antler	dry moist.	Alb.	5	20	
62. antler	dry moist.	Kur.	10	30	
63. antler	dry moist.	Kur.	60	45	
64. antler	dry moist.	Mot.	12	20	Hafted
65. antler	dry moist.	Mor.	42	25	Hafted
66. antler	dry soaked	Alb.	3	27	
67. antler	dry soaked	Ber.	20+15	28	
68. antler	dry soaked	Ber.	20	18	
69. antler	dry soaked	Ber.	10	30	
70. antler	dry soaked	Ber.	15	24	
71. antler	dry soaked	Ber.	5	25	
72. antler	fresh	Kur.	60	50	
73. bone	cooked	Kur.	20	39	
74. bone	cooked	Kur.	15	37	
75. bone	dry moist.	Kur.	60	44	
76. bone	dry soaked	Alb.	1...7	54	
77. bone	dry soaked	Alb.	1...7	43	
78. bone	fresh	Kur.	60	37	
79. bone	fresh	Kur.	X	35, 20	
80. bone	fresh	Ber.	15	18	
81. bone	fresh	Ber.	45	22	
82. bone	fresh	Ber.	20	45	
83. bone	fresh	Ber.	40	15, 20	
84. bone	fresh	Ber.	30	30	
85. vegetal	fresh	Ber.	8	15	Willow
86. vegetal	fresh	Ber.	8	15	Willow
87. wood	dry	Kur.	1...2	40	Hazeltree
88. wood	dry	Kur.	3	17	Oak
89. wood	dry	Kur.	30	20	Oak
90. wood	dry	Kur.	60	26	Oak
91. wood	dry moist.	Kur.	60	34	Oak
92. wood	dry moist.	Kur.	60	24	Oak
93. wood	fresh	Kur.	15	25	Oak
94. wood	fresh	Kur.	30	20	Oak
95. wood	fresh	Kur.	60	30	Oak
96. wood	fresh	Kur.	7	14	Oak
97. wood	dry	Kur.	60	47	Pine
98. wood	dry	Kur.	60	42	Pine
99. wood	fresh	Kur.	1...15	65	Pine
100.wood	fresh	Kur.	60	29	Pine
SCRAPING; LOW WORKING ANGLE; MOTION IN ONE SENSE					
101.antler	dry	Alb.	25		
102.antler	dry moist.	Alb.	45	90	Hafted.
103.antler	dry moist.	Kur.	11	76	
104.antler	dry moist.	Kur.	60	82	
105.antler	dry moist.	Kur.	60	35	
106.antler	dry soaked	Ber.	3/10	32/30	
107.antler	dry soaked	Alb.	25	72	Hafted
108.antler	dry soaked	Alb.	35	62	
109.antler	dry soaked	Alb.	4	22	
110.antler	dry soaked	Alb.	60	81	
111.antler	dry soaked	Alb.	150/25	70/82	
112.antler	dry soaked	Ber.	5/15	80/75	
113.antler	dry soaked	Syr.	9	46	
114.antler	dry soaked	Ber.	8	90	
115.antler	dry soaked	Ber.	25	88	
116.antler	dry soaked	Alb.	20	96	
117.antler	dry soaked	Syr.	10	84	
118.antler	fresh	Kur.	3	64	
119.antler	fresh	Kur.	5	82	
120.antler	fresh	Kur.	60	100	
121.antler	fresh	Kur.	6	30	
122.bark	partly dry	Kur.	25	32	Oak
123.bark	soaked	Syr.	30	26	Willow.
124.bark	fresh	Kur.	20	28	Willow.
125.bone	dry soaked	Kur.	10	66	Willow
126.bone	partly dry	Ber.	25	36	
127.bone	fresh	Alb.	15	70	
128.bone	fresh	Kur.	60	90	
129.bone	fresh	Kur.	60	48	
130.bone	fresh	Kur.	60	64	
131.bone	fresh	Kur.	30	80	
132.bone	fresh	Kur.	18	34	
133.bone	fresh	Kur.	20	78	
134.bone	fresh	Kur.	10	31	
135.bone	fresh	Kur.	5	24	
136.bone	fresh	Kur.	10	23	
137.fish	fresh	Kur.	60	95	
138.hide	dry	Kur.	20	85	
139.hide	dry	Alb.	22	72	Hafted
140.hide	dry	Kur.	60	76	
141.hide	dry	Syr.	15	22	Hafted.
142.hide	dry	Syr.	15	24	Hafted.
143.hide	dry	Syr.	15	25	Hafted.
144.hide	dry moist.	Alb.	20	79	Hafted
145.hide	dry moist.	Kur.	24	80	
146.hide	dry moist.	Kur.	60	76	
147.hide	dry moist.	Kur.	60	105	
148.hide	dry moist.	Kur.	23	83	Hafted
149.hide	dry moist.	Alb.	25	74	Hafted
150.hide	dry moist.,	Kur.	6	78	Hafted
151.hide	dry moist.,	Kur.	60	80	
152.hide	dry moist.,	Kur.	24	75	Hafted
153.hide	dry tallowed	Kur.	16	88	Hafted
154.hide	dry &ocher	Alb.	20	26	
155.hide	dry &ocher	Kur.	15	80	Hafted
156.hide	dry &ocher,	Alb.	40	30	Hafted
157.hide	dry, hair side	Kur.	25	79	Hafted
158.hide	fresh	Alb.	35	70	Hafted
159.hide	fresh	Kur.	25	66	Hafted
160.hide	fresh	Kur.	25	74	Hafted

Worked material	State	Flint .	Time of use	Edge Angle	
161.hide	fresh	Kur.	25	75	Hafted
162.hide	fresh	Kur.	40	65	Hafted
163.hide	fresh	Kur.	60	54	
164.hide	fresh	Kur.	60	75	
165.hide	fresh	Kur.	8	60	Hafted
166.hide	fresh, hair	Alb.	35	50	on wood
167.hide	fresh, hair s.	Alb.	30	52	
168.hide	fresh, hair s.	Alb.	50	85	Hafted.
169.hide	p.dry &ocher	Vau.	20	24	
170.hide	tanned	Syr.	65	25	on wood
171.hide	tanned moist.	Alb.	45	54	on stone
172.hide	tanned moist.	Syr.	15	74	&ocher
173.hide	tanned tallow.	Syr.	30	41	&ocher
174.Ocher		Alb.	7	70	
175.Ocher	moist.	Kur.	10	81	
176.tendon	fresh	Kur.	25	84	
177.vegetal	fresh	Mar.	65	20	Bramble
178.vegetal	fresh	Vau.	70	27	Bramble
179.vegetal	dry	Kur.	60	46	Cereal
180.vegetal	dry	Vau.	75	32	Reed
181.vegetal	fresh	Vau.	60	42	Esparto
182.vegetal	fresh	Neg.	90	66	Esparto
183.vegetal	fresh	Mar.	120	30	Willow
184.vegetal	fresh	Vau.	120	30	Willow
185.wood	fresh	Alb.	30	80	Acacia
186.wood	fresh	Kur.	3	64	Boxwood
187.wood	fresh	Kur.	4	58	Boxwood
188.wood	fresh	Kur.	4	82	Boxwood
189.wood	dry	Alb.	10	74	Hazeltree
190.wood	dry	Alb.	35	80	Hazeltree
191.wood	dry	Vau.	60	28	Hazeltree
192.wood	dry	Syr.	15	24	Hazeltree
193.wood	dry	m-n	5	81	Hazeltree
194.wood	dry	Kur.	3	39	Hazeltree
195.wood	dry	m-n	12	14	Hazeltree
196.wood	dry	Kur.	60	51	Oak
197.wood	dry moist.	Kur.	60	35	Oak
198.wood	fresh	Alb.	40	75	Oak
199.wood	fresh	Kur.	5	23	Oak
200.wood	fresh	Kur.	60	18	Oak
201.wood	fresh	Kur.	60	50	Oak
202.wood	fresh	Alb.	30	84	Pine
203.wood	fresh	Kur.	14	82	Pine
204.wood	dry	Kur.	60	27	Pine
205.wood	fresh	Kur.	60	22	Pine
206.wood	dry	Alb.	20	55	Yew
207.wood	dry	Alb.	30	18	Yew

SCRAPING, HIGH WORKING ANGLE; MOTION IN BOTH SENSES

Worked material	State	Flint .	Time of use	Edge Angle	
208.antler	dry moist.	Kur.	60	77	
209.antler	dry moist.	Alb.	10	66	
210.bone	dry soaked	Syr.	5	32	
211.bone	fresh	Kur.	60	90	
212.hide	dry	Kur.	60	75	
213.hide	dry moist.	Kur.	60	82	
214.hide	fresh	Alb.	30	65	Hafted
215.hide	fresh	Kur.	60	94	
216.vegetal	dry	Alb.	60	29	Reed
217.vegetal	fresh	Syr.	75	17	Willow
218.wood	fresh	Kur.	60	67	Oak
219.wood	dry moist.	Kur.	60	75	Pine
220.antler	dry moist.	Kur.	60	76	
221.antler	dry moist.	Kur.	60	31	
222.bone	dry soaked	Syr.	3	84	
223.bone	dry moist.	Kur.	60	50	
224.bone	fresh	Kur.	60	29	
225.bone	fresh	Kur.	60	98	
226.hide	dry	Kur.	60	94	
227.hide	dry	Syr.	120	40	on wood
228.hide	tanned moist.	Syr.	120	51	Hafted
229.hide	dry moist.	Kur.	60	84	
230.hide	fresh	Kur.	60	91	
231.wood	dry	Kur.	60	52	Oak
232.wood	dry moist.	Kur.	60	38	Oak
233.wood	fresh	Kur.	15	32	Oak
234.wood	fresh	Kur.	60	22	Oak
235.wood	fresh	Kur.	7	21	Oak
236.wood	fresh	Kur.	60	76	Oak
237.wood	burned	Kur.	60	40	Pine
238.wood	dry	Kur.	30	46	Pine
239.wood	dry	Kur.	30	38	Pine
240.wood	dry	Kur.	30	60	Pine,
241.wood	dry	Ber.	40	39	Hafted
242.wood	dry	Ber.	45	46	Hafted

ENGRAVING

Worked material	State	Flint .	Time of use	Edge Angle	
243.antler	dry	Ber.	35	x	
244.antler	dry moist.	Kur.	60	x	
245.antler	dry moist.	Kur.	7	x	
246.bone	dry soaked	Kur.	20	x	
247.bone	fresh	Kur.	60	x	
248.bone	fresh	Kur.	X	x	
249.bone	fresh	Ber.	30	x	
250.bone	fresh	Ber.	5	x	
251.bone	fresh	Ber.	20	x	
252.bone	fresh	Ber.	30	x	
253.hide	tanned	Kur.	60	x	
254.stone		Kur.	60	x	limestone
255.stone		Kur.	60	x	Slate
256.wood	fresh	Kur.	60	x	Oak
257.wood	fresh	Alb.	30	x	Pine
258.wood	dry	Kur.	60	x	Pine
259.antler	dry moist.	Kur.	45	x	
260.antler	dry moist.	Kur.	60	x	
261.antler	dry soaked	Alb.	45	x	
262.antler	dry soaked	Alb.	40	x	
263.antler	dry soaked	Alb.	80	x	

Worked material	State	Flint .	Time of use	Edge Angle	
264.antler	dry soaked	Alb.	20, 70	x	
265.antler	dry soaked	Ber.	30	x	
266.antler	dry soaked	Alb.	5	x	
267.antler	dry soaked	Ber.	60	x	
268.antler	dry soaked	Syr.	x	x	
269.bone	dry soaked	Kur.	30	x	
270.bone	fresh	Kur.	60	x	
271.stone		Kur.	60	x	Slate
272.wood	dry	Kur.	10	x	Oak
273.wood	dry soaked	Kur.	20	x	Oak
274.wood	fresh	Kur.	60	x	Oak
275.wood	dry	Kur.	60	x	Pine
276.antler	dry moist.	Kur.	15	x	
277.antler	dry moist.	Kur.	60	x	
278.antler	dry soaked	Kur.	25	x	
279.bone	fresh	Kur.	60	x	
280.stone		Kur.	60	x	Sandst.
281.stone		Kur.	60	x	Slate
282.wood	fresh	Kur.	60	x	Oak
283.wood	dry	Alb.	25	x	Hazeltree

BORING, ONE SENSE

Worked material	State	Flint .	Time of use	Edge Angle	
284.antler	dry moist.	Kur.	60	x	
285.bone	fresh	Kur.	60	x	
286.hide	dry	Alb.	8	x	
287.hide	dry	Kur.	15	x	
288.shell		Kur.	60	x	
289.wood	fresh	Kur.	60	x	Oak

BORING, TWO SENSES

Worked material	State	Flint .	Time of use	Edge Angle	
290.antler	dry moist.	Kur.	15	x	
291.antler	dry moist.	Kur.	60	x	
292.antler	fresh	Kur.	60	x	
293.bone	dry moist.	Kur.	60	x	
294.bone	dry soaked	Alb.	25	x	
295.bone	fresh	Kur.	60	x	
296.hide	dry	Kur.	60	x	
297.hide	dry moist.	Alb.	25	x	
298.hide	dry moist.	Kur.	60	x	
299.hide	dry moist.	Kur.	60	x	
300.hide	fresh	Kur.	60	x	
301.hide	tanned	Kur.	60	x	
302.shell		Kur.	60	x	
303.stone	moistened	Mar.	35	x	limestone
304.wood	dry	Kur.	10	x	Hazeltree
305.wood	dry	Kur.	60	x	Oak
306.wood	dry moist.	Kur.	60	x	Oak
307.wood	fresh	Kur.	60	x	Oak
308.wood	fresh	Kur.	60	x	Oak
309.wood	dry	Kur.	60	x	Oak
310.wood	fresh	Kur.	1...7	x	Oak
311.wood	fresh	Kur.	10	x	Oak
312.wood	fresh	Kur.	60	x	Oak

PROJECTILES

Worked material	State	Flint .	Time of use	Edge Angle
313.goat		Alb.	2 hits	Javelin
314.goat		Alb.	3 hits	Pike
315.goat		Kur.	1 hit	Javelin
316.goat		Kur.	1 hit	Pike
317.goat		Kur.	1 hit	Arrow
318.goat		Kur.	1 hit	Pike
319.goat		Kur.	1 hit	Arrow
320.goat		Kur.	1 hit	Arrow
321.goat		Kur.	1 hit	Arrow
322.goat		Kur.	1 hit	Pike
323.goat		Kur.	1 hit	Javelin
324.goat		Kur.	1 hit	Javelin
325.goat		Kur.	1 hit	Arrow
326.goat		Kur.	1 hit	Javelin
327.goat		Kur.	1 hit	Pike
328.goat		Kur.	1 hit	Arrow
329.goat		Kur.	2 hits	Pike
330.goat		Kur.	2 hits	Arrow
331.goat		Kur.	2 hits	Javelin
332.goat		Kur.	2 hits	Arrow
333.goat		Kur.	2 hits	Arrow
334.goat		Kur.	2 hits	Javelin
335.goat		Kur.	2 hits	Arrow
336.goat		Kur.	2 hits	Arrow
337.goat		Kur.	2 hits	Arrow
338.goat		Kur.	2 hits	Javelin
339.goat		Kur.	2 hits	Arrow
340.goat		Kur.	3 hits	Pike
341.goat		Kur.	3 hits	Pike
342.goat		Kur.	3 hits	Pike
343.goat		Kur.	3 hits	Arrow
344.goat		Kur.	3 hits	Javelin
345.goat		Kur.	3 hits	Arrow
346.goat		Kur.	3 hits	Arrow
347.goat		Kur.	3 hits	Pike
348.goat		Kur.	3 hits	Javelin
349.goat		Kur.	3 hits	Javelin
350.goat		Kur.	3 hits	Arrow
351.goat		Kur.	3 hits	Arrow
352.goat		Kur.	3 hits	Arrow
353.goat		Kur.	5 hits	Javelin
354.goat		Kur.	6 hits	Javelin
355.goat		Kur.	8 hits	Pike
356.goat		Kur.	8 hits	Javelin
357.goat		Kur.	9 hits	Pike
358.goat		Kur.	9 hits	Pike
359.goat		Kur.	9 hits	Pike
360.goat		Alb.	3 hits	Barb
361.goat		Kur.	3 hits	Barb
362.goat		Kur.	3 hits	Barb
363.goat		Kur.	3 hits	Barb

DIRECT PERCUSSION,: ADZE

Worked material	State	Flint	Time of use	Edge Angle	
364.antler	dry	Ber.	20	46	Hafted
365.hide	fresh	Kur.	45	82	Hafted
366.wood	dry	Ber.	13	47	Hazeltree
367.wood	fresh	Kur.	60	38-75	Oak
368.wood	fresh	Kur.	60	40	Oak
369.wood	dry	Ber.	30	24	Yew
370.wood	dry	Ber.	90	34	Yew

DIRECT PERCUSSION, AXE

Worked material	State	Flint	Time of use	Edge Angle	
371.antler	dry	Alb.	5	62	Hafted
372.antler	dry	Par.	7	29	Hafted
373.antler	dry	Ber.	10	75	Hafted
374.antler	fresh	Cuc.	2 blow	44	
375.bone	dry soaked	Alb.	5	25	Hafted
376.earth		Kur.	60	21	
377.goat	fresh	Kur.	20	50	Butchery
378.goat	fresh	Kur.	25	43	Butchery

INDIRECT PERCUSSION, WEDGE

Worked material	State	Flint	Time of use	Edge Angle	
379.antler	dry	Syr.	10	40	
380.antler	dry	Syr.	5	60	
381.antler	dry	Ber.	15	55	
382.antler	fresh	Kur.	60	36	
383.bark	fresh	Ber.	40	32	Willow
384.bone	dry	Ber.	15	50	
385.bone	dry	Ber.	15	60	Hafted
386.bone	dry	Ber.	25	70	Hafted
387.bone	fresh	Kur.	3	42	
388.bone	fresh	Kur.	3	67	
389.bone	fresh	Kur.	5	41	
390.bone	fresh	Kur.	6	27	
391.bone	fresh	Kur.	60	40	
392.bone	fresh	Kur.	3	25	
393.bone	fresh	Kur.	5	38	
394.bone	fresh	Kur.	8	40	
395.bone	fresh	Kur.	10	33	
396.wood	fresh	Alb.	5	60	Hafted
397.wood	fresh	Alb.	5	24	Hafted
398.wood	dry	Kur.	60	53	
399.wood	dry moist.	Kur.	60	52	
400.wood	fresh	Kur.	15	36	
401.wood	fresh	Kur.	60	58	
402.wood	fresh	Kur.	7	26	
403.wood	fresh	Kur.	6	29	
404.wood	fresh	Kur.	7	70	
405.wood	fresh	Kur.	60	50	
406.wood	dry	Ber.	2	25	Yew
407.wood	dry	Ber.	5	30	Yew

APPENDIX 3

PHOTOGRAPHS OF EXPERIMENTAL USE TRACES

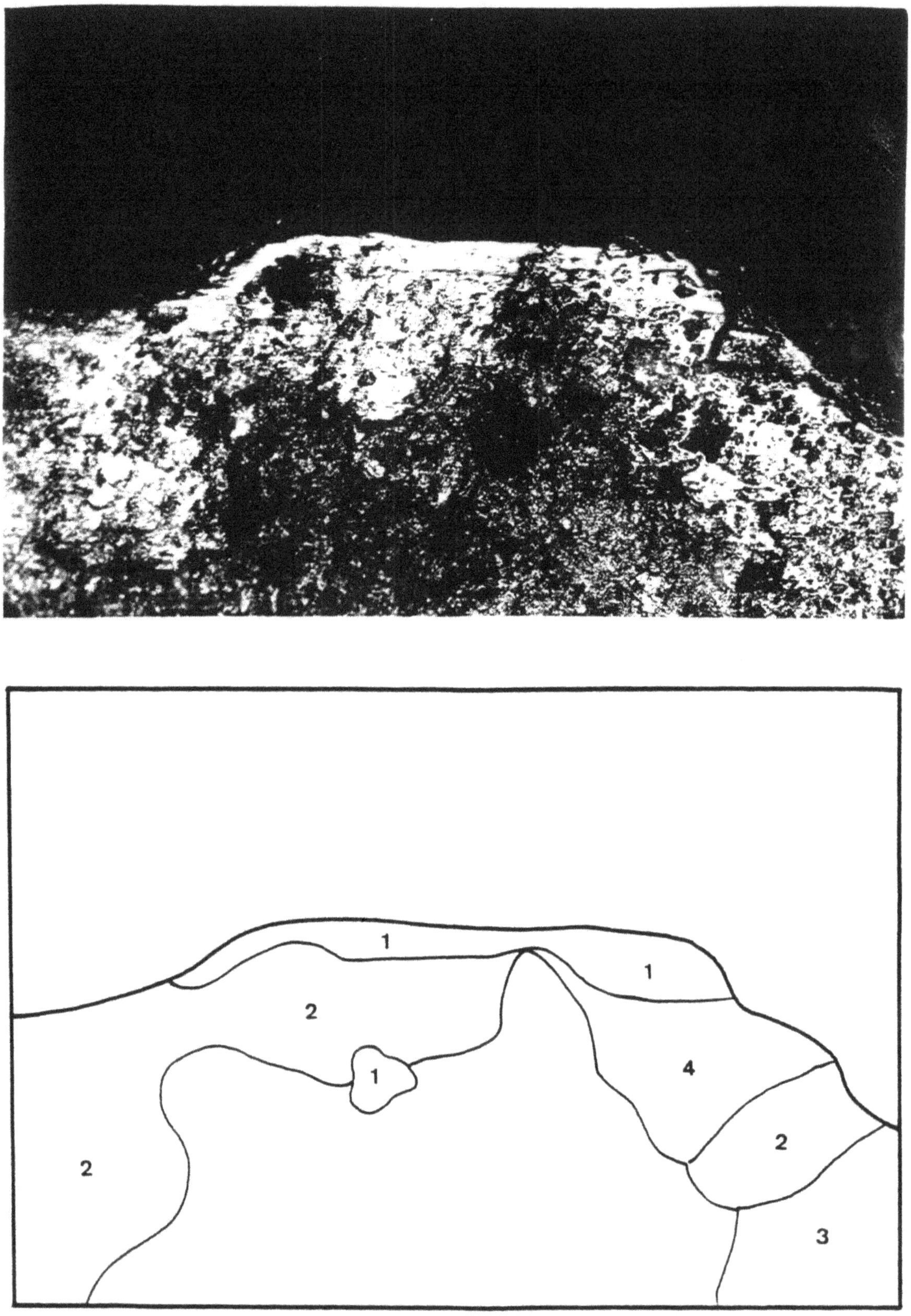

Photo 1. Polish from cutting wood. 100 X. Association of patterns: 1. open; 2. semi-closed; 3. closed; 4. compact.

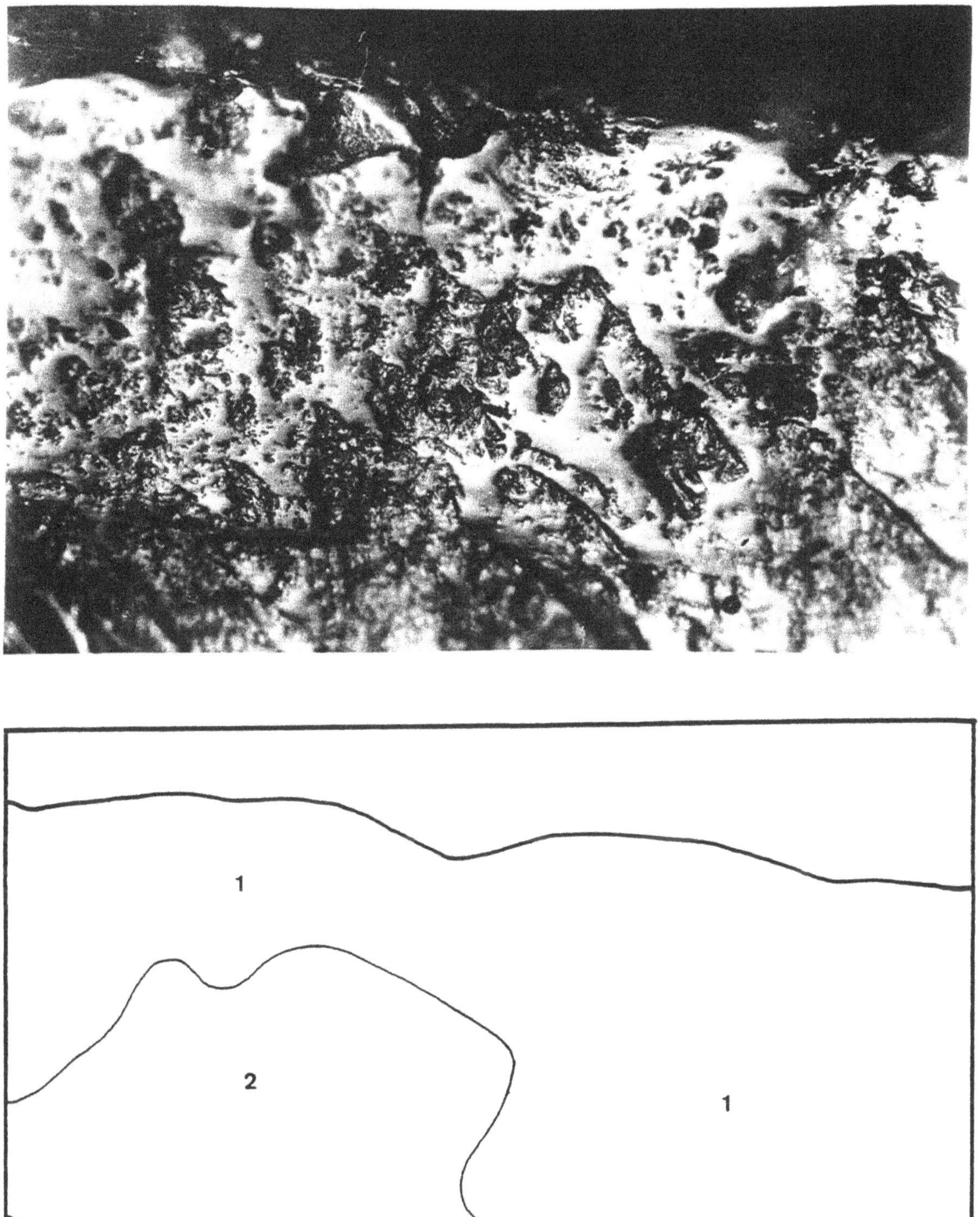

Photo 2. Polish from cutting wood. 200 X. Association of patterns: 1. closed; 2. semi-closed. Network: wide. Topography: undulating.

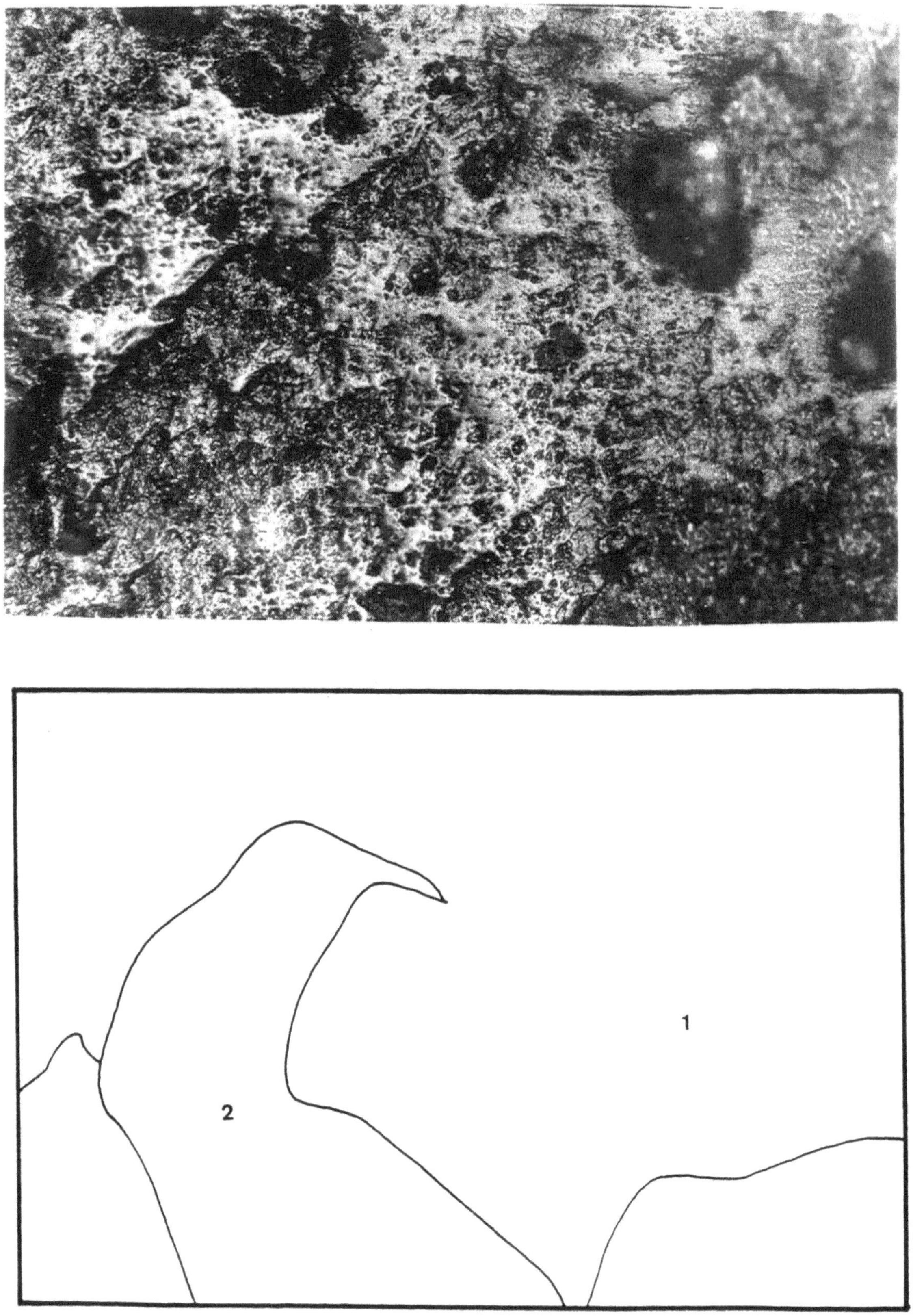

Photo 3 Polish from cutting bone, 200 X. Pattern: closed. Network: narrow. Topography: 1. smooth; 2. irregular

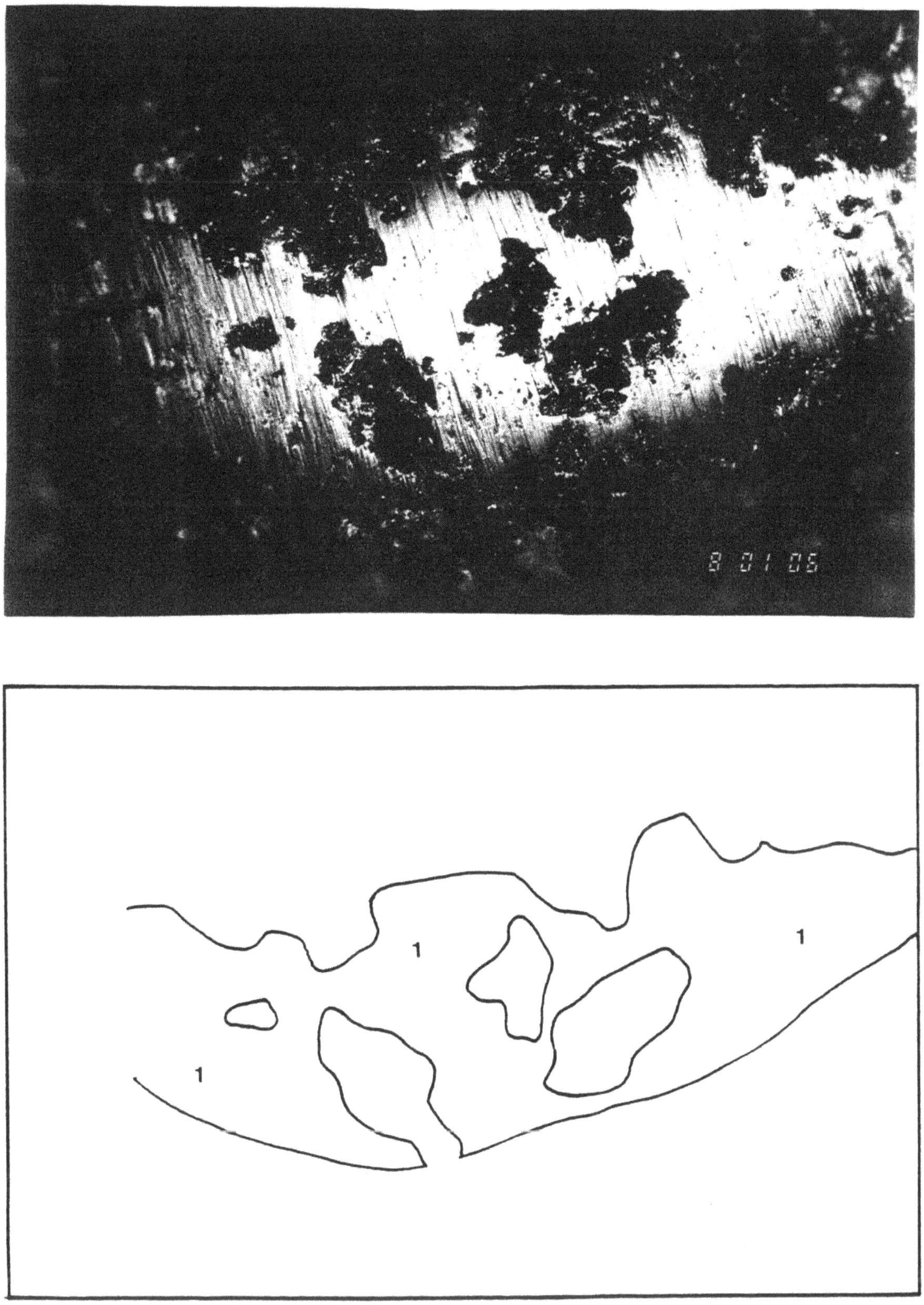

Photo 4. Polish from scraping limestone. 200 X. Pattern: compact. Topography: smooth.

Photo 5. Polish from cutting wood. 200 X. Association of patterns: 1. compact; 2. closed; 3. semi-closed. Topography: undulating.

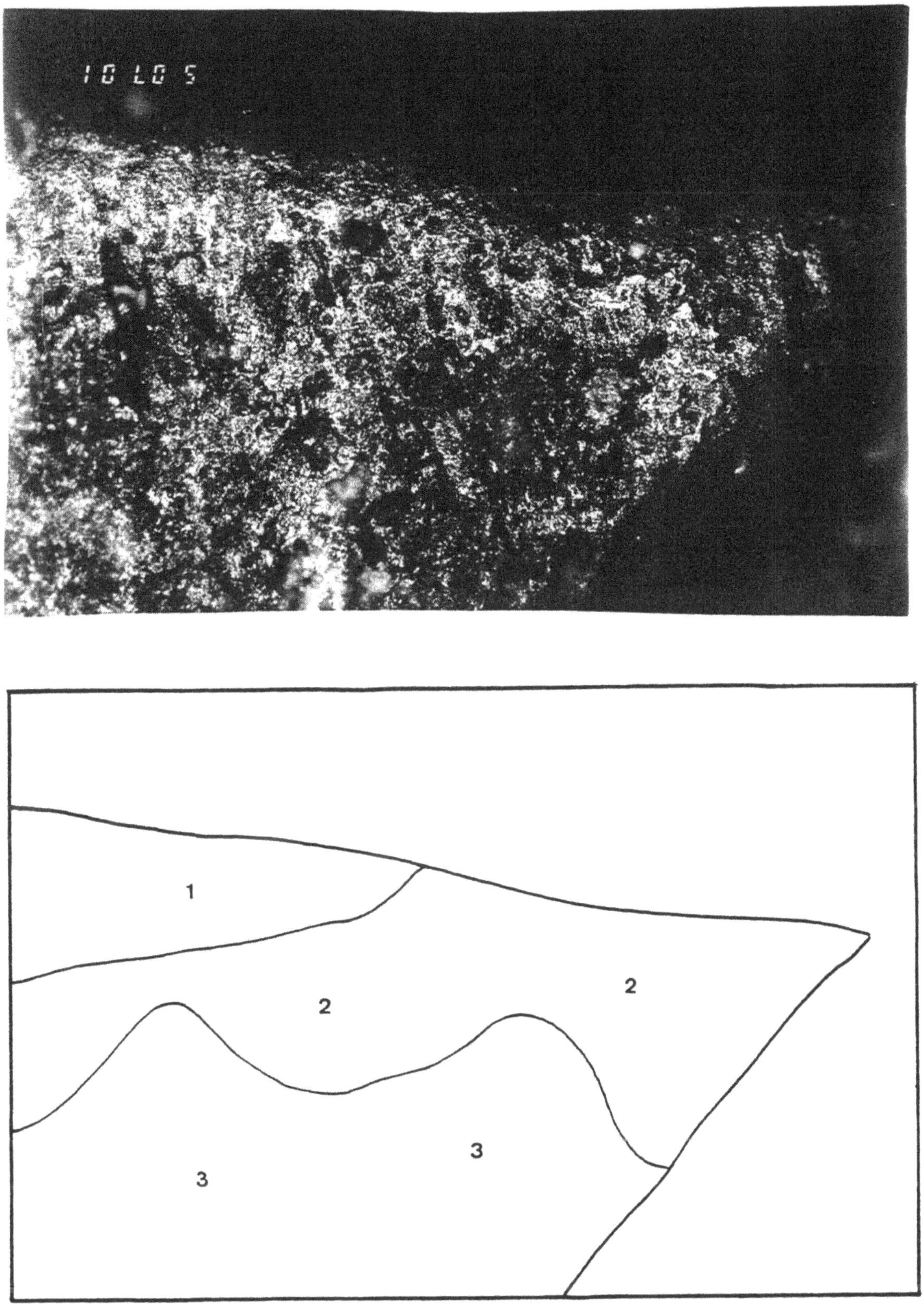

Photo 6. Polish from scraping dry hide. 200 X. Association of patterns: 1. closed; 2. semi-closed;; 3. open. Association of topographies: 1. undulating/irregular; 2 and 3: irregular.

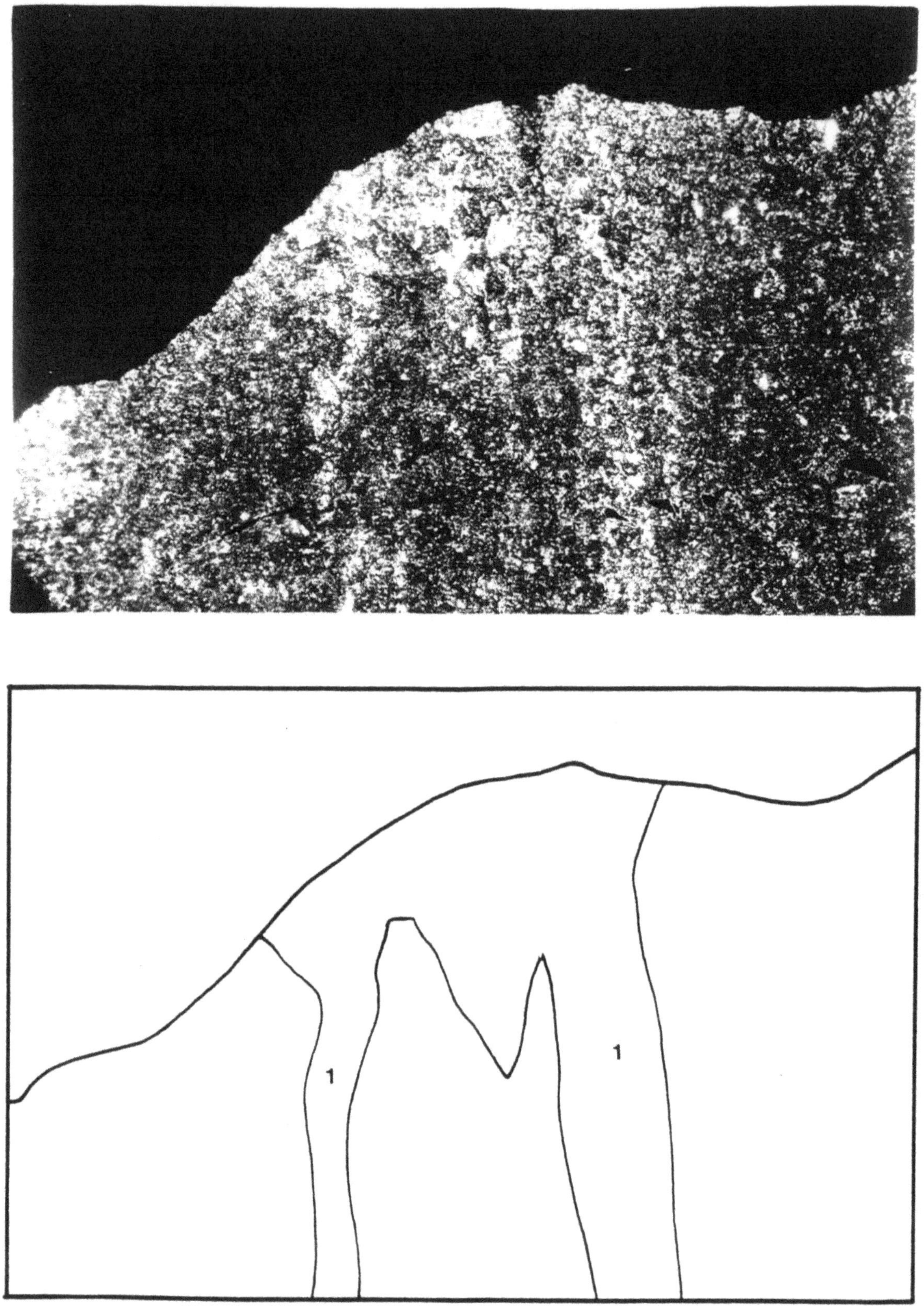

Photo 7. Polish with linear disposition from scraping wood. 200 X.

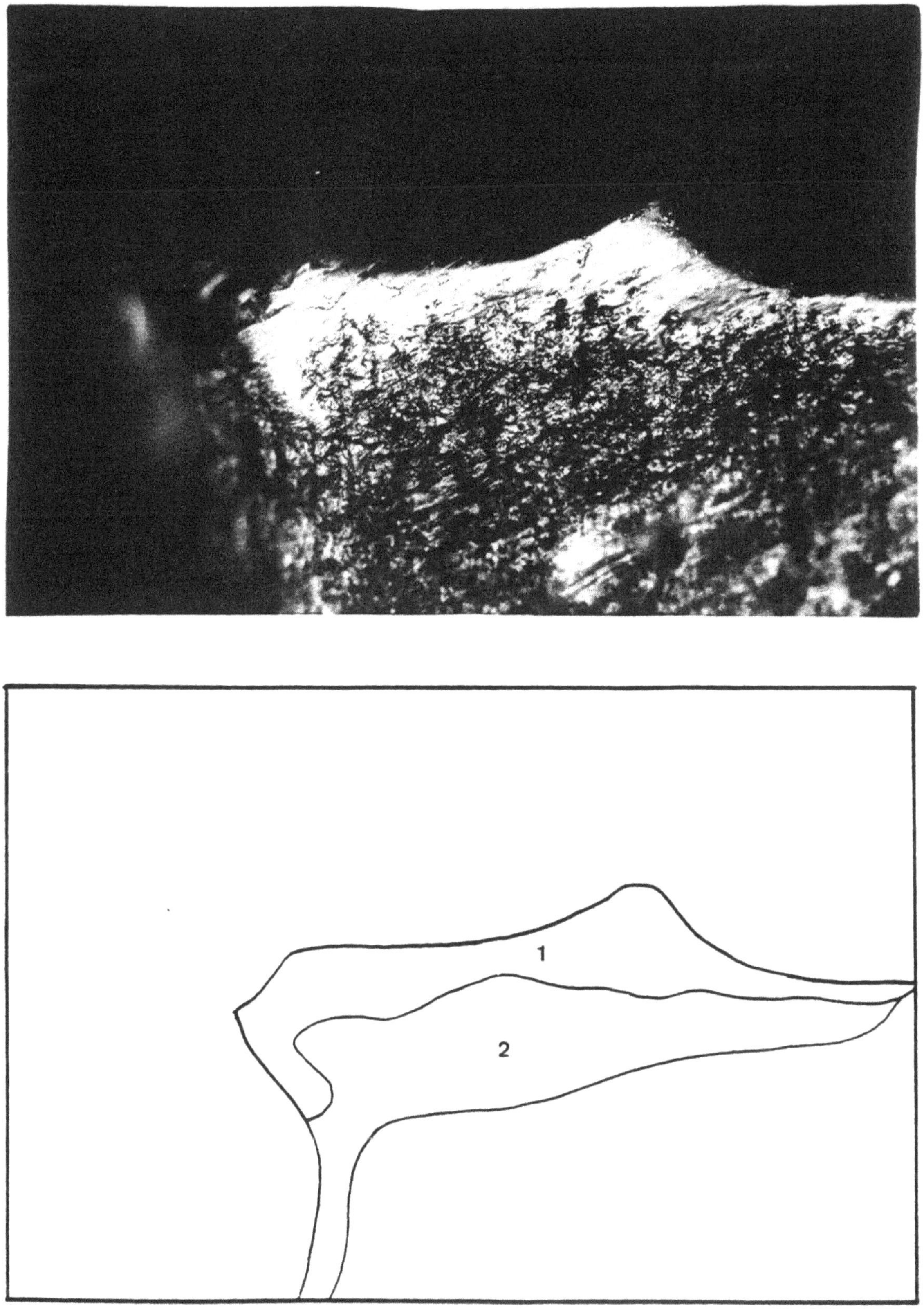

Photo 8. Polish from cutting bone. 200 X. Linear components. Association of patterns: 1. compact; 2. closed. Association of topographies: 1. smooth; 2. irregular.

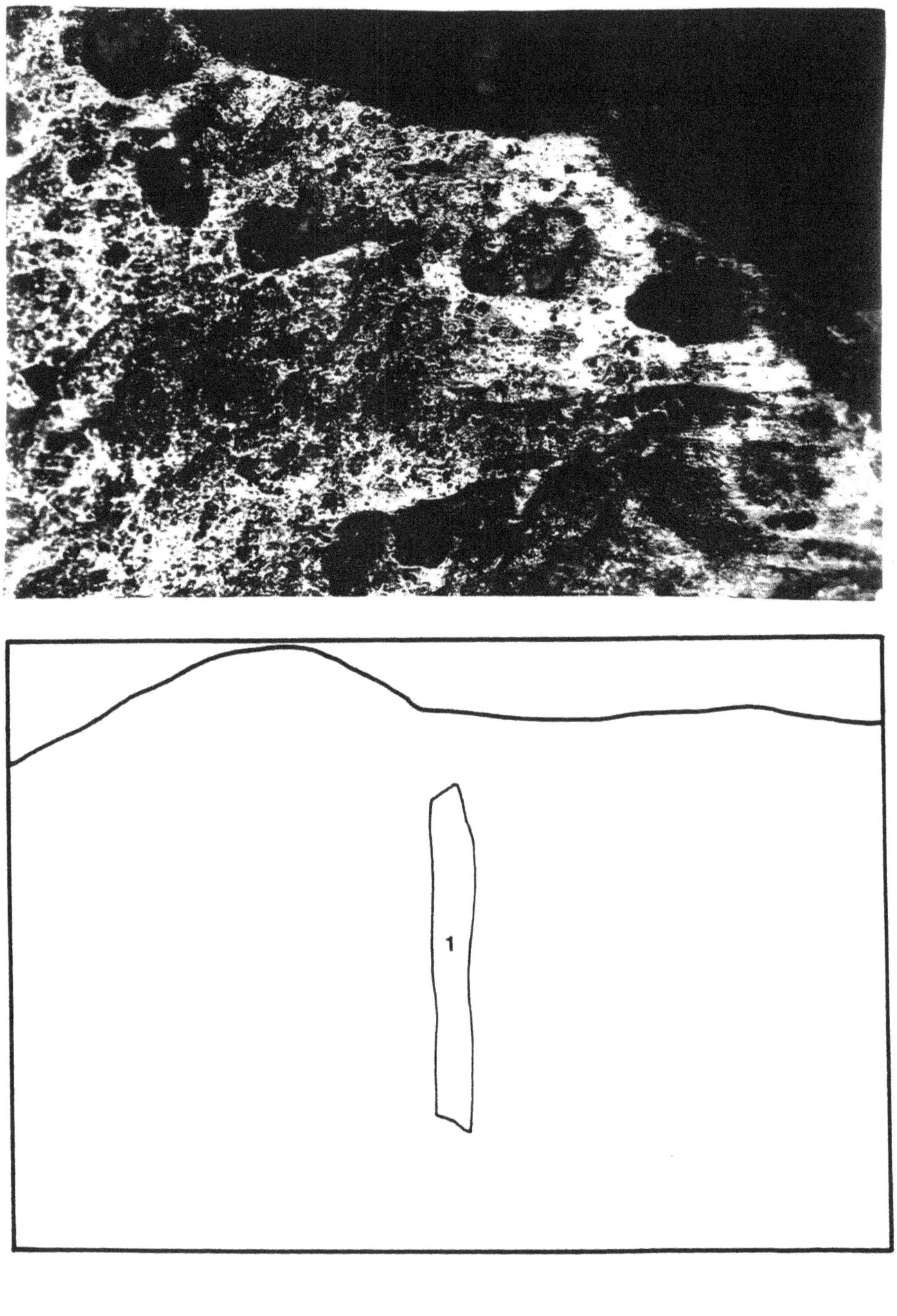

Photo 9. Striation from impact on a projectile point. 100 X.

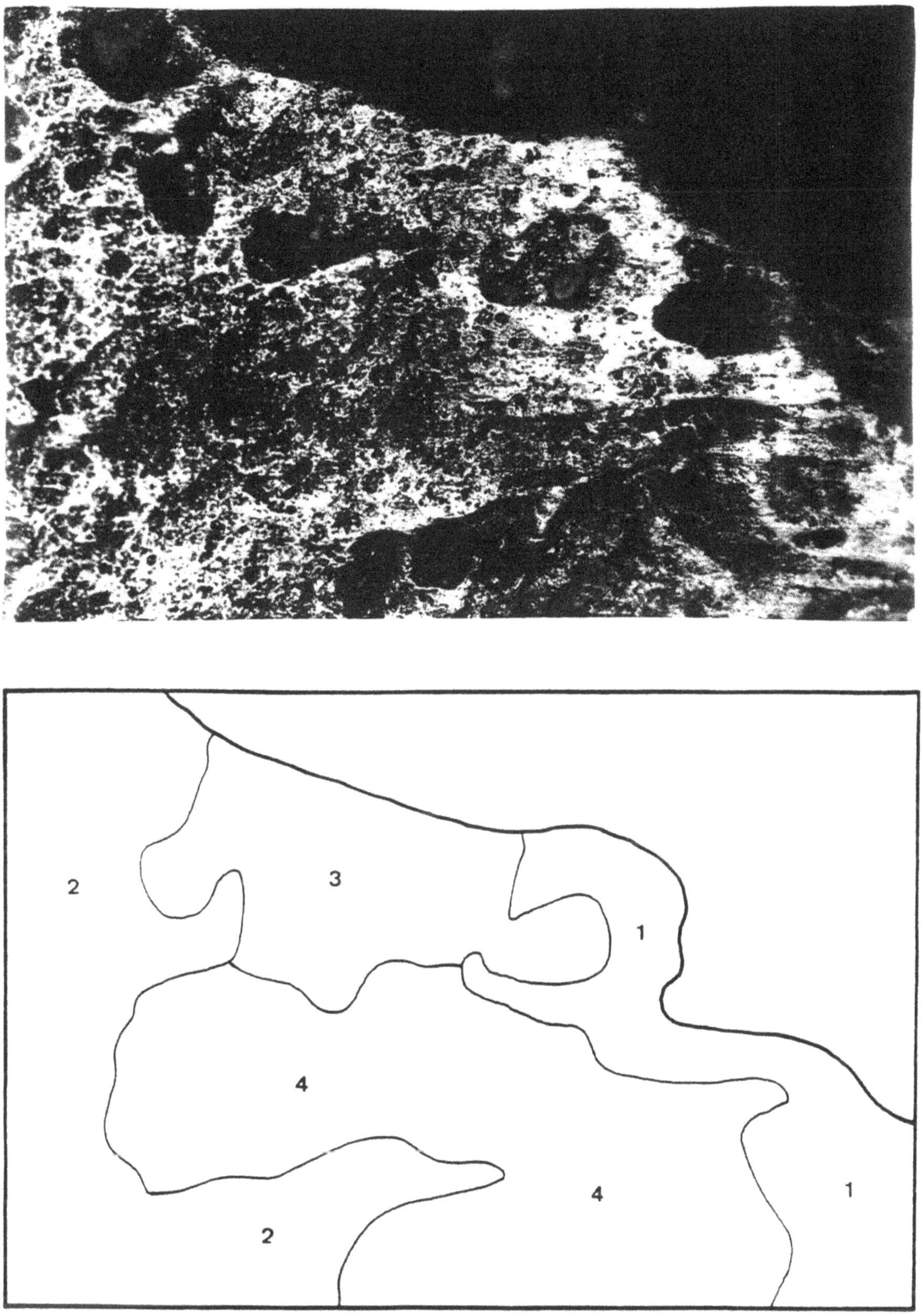

Photo 10. Polish from cutting wood. 100 X. Association of patterns: 1. compact; 2. closed; 3. semi-closed. Topography: undulating.

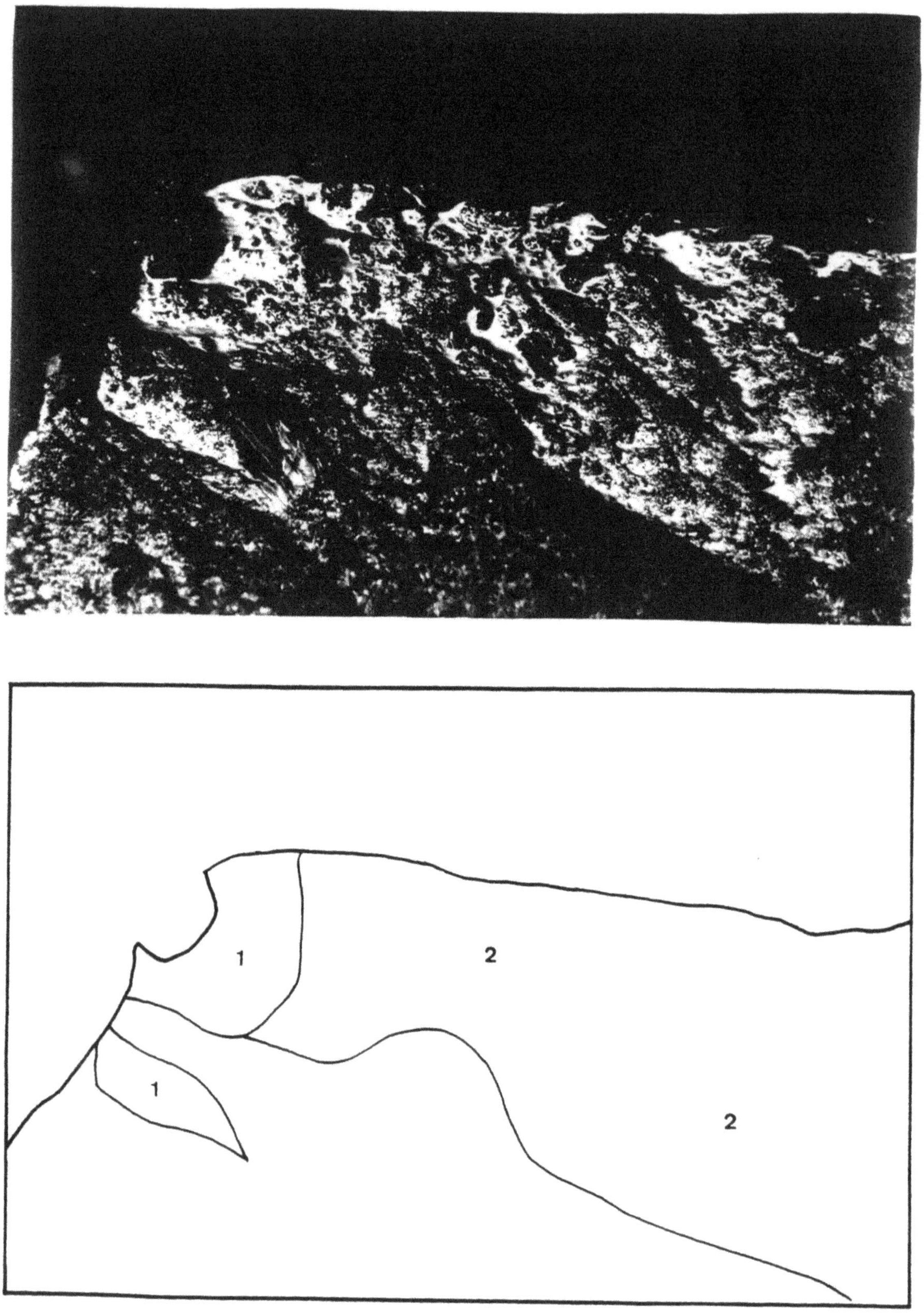

Photo 11. Polish from cutting wood. 100 X. Association of patterns: 1. closed; 2. semi-closed.

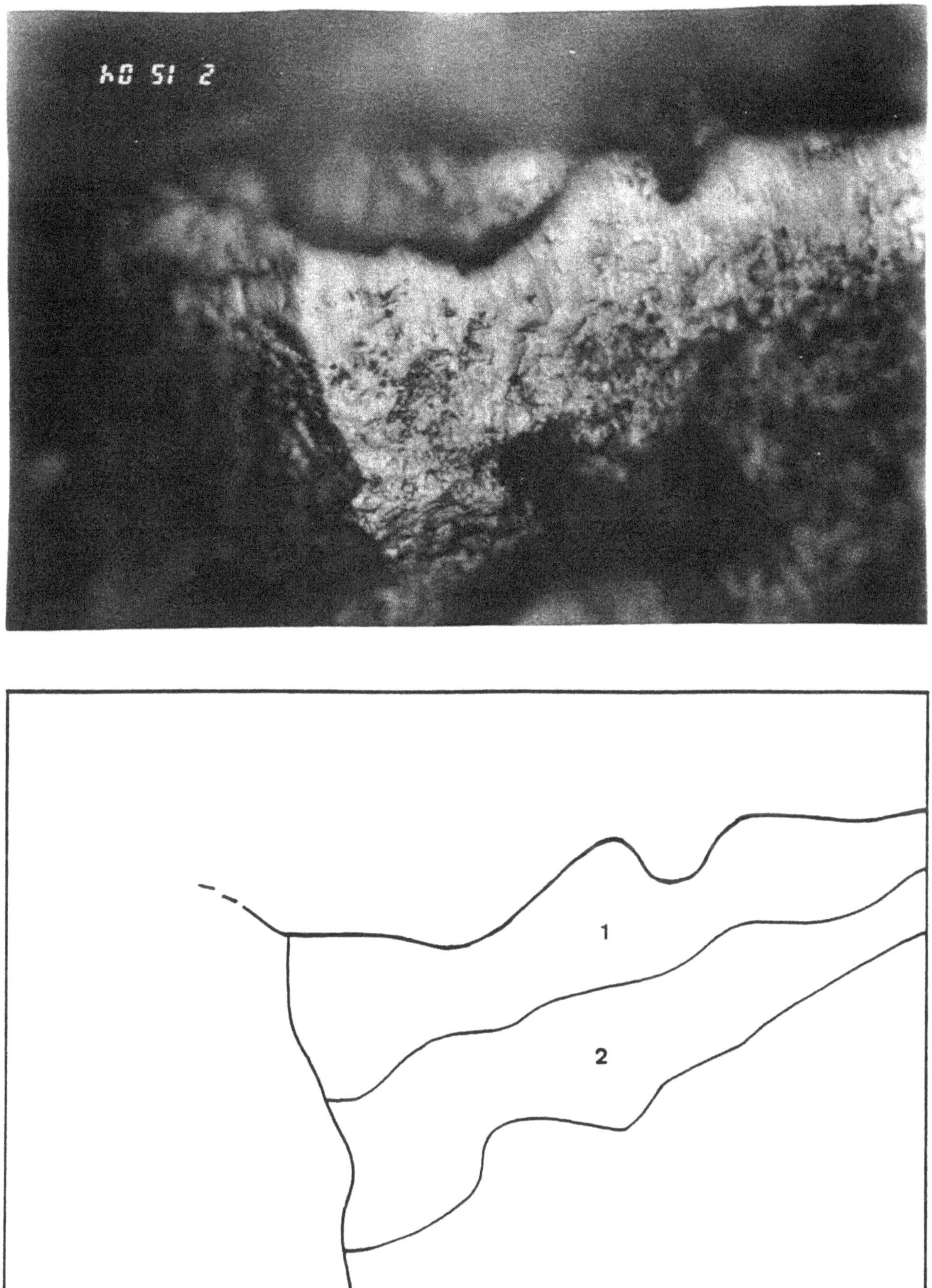

Photo 12. Polish from scraping bone. 200 X. Association of patterns: 1. compact; 2. closed. Association of topographies: 1. smooth; 2. irregular.

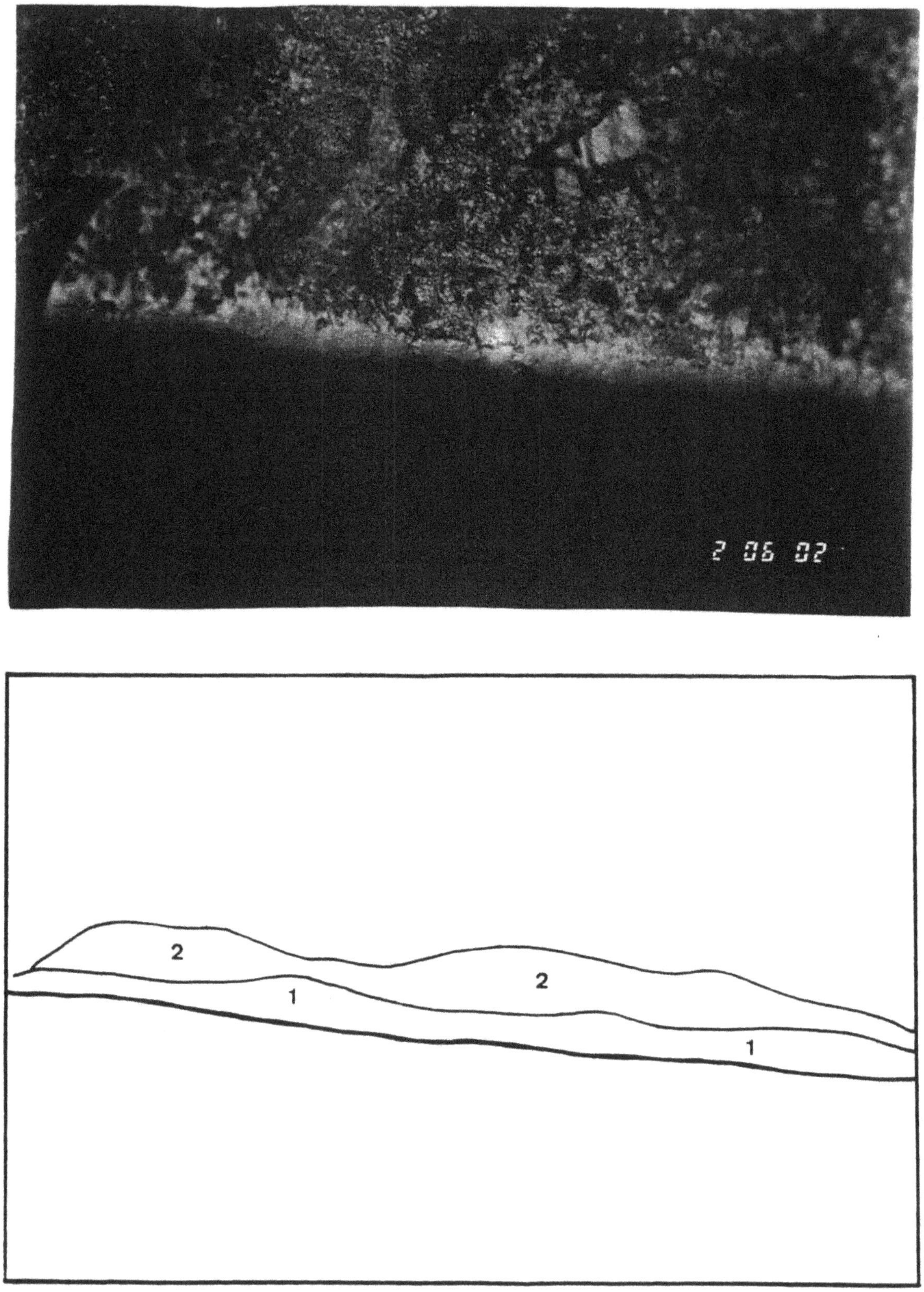

Photo 13 Polish from scraping bone. 200 X. Association of patterns: 1. compact; 2. closed.

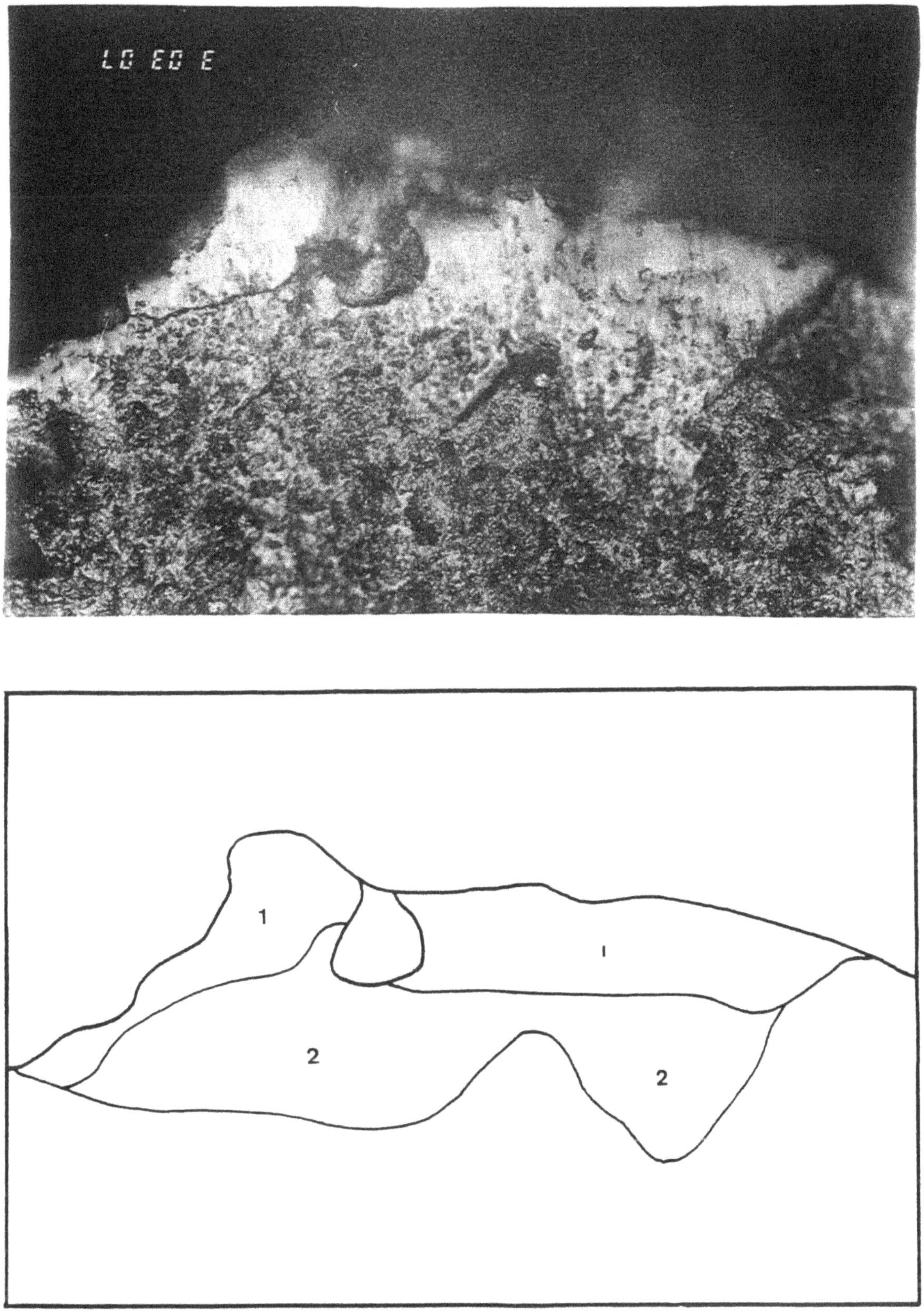

Photo 14. Polish from scraping antler. 200 X. Association of patterns: 1. compact; 2. closed. Association of topographies: 1. smooth; 2. smooth/undulating.

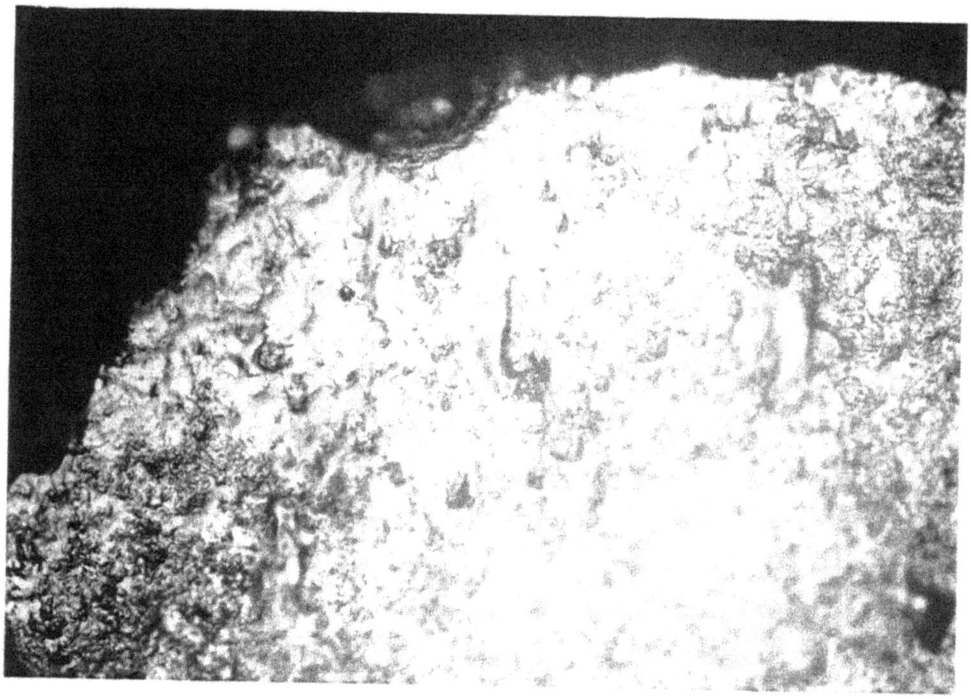

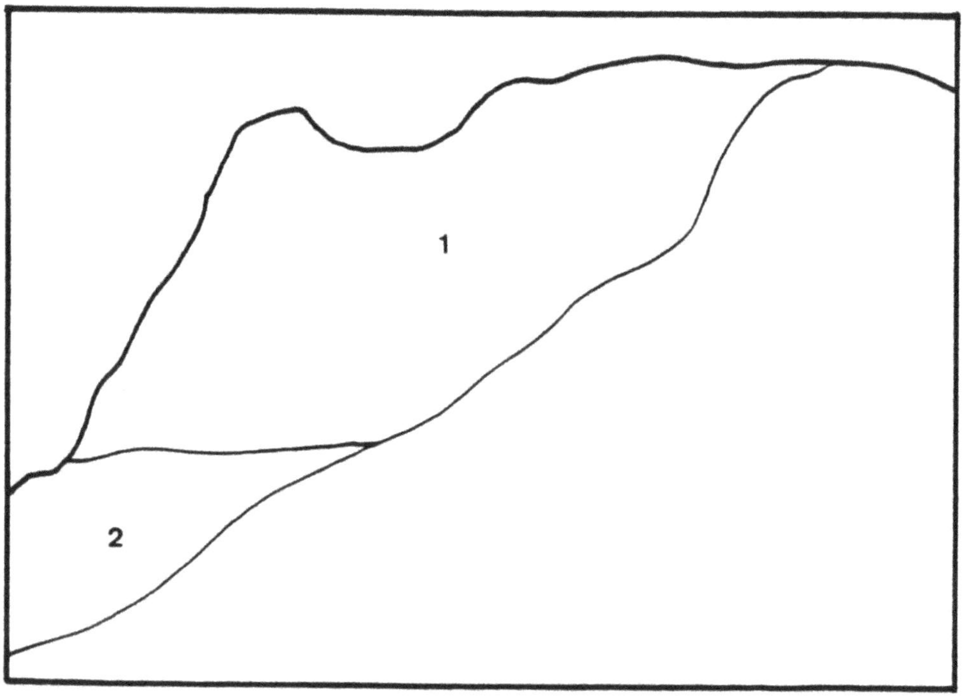

Photo 15. Polish from scraping antler. 200 X. Pattern: compact. Topography: smooth/undulating.

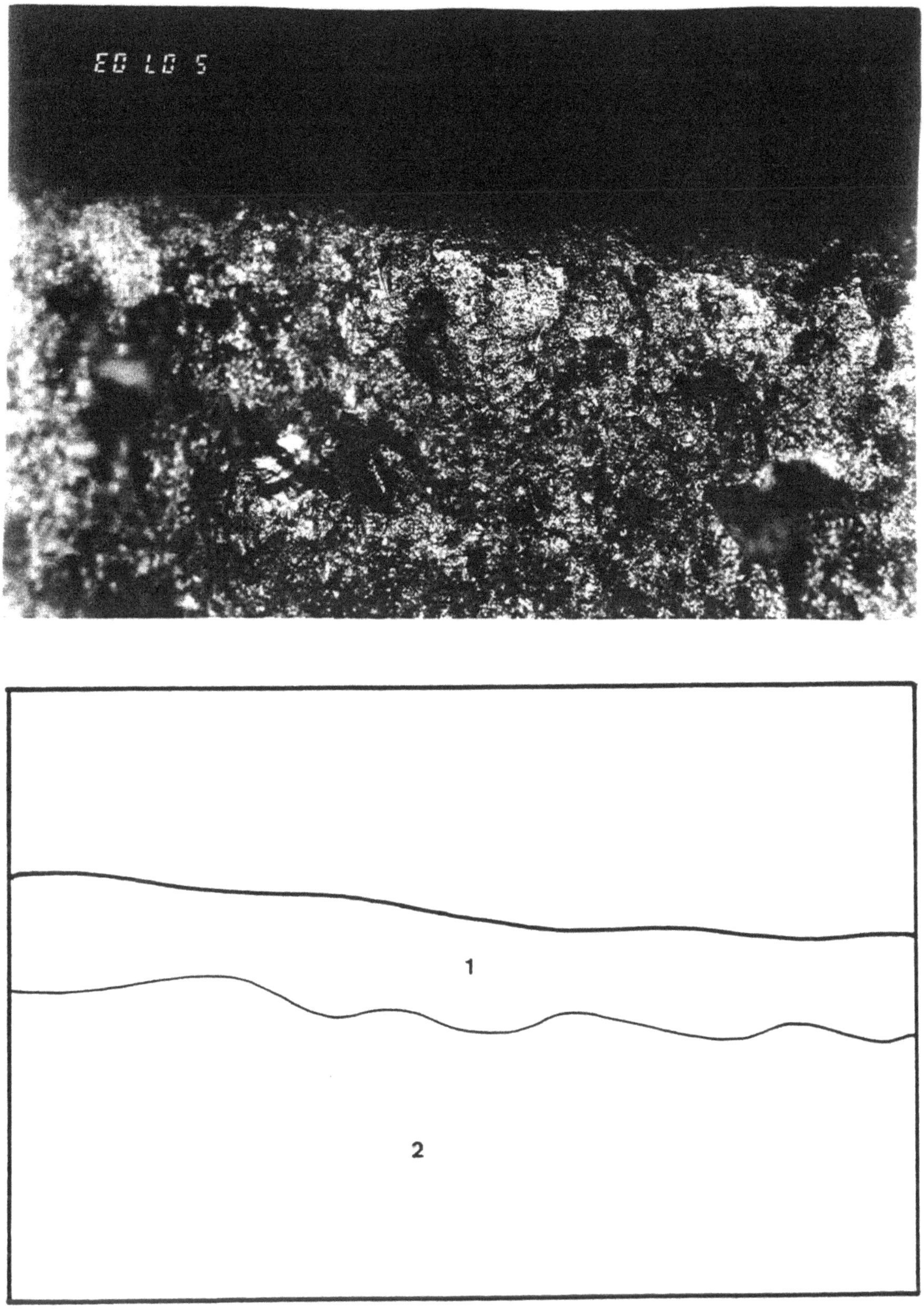

Photo 16. Polish from scraping dry hide. 200 X. Association of patterns: 1. closed; 2. semi-closed. Association of topographies: 1. undulating/irregular; 2. irregular.

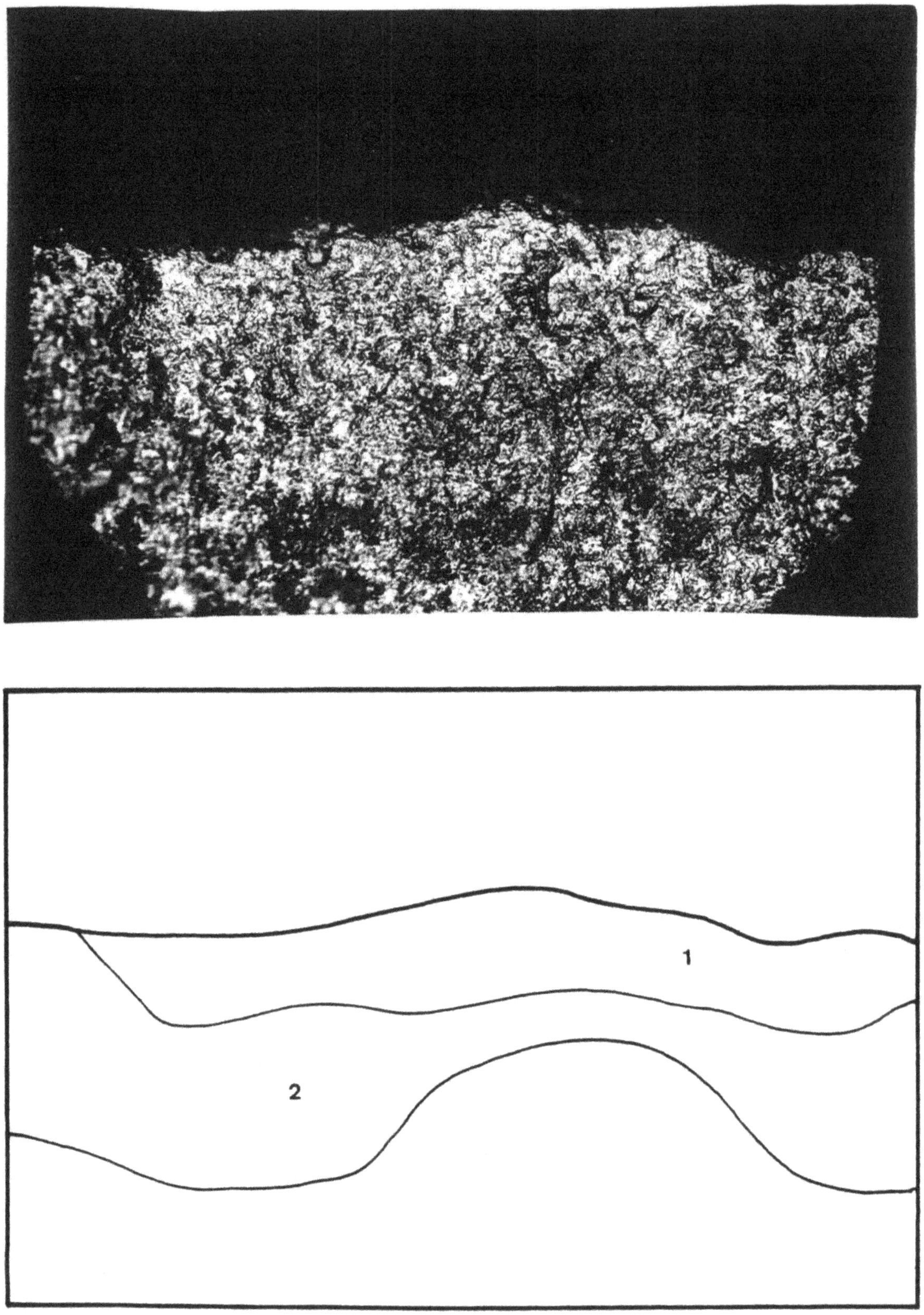

Photo 17. Polish from scraping fresh hide. 200 X. Association of patterns: 1. closed; 2. semi-closed. Topography: irregular.

APPENDIX 4

DIAGRAMS OF EXPERIMENTAL USE TRACES

In these diagrams the distribution of use-wear traces on some experimental tools is shown. The different surfaces of the active zone (i.e. dorsal and ventral surface of the edge) and the microscarring from use are drawn. Use polish is represented by means of screens of different intensities, which are related to the degree of polish development. The sketches showing the working motion explain the activity carried out with the tool. One of the surfaces in the sketch is identified with the corresponding surface in the diagram. This allows to relate the distribution of traces (diagram) with respect to the working motion (sketch). All these tools were used for 60 minutes

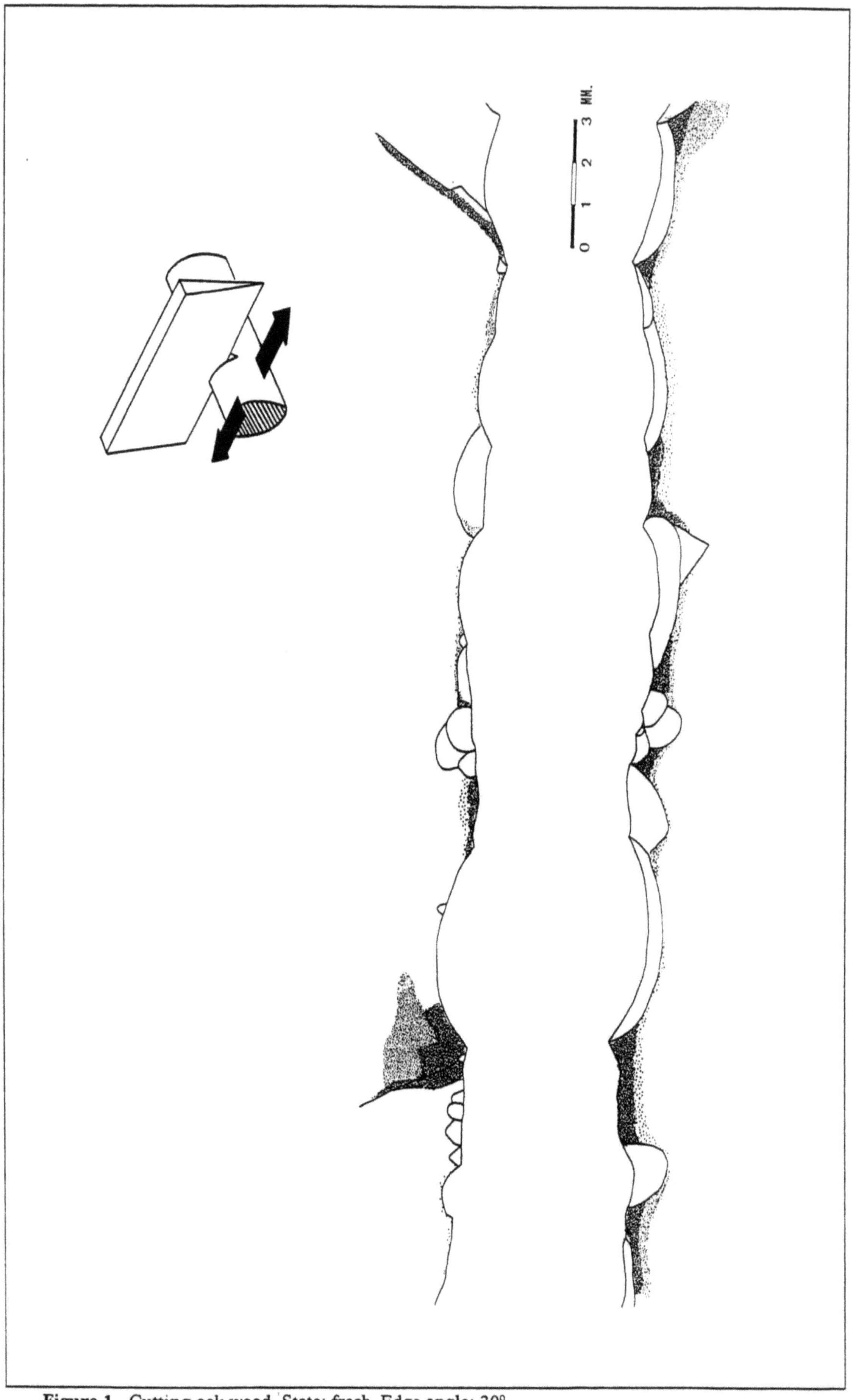

Figure 1. Cutting oak wood. State: fresh. Edge angle: 30°.

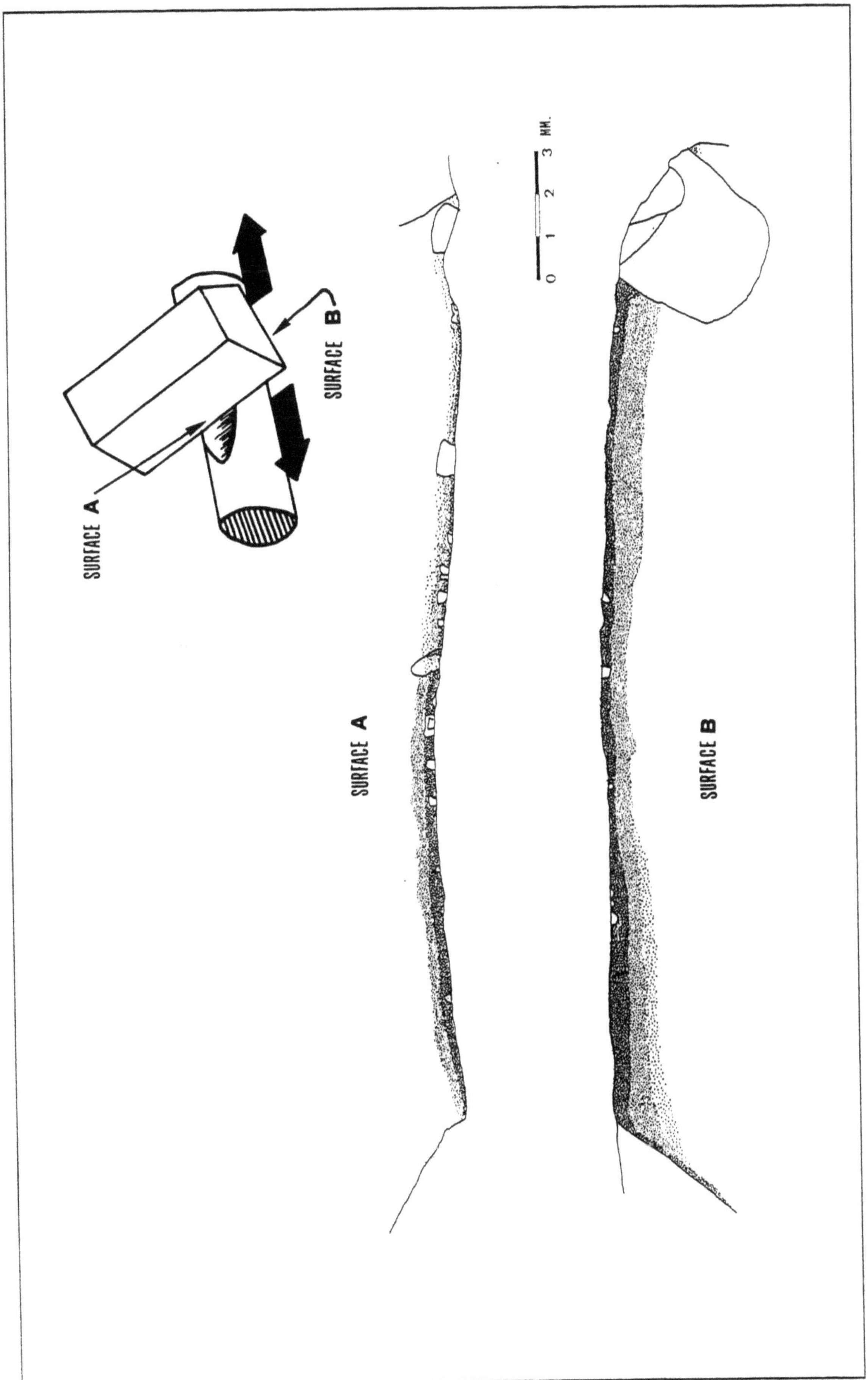

Figure 2. Scraping oak wood. State: fresh. Edge angle: 76°.

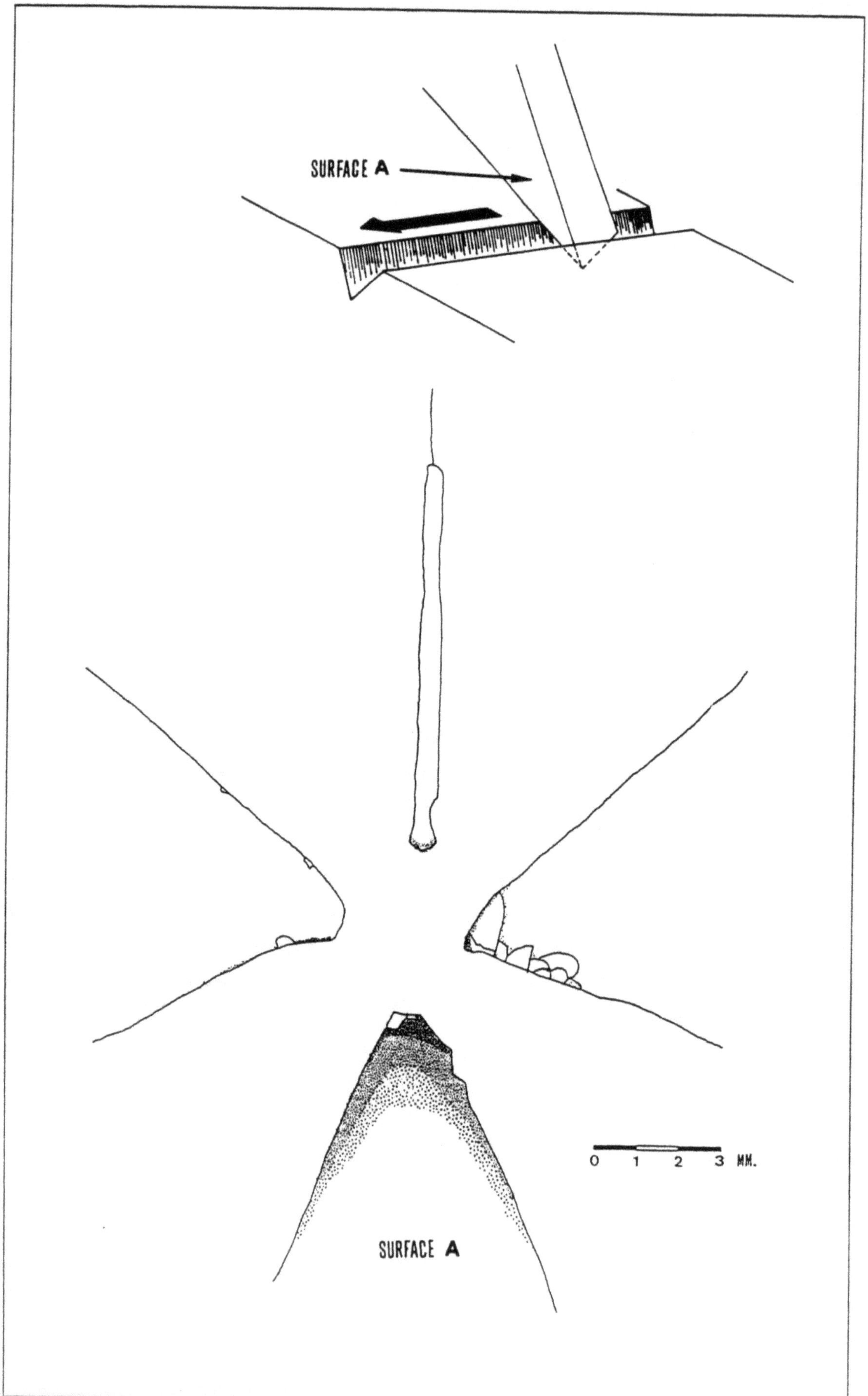

Figure 3. Engraving oak wood. State: fresh.

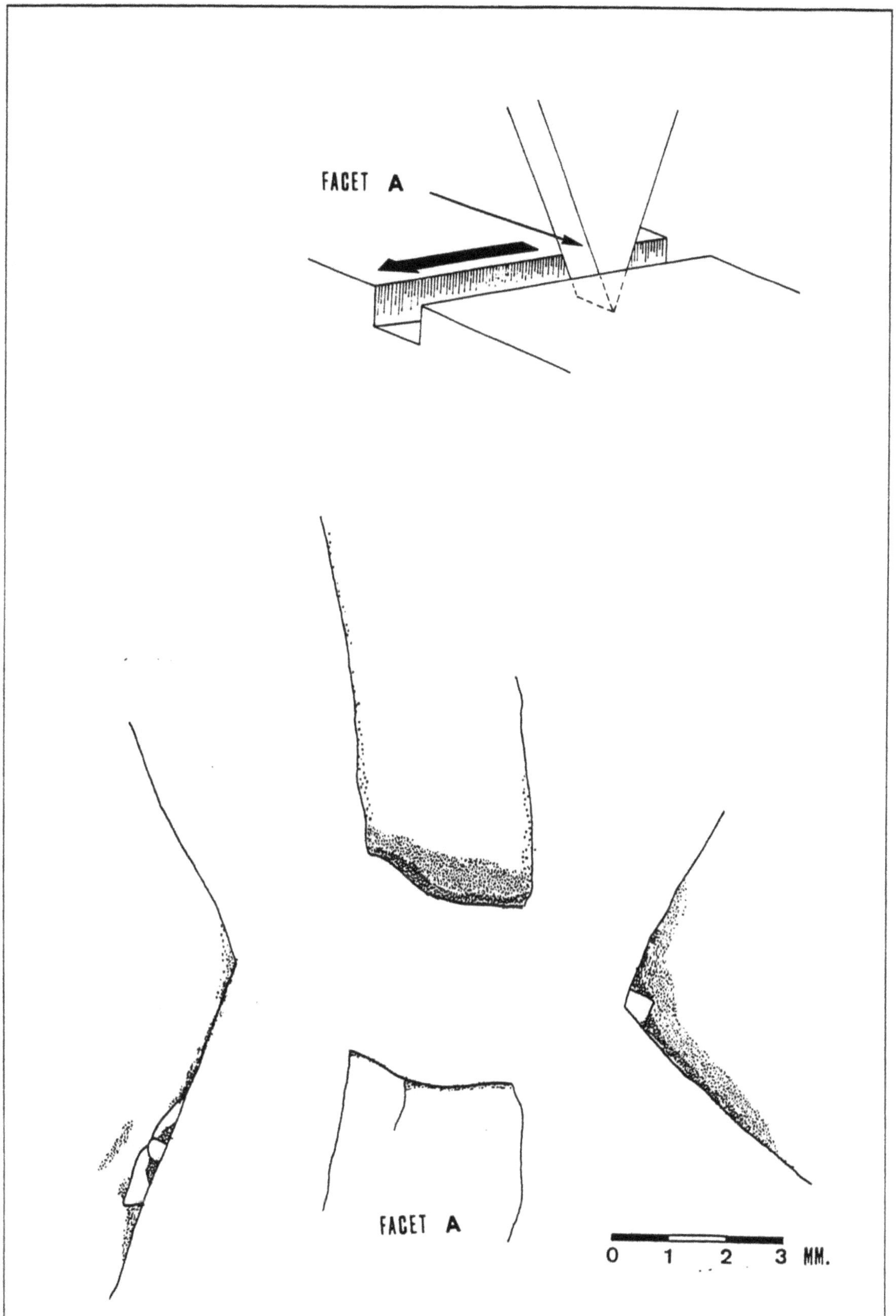

Figure 4. Engraving aok wood. State: fresh.

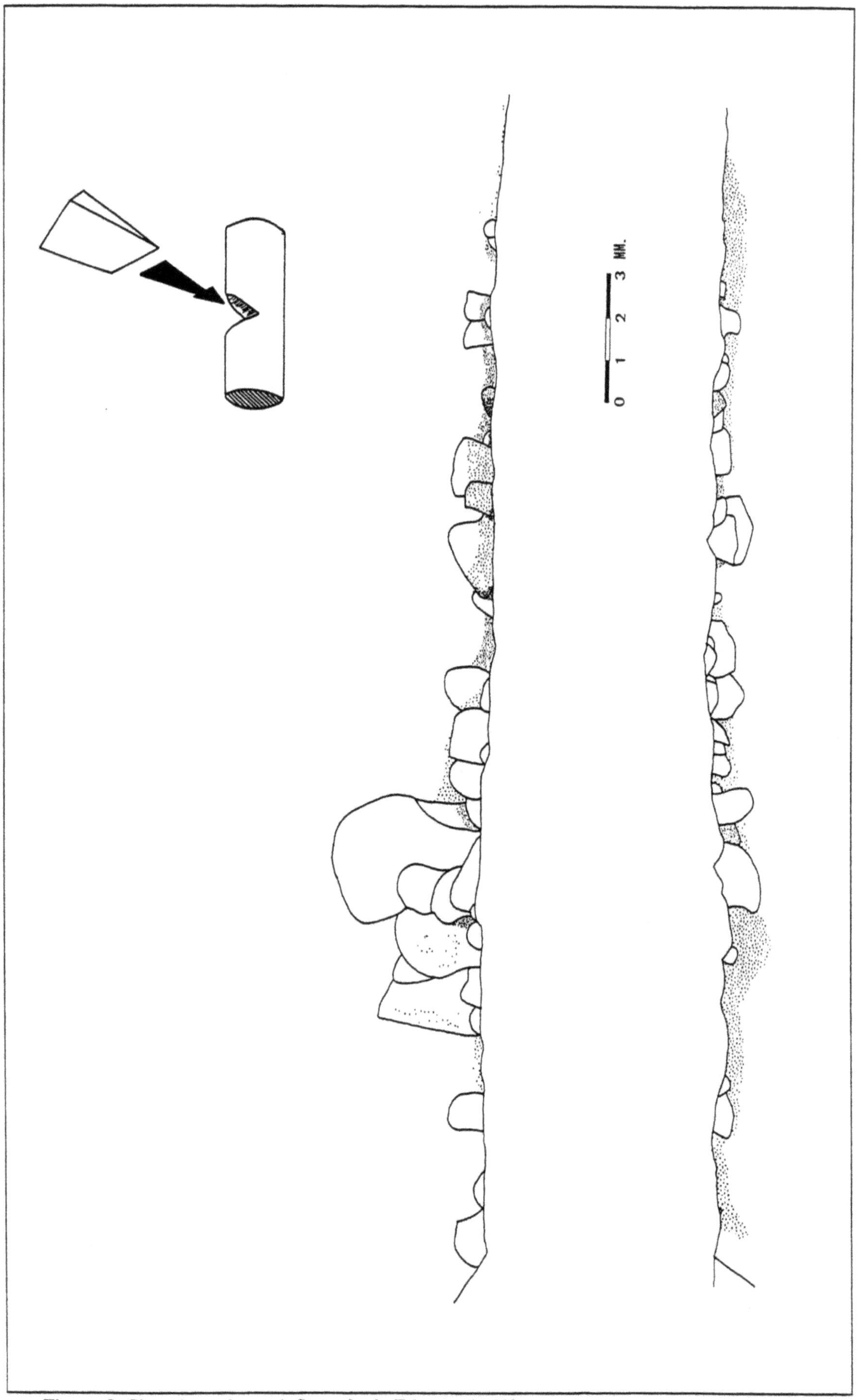

Figure 5. Chopping oak wood. State: fresh. Edge angle 58°.

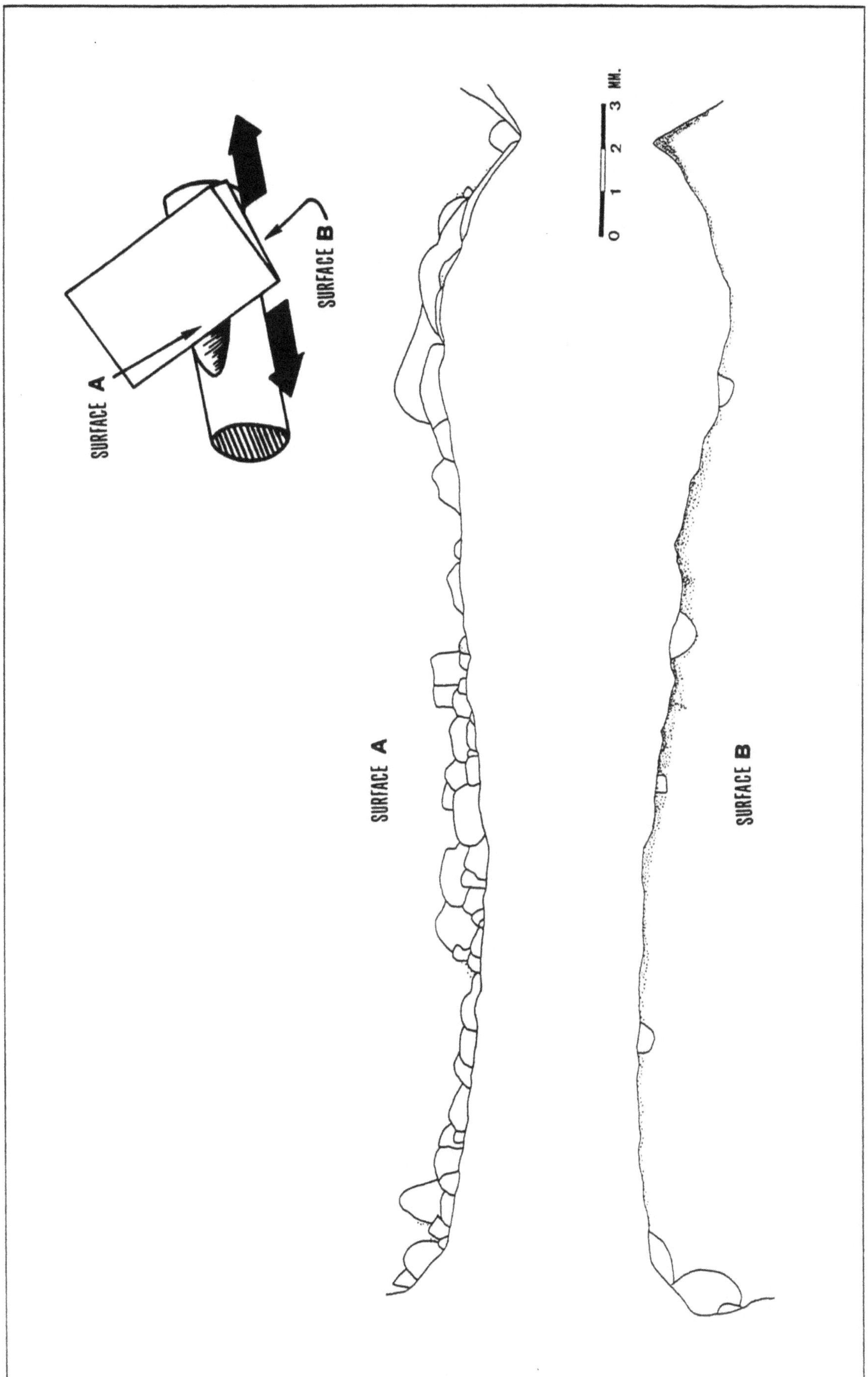

Figure 6. Scaping oak wood. State: fresh. Edge angle: 52°.

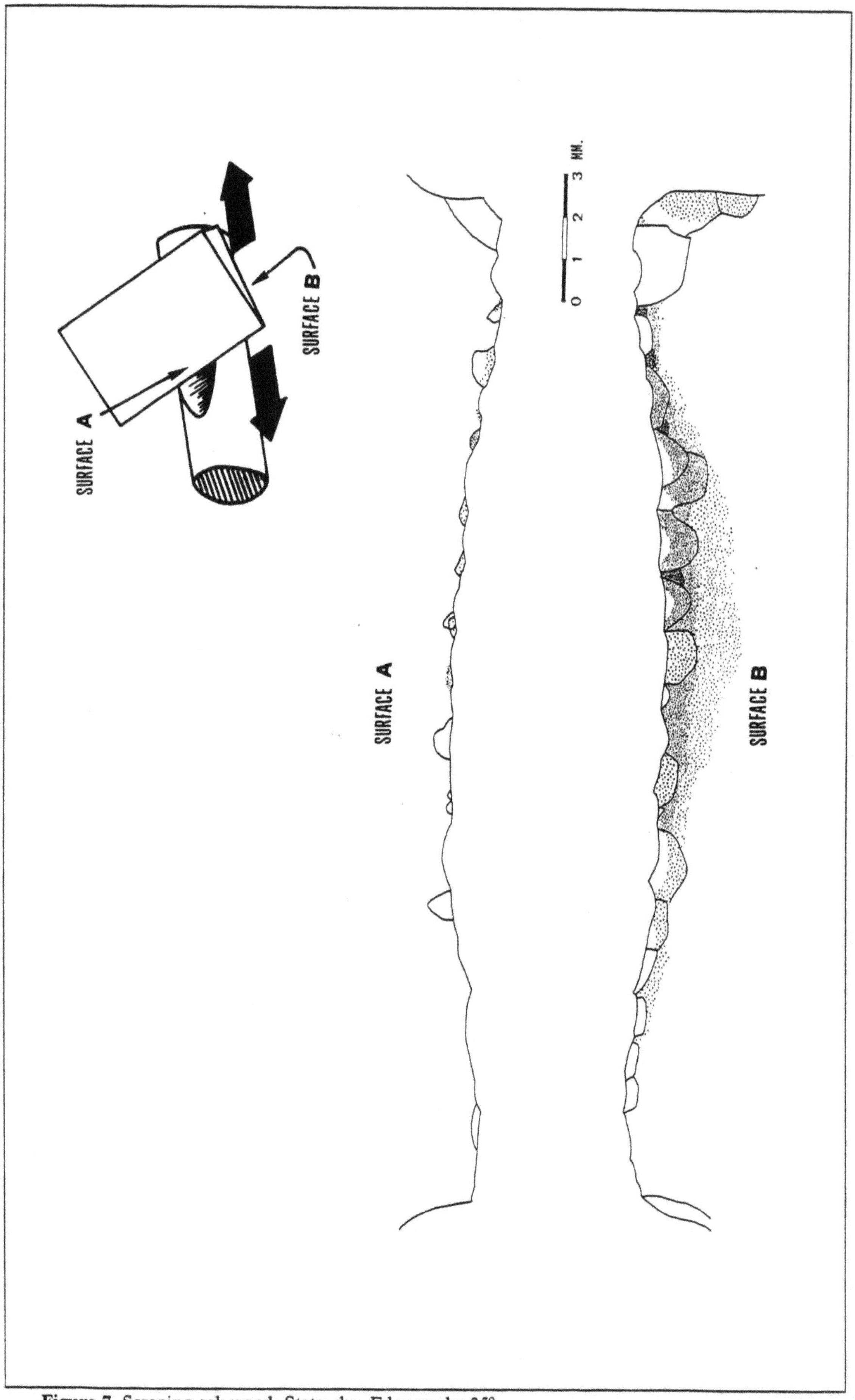

Figure 7. Scraping oak wood. State: dry. Edge angle: 35°.

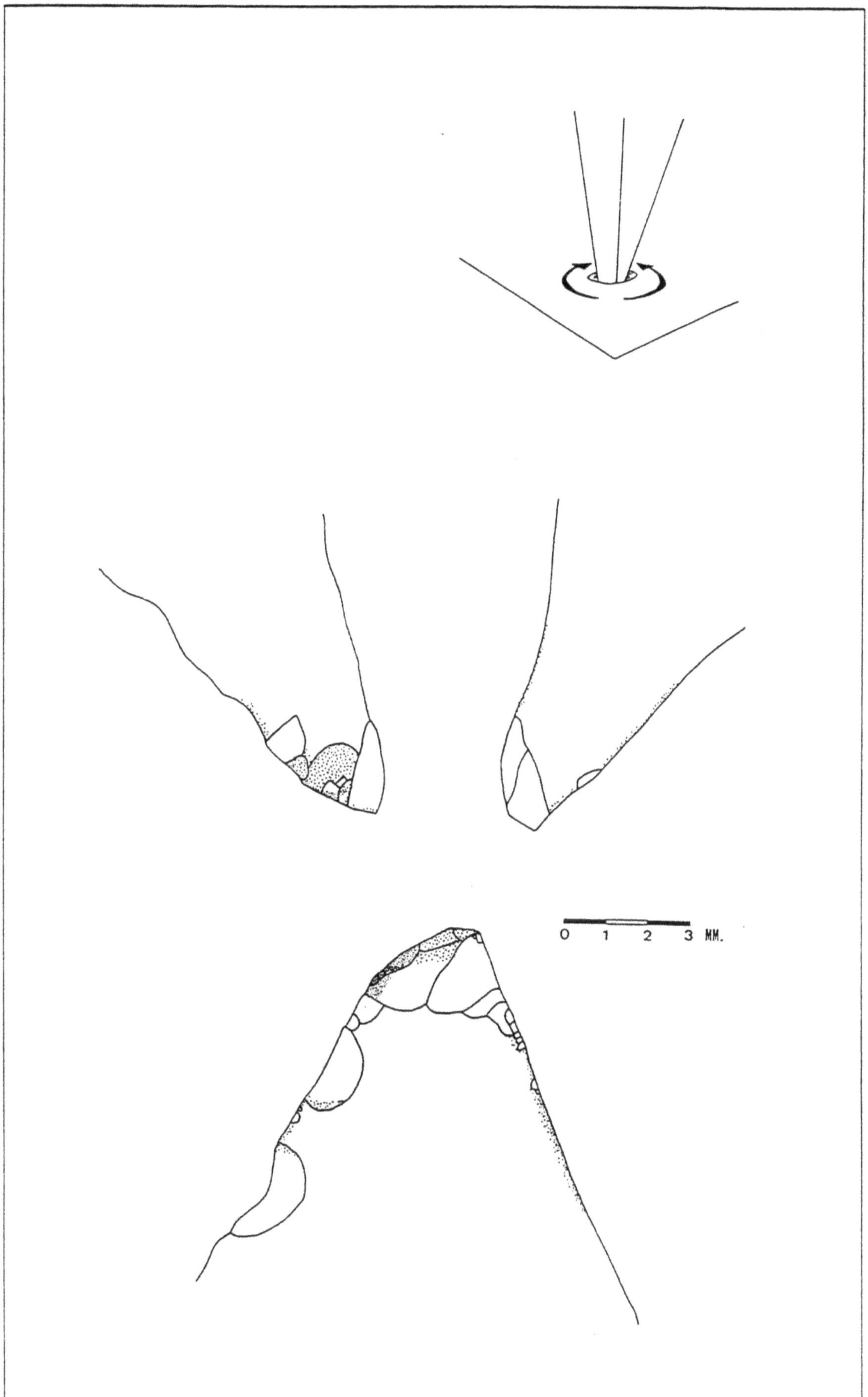

Figure 8. Boring pine wood. State: fresh.

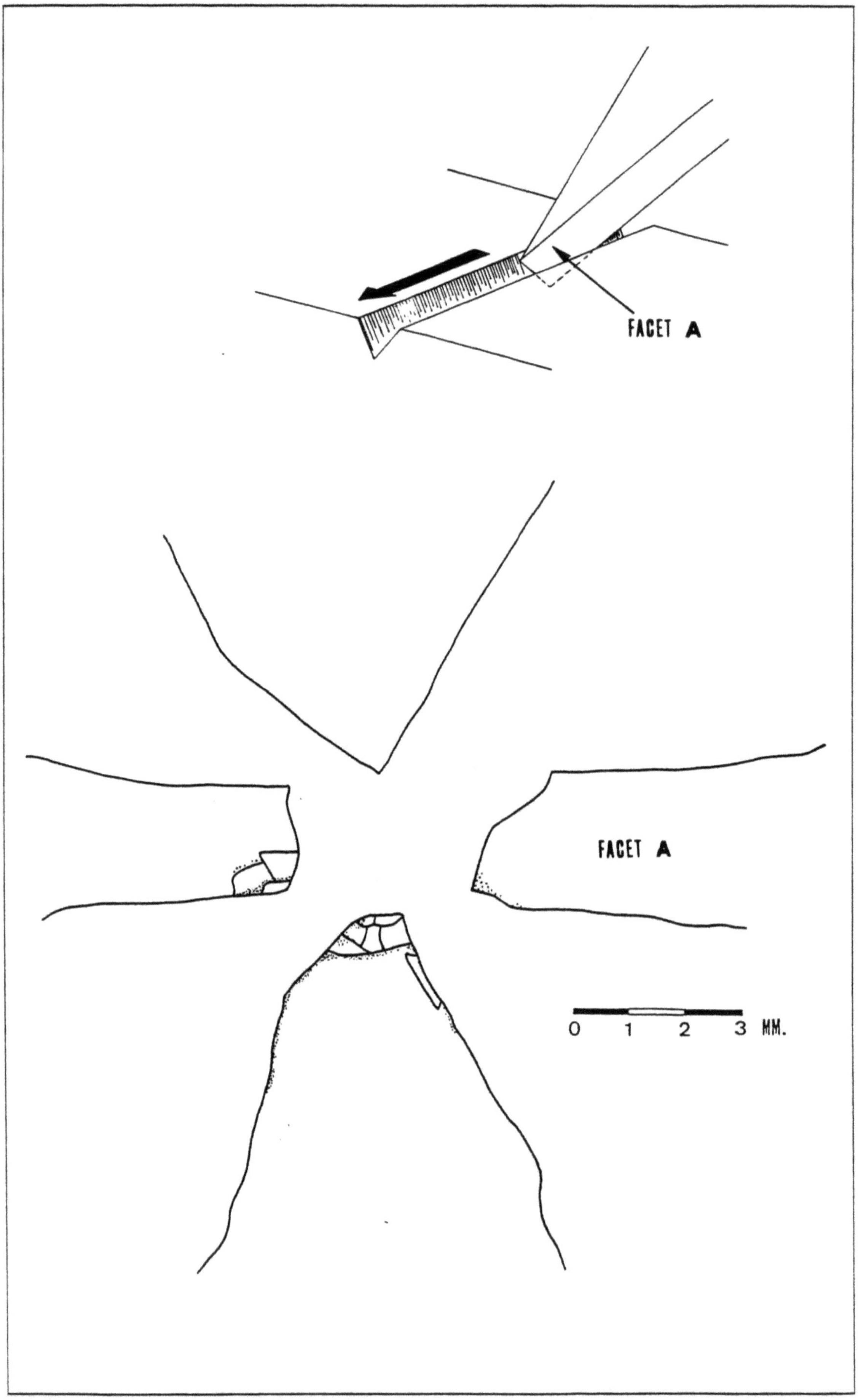

Figure 9. Engraving boxwood. State: dry.

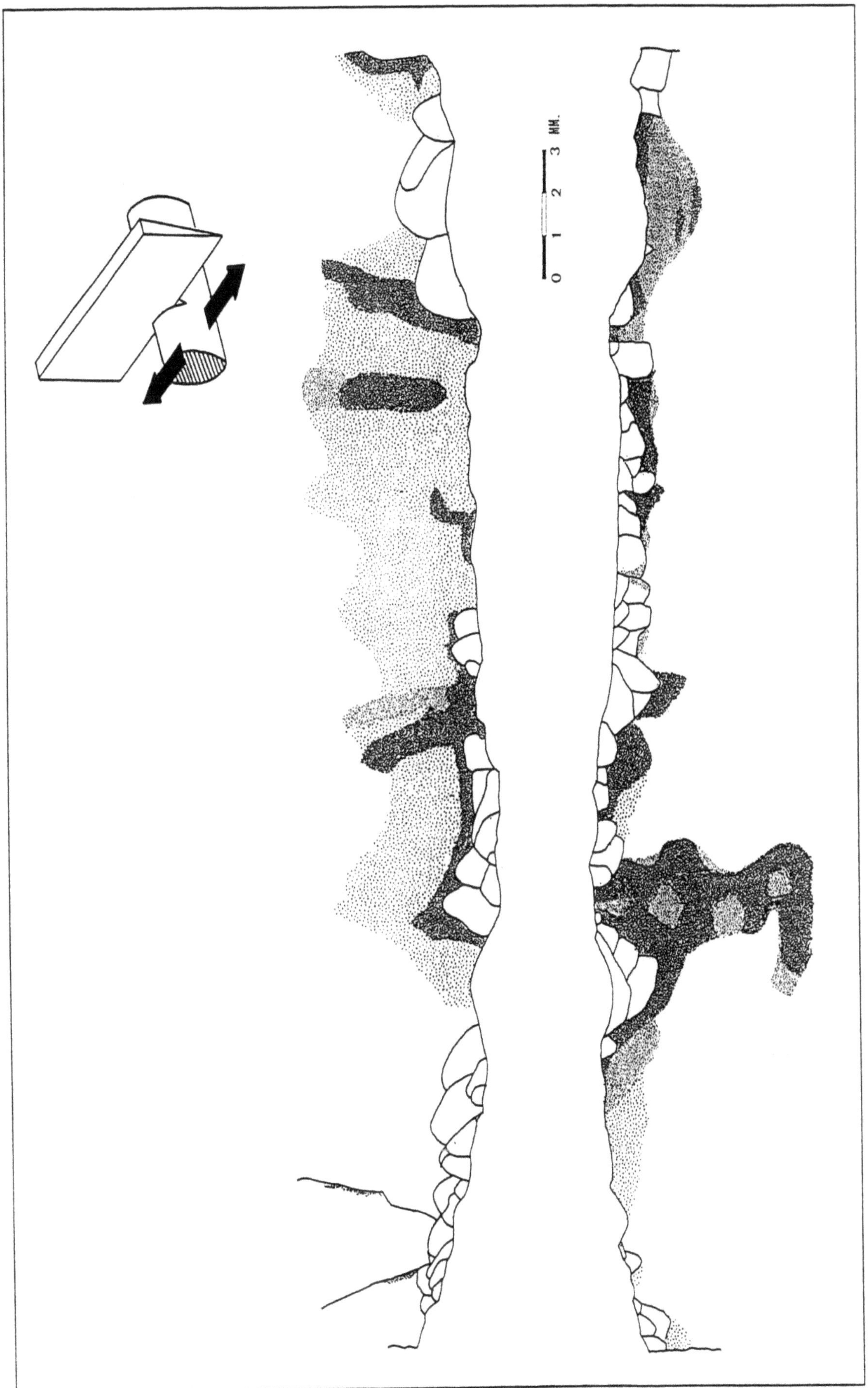

Figure 10. Cutting bone. State: soaked. Edge angle: 44°.

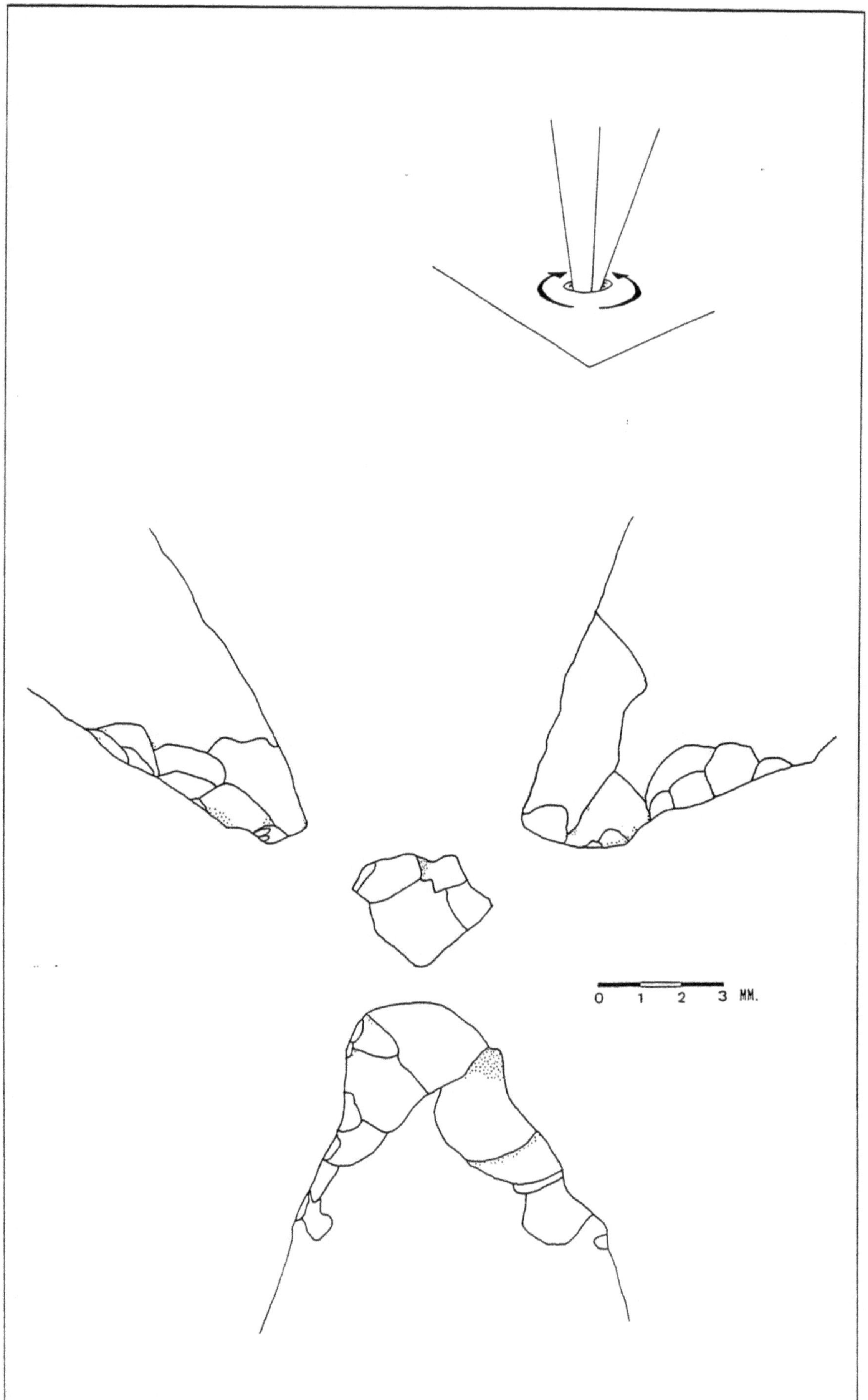

Figure 11. Boring bone. State: fresh.

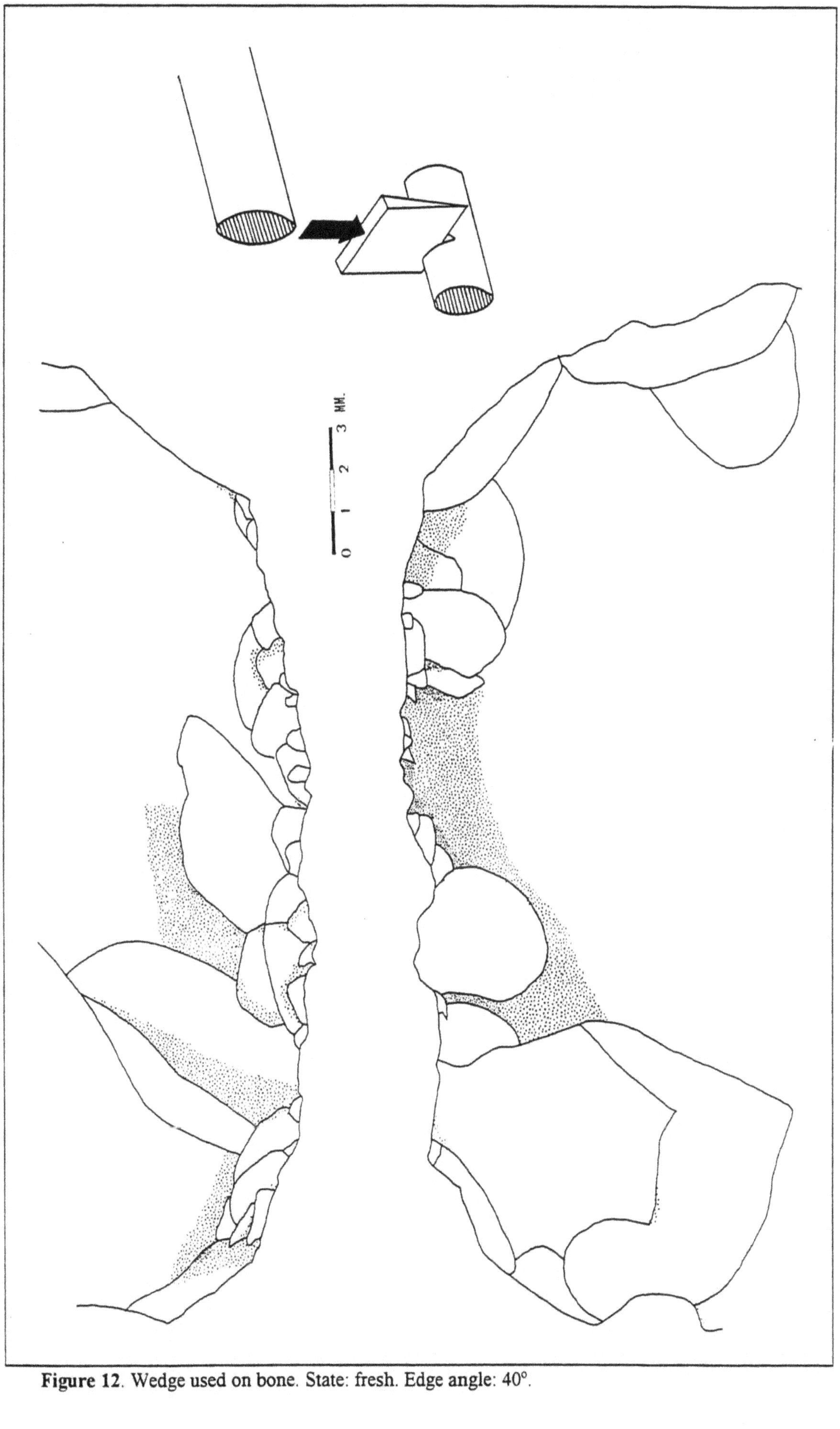

Figure 12. Wedge used on bone. State: fresh. Edge angle: 40°.

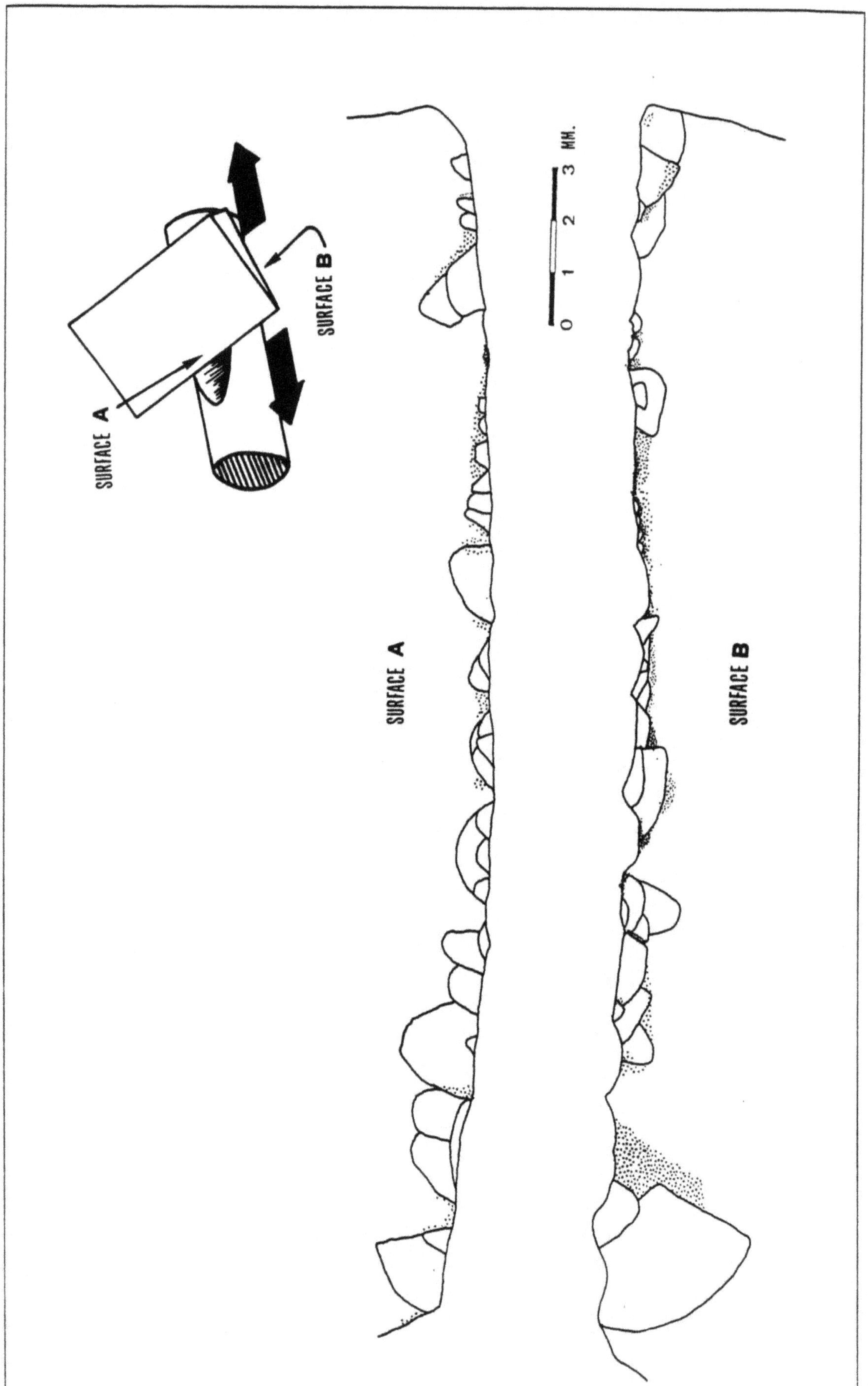

Figure 13. Scraping bone. State: soaked. Edge angle: 50°.

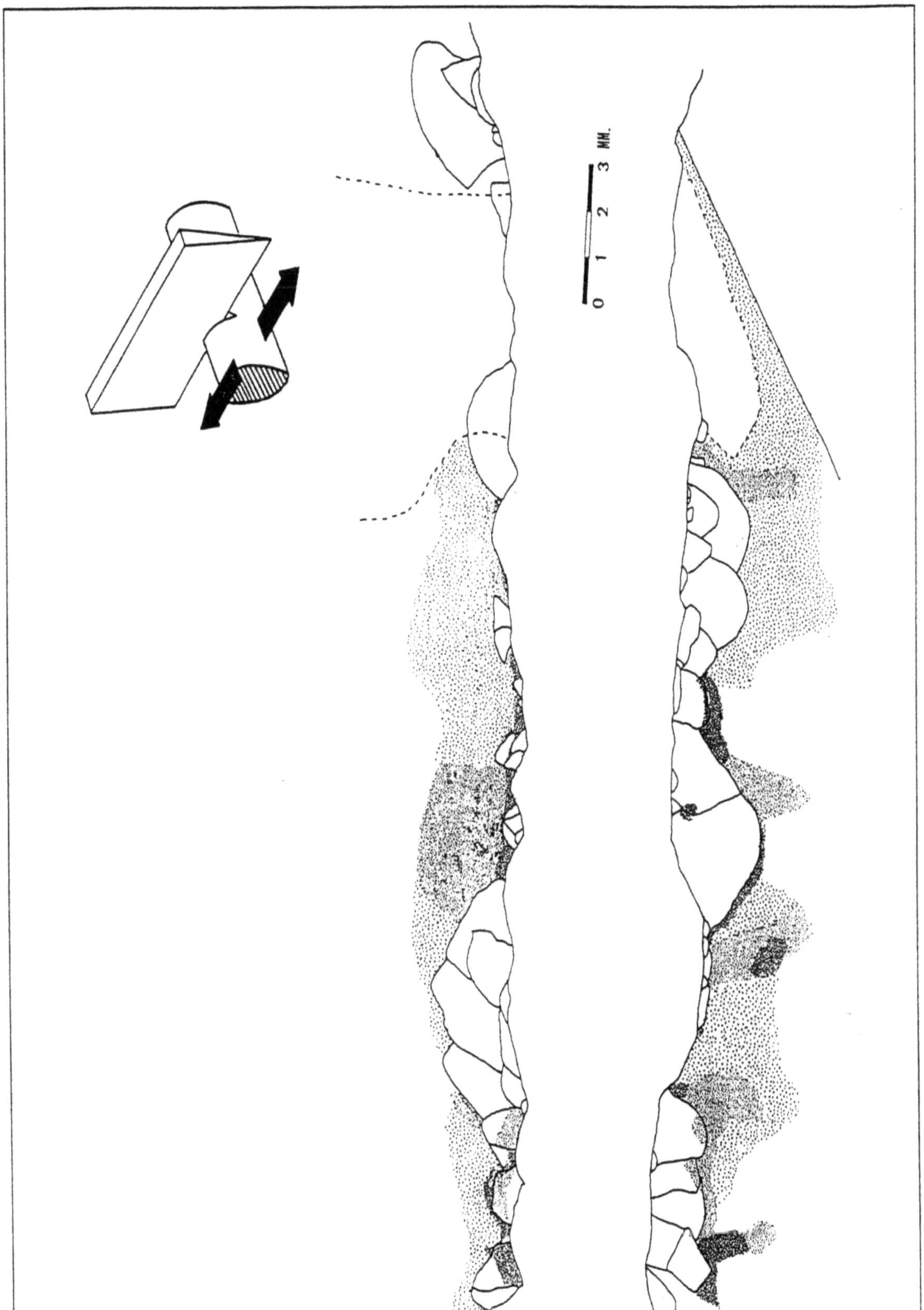

Figure 14. Cutting antler. State: soaked. Edge angle: 45.

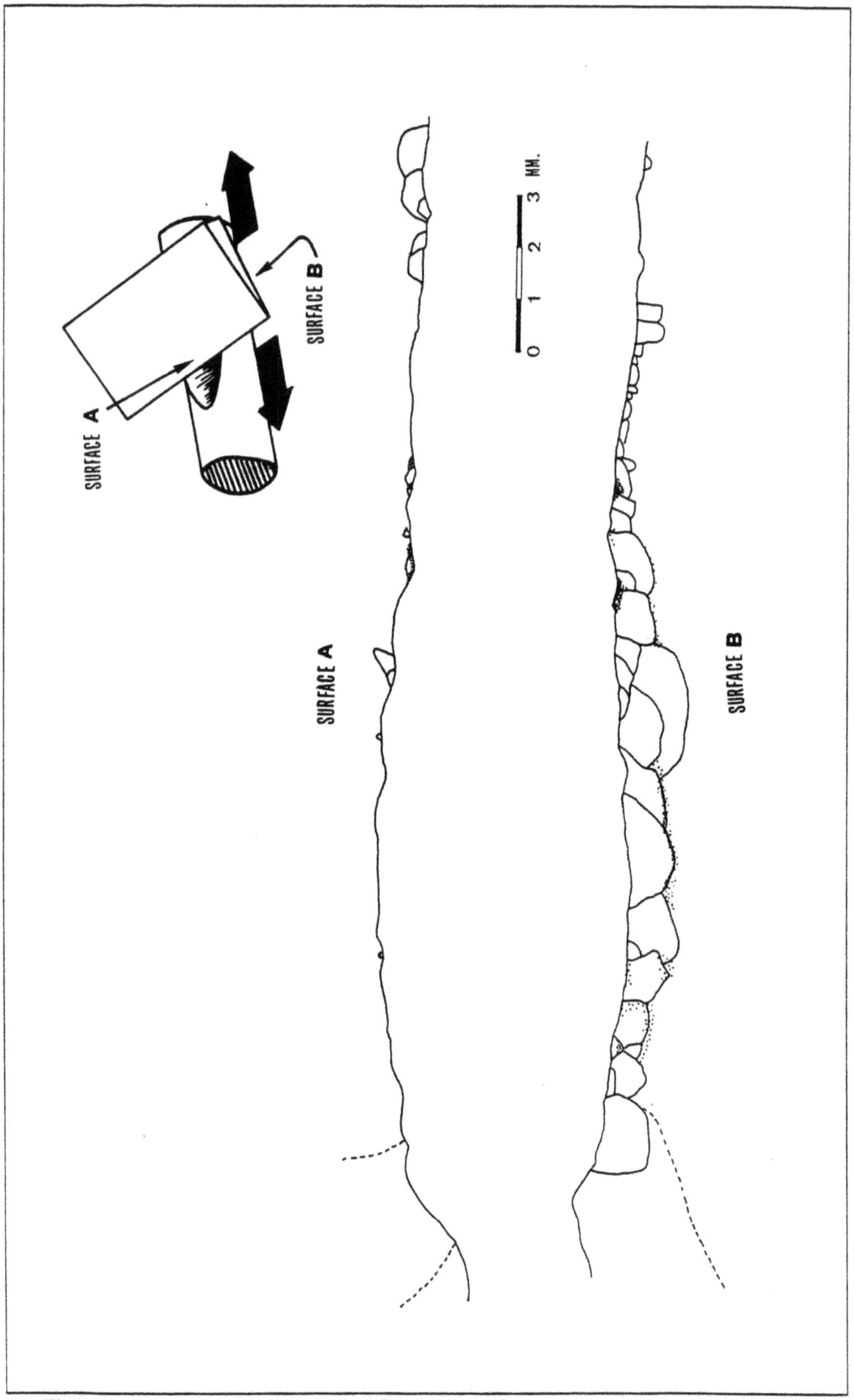

Figure 15. Scraping antler. State: soaked. Edge angle: 45°.

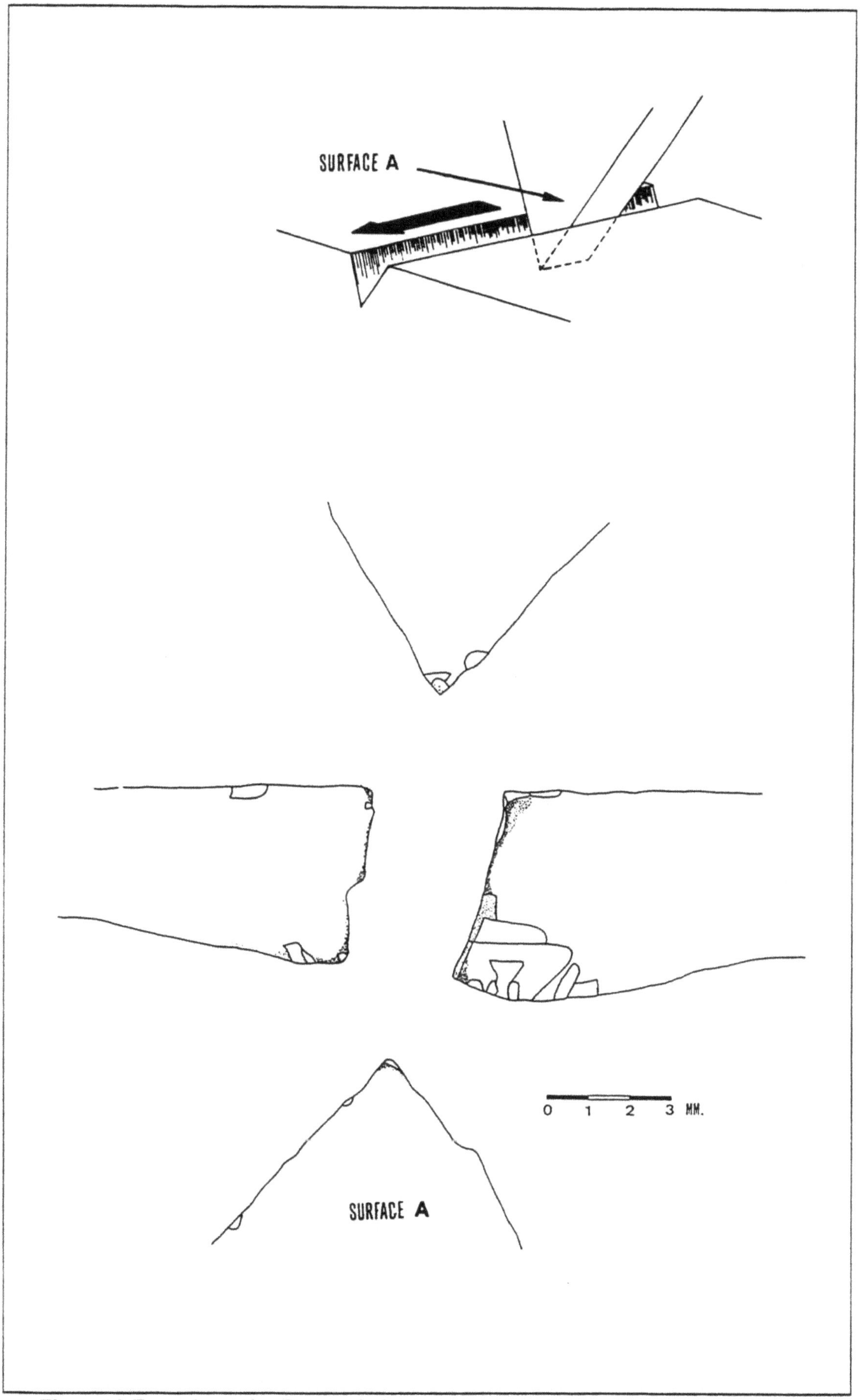

Figure 16. Engraving antler. State: soaked.

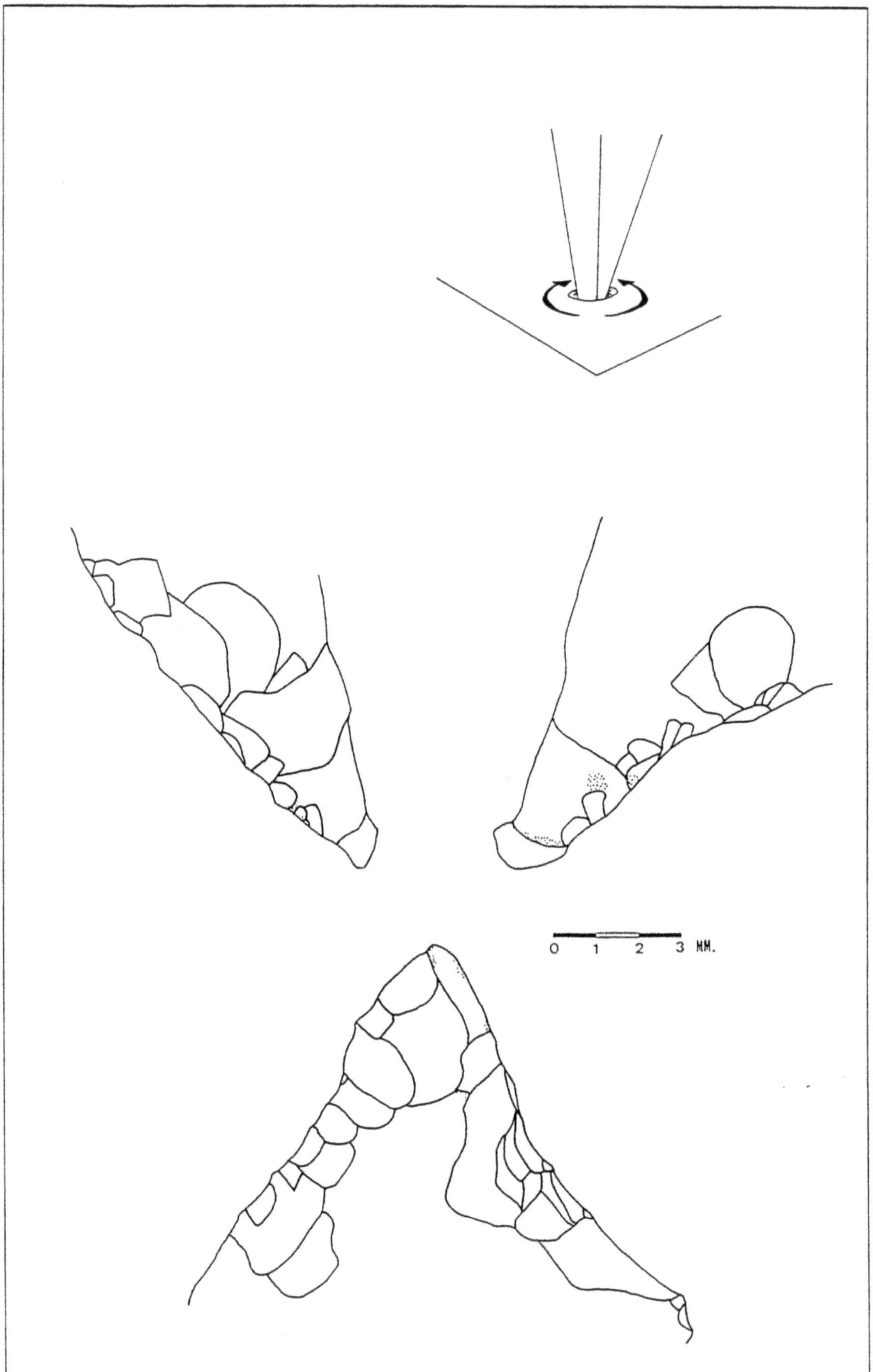

Figure 17. Boring antler. State: soaked.

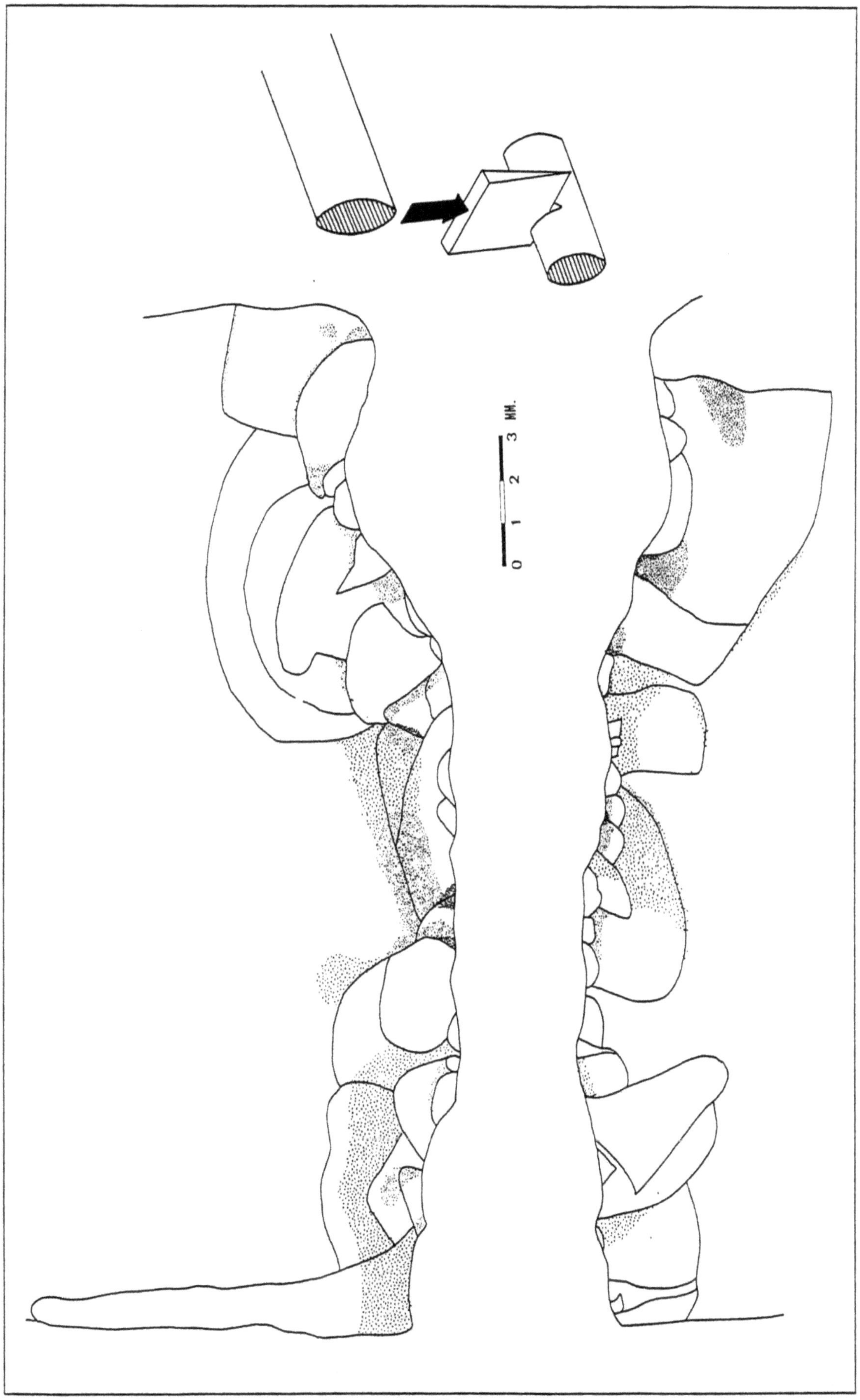

Figure 18. Wedge used on antler. State: soaked. Edge angle: 36°.

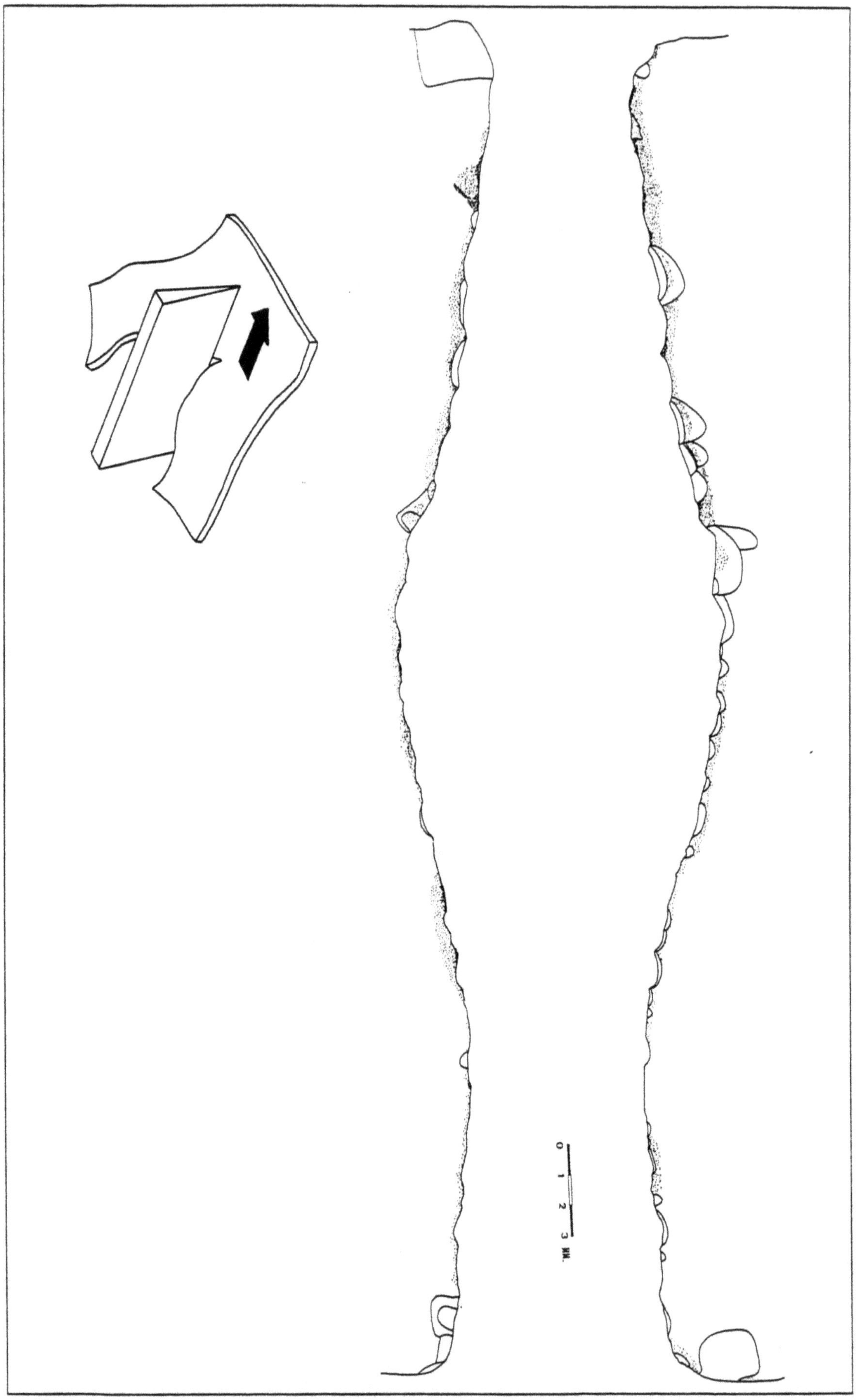

Figure 19. Cutting hide. State: dry. Edge angle: 33°.

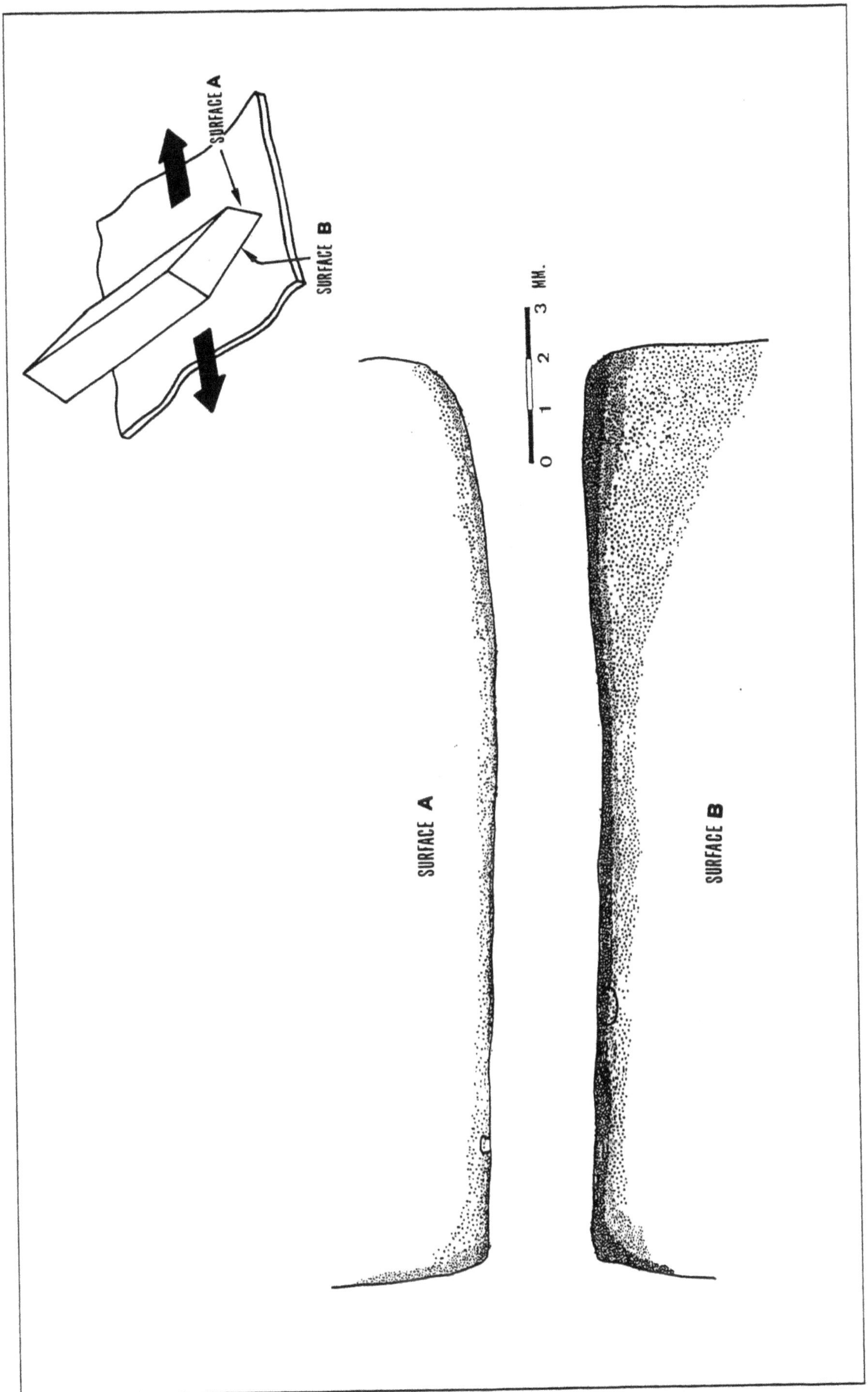

Figure 20. Scraping hide. State: dry. Edge angle: 94°.

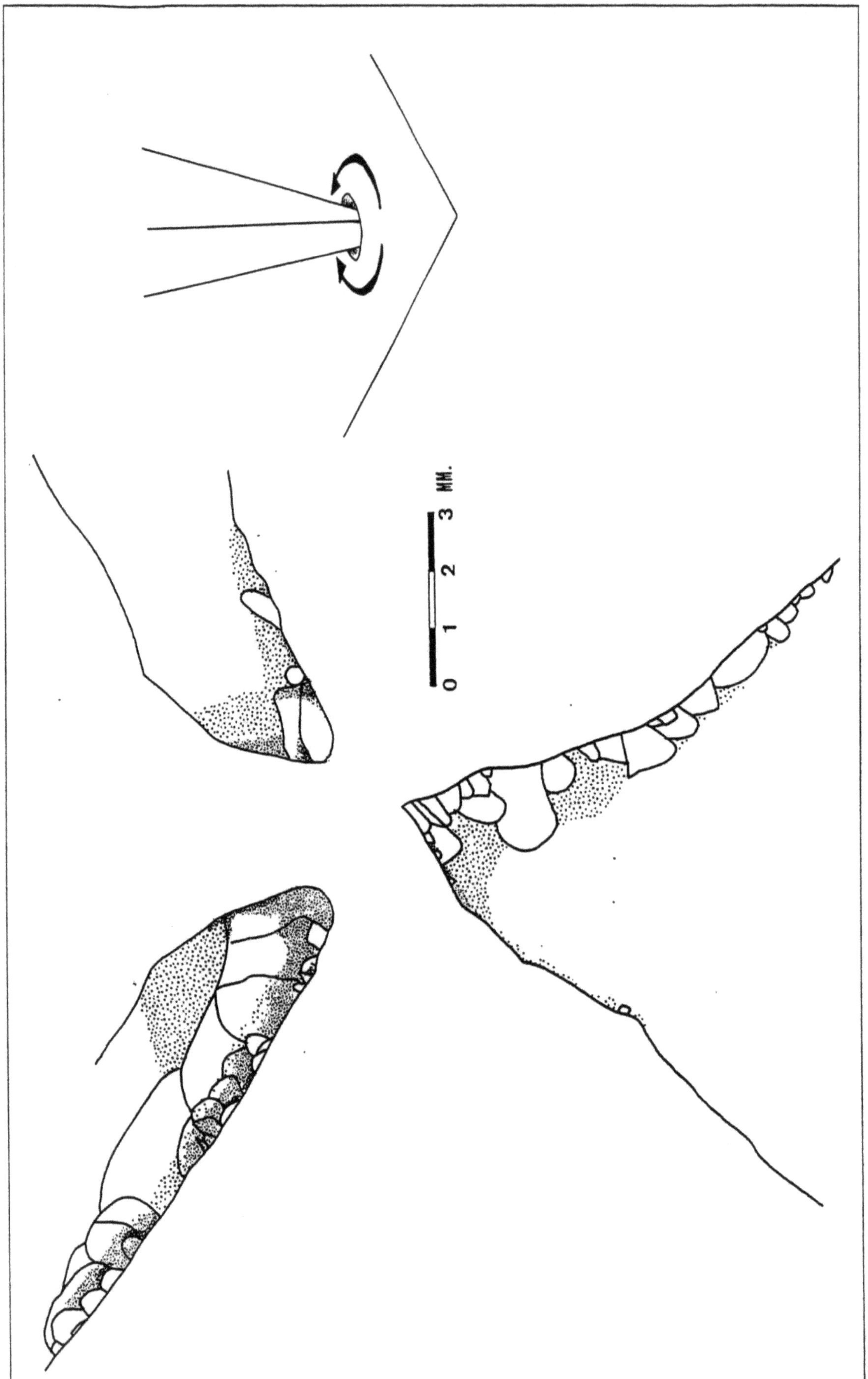

Figure 21. Boring hide. State: dry.

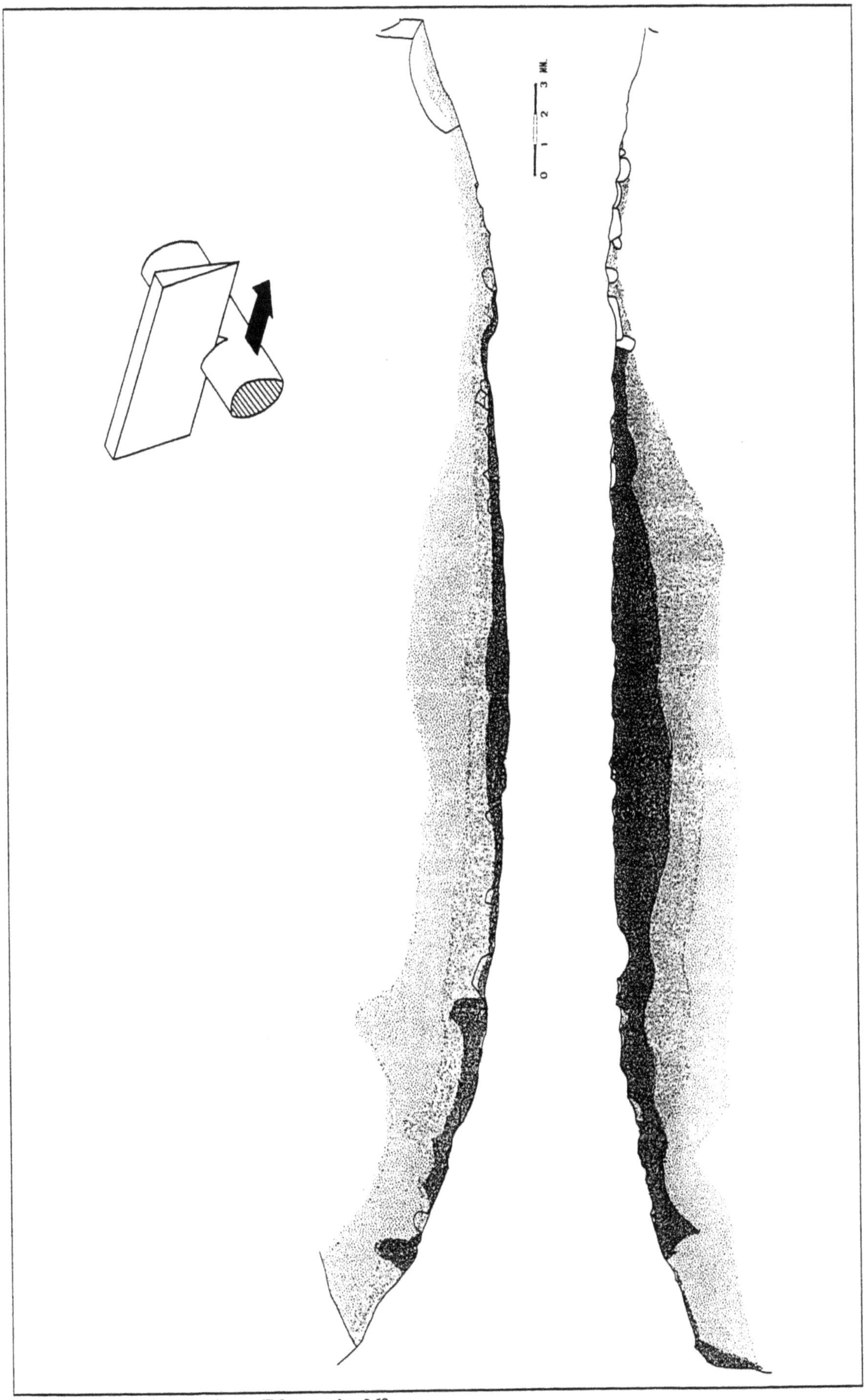

Figure 22. Cutting ferns. Edge angle: 36°.

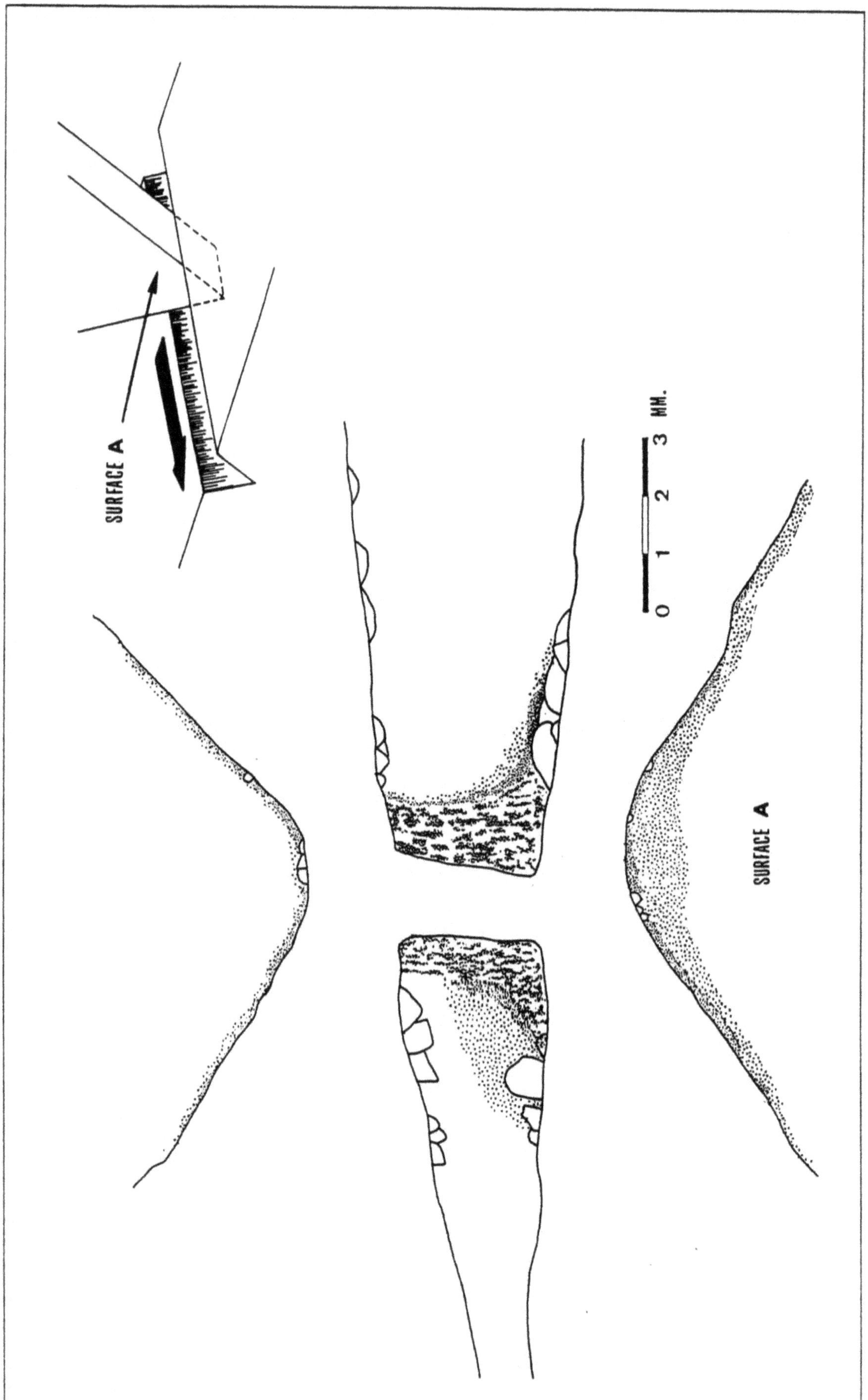

Figure 23. Engraving limestone. State: dry.

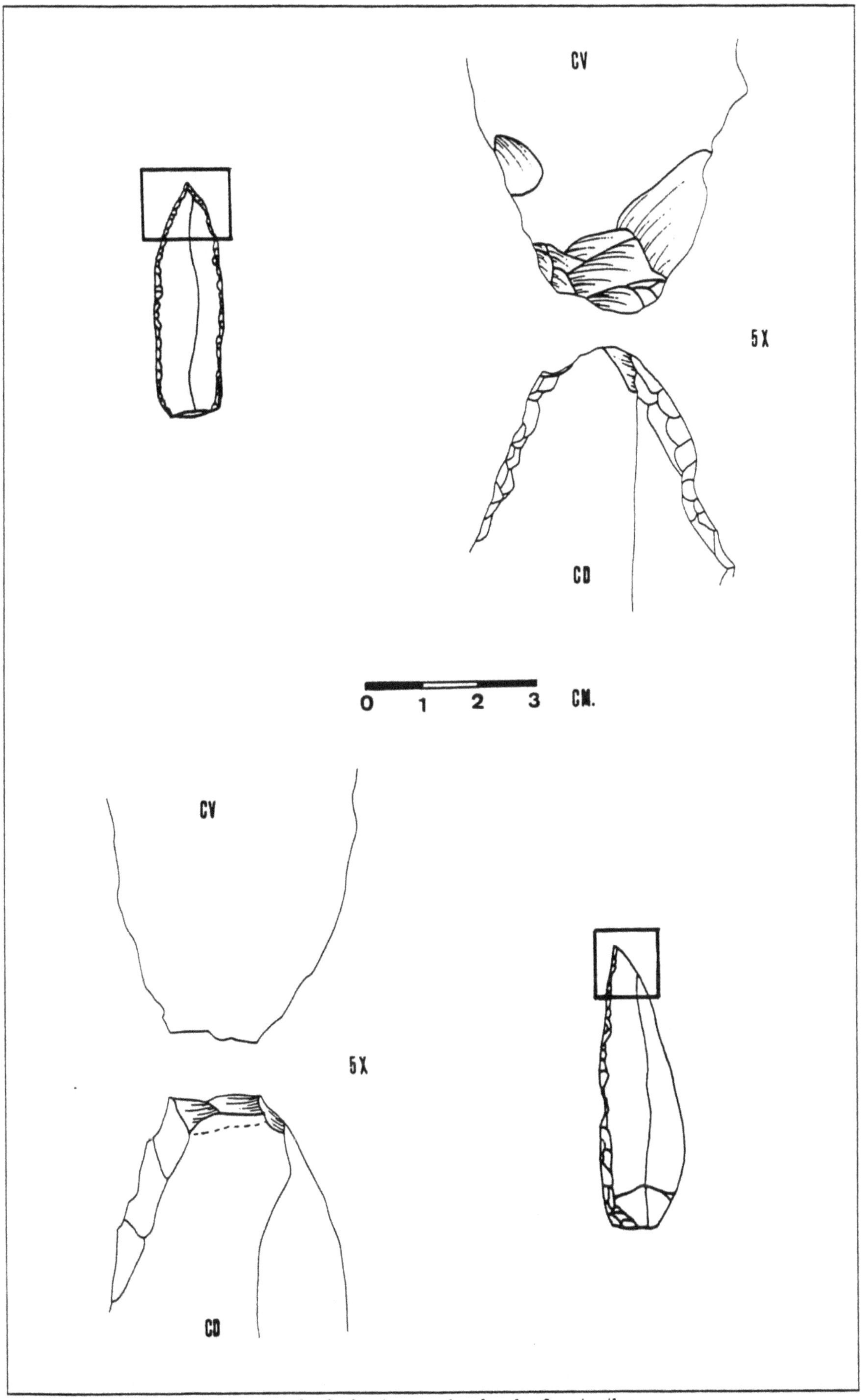

Figure 24: Impact fractures on backed points used as heads of projectiles.

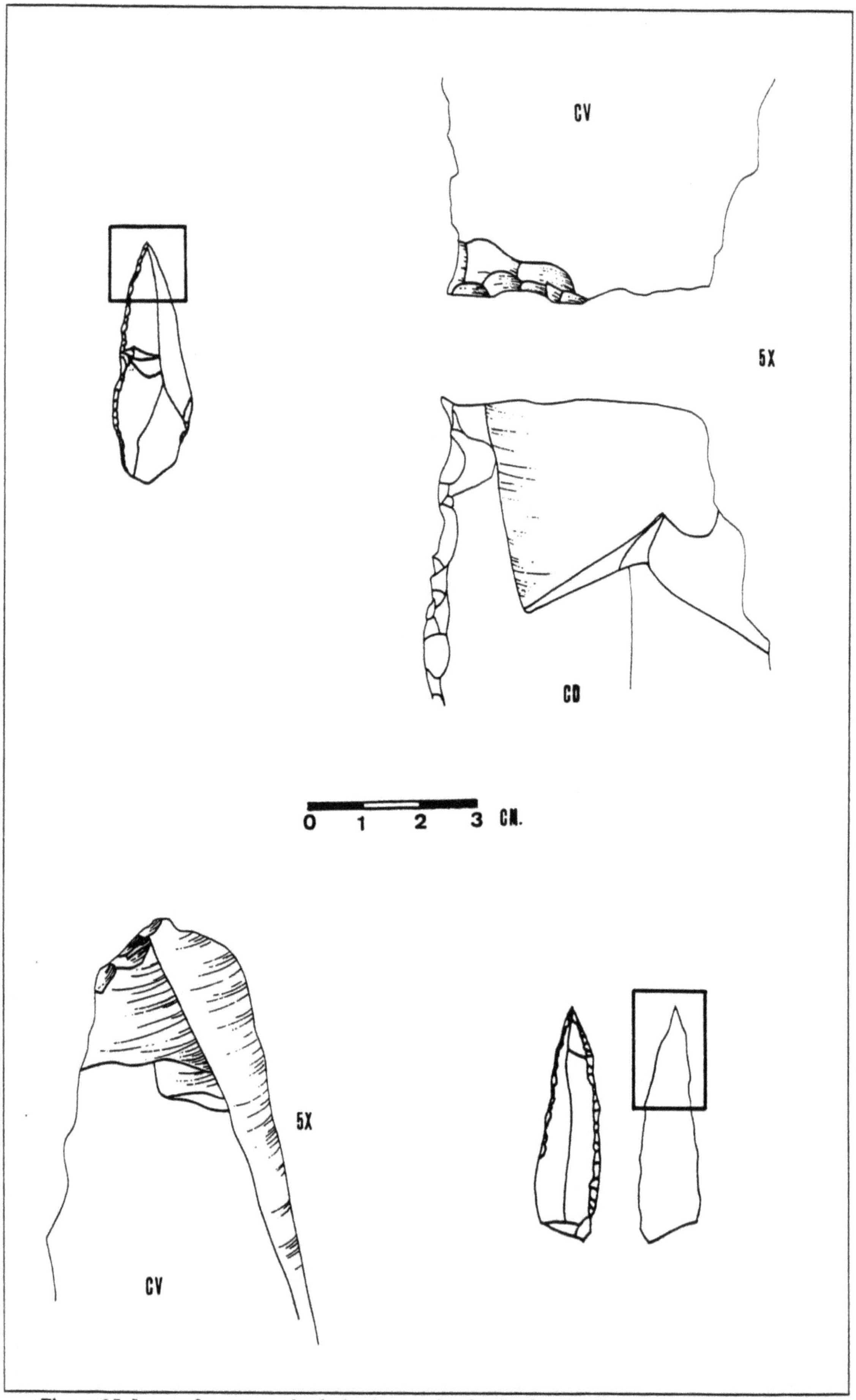

Figure 25: Impact fractures on backed points used as heads of projectiles.

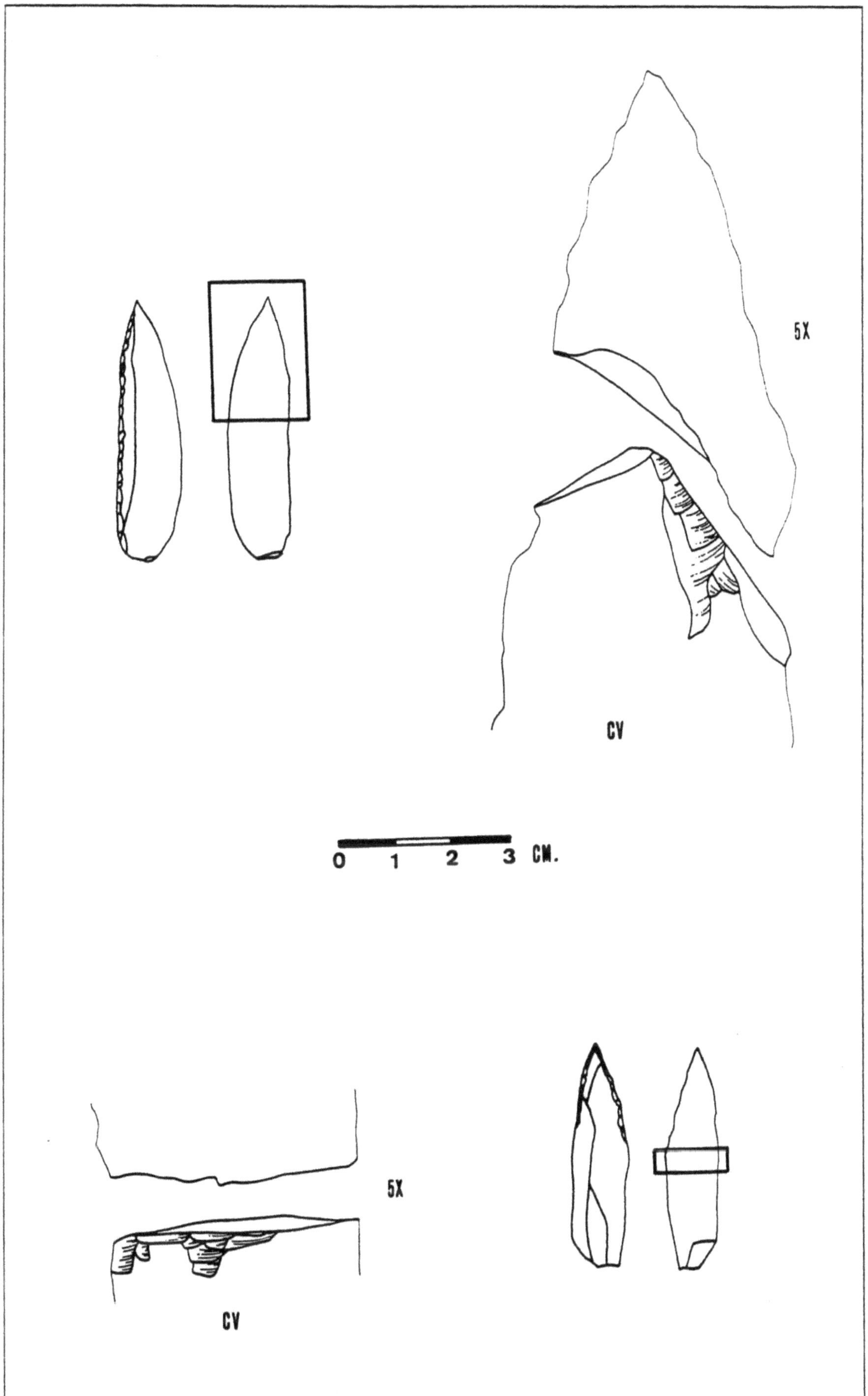

Figure 26: Impact fractures on backed points used as heads of projectiles.

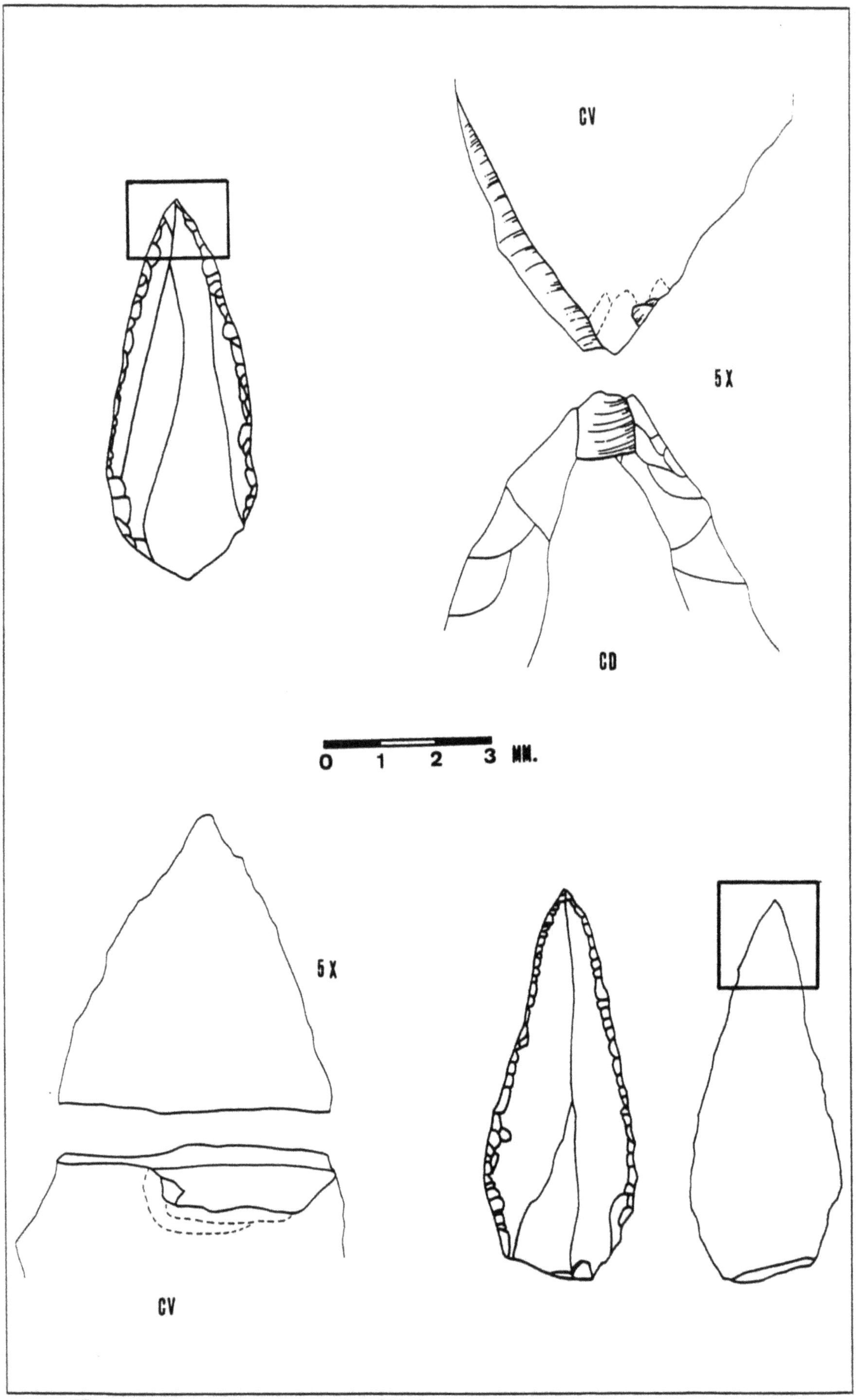

Figure 27: Impact fractures on pointed blades.

APPENDIX 5

RESULTS OF THE USE-WEAR ANALYSIS

	Piece num.	Activity	Worked Material	Edge Angle
	END-SCRAPERS			
	Berniollo			
1.	A10-102 88	Scraping	Antler	90
2.	A10-102 88	Scraping	Antler	83
3.	A10-102 88	Scraping	Antler	90
4.	A10-102 88	Scraping	Antler	100
5.	A11 -80 53	Scraping	Hide	71
6.	A11 -87 75	Scraping	Wood	90
7.	A11 -87 75	Scraping	Wood	80
8.	A11 -87 75	Scraping	Wood	83
9.	A12 -93 32	Scraping	Hide	90
10.	A12 -93 32	Scraping	Hide	68
11.	A12 -93 32	Scraping	Hide	76
12.	A12 -93 32	Scraping	Hide	72
13.	A15 -72 11	Scraping	Dry Hide	63
14.	A15 -72 11	Scraping	Dry Hide	50
15.	A2 -91 101	Scraping	Hide	85
16.	A2 -91 101	Scraping	Indeterm.	80
17.	A4 -93 143	Scraping	Hide	72
18.	A4 -93 143	Cutting	Hide	62
19.	A4 -93 143	Scraping	Hide	90
20.	A4 -93 143	Cutting	Hide	55
21.	A4 -93 143	Engraving	Hard Mat.	-
22.	B4 -89 94	Scraping	Hide	65
23.	B4 -89 94	Cutting	Hard Mat.	47
24.	B4 -89 94	Cutting	Hard Mat.	57
25.	B6 -90 27	Indeterm.	Indeterm.	-
26.	B7 -69 30	Scraping	Wood	29

	Piece num.	Activity	Worked Material	Edge Angle
27.	B7 -69 30	Scraping	Bone	83
28.	B9 -81 20	Scraping	Ant./Bone	-
29.	B9 -81 20	Scraping	Ant./Bone	-
30.	C1 -73 23	Scraping	Indeterm.	70
31.	C1 -73 23	Scraping	Indeterm.	80
32.	C11-102 27	Cutting	Dry Hide	47
33.	C11-102 27	Scraping	Dry Hide	80
34.	C11-102 27	Scraping	Dry Hide	33
35.	C11-102 27	Scraping	Dry Hide	47
36.	C11-102 27	Scraping	Dry Hide	88
37.	C2 -80 76	Scraping	Dry Hide	50
38.	C2 -80 76	Cutting	Dry Hide	29
39.	C7 -100 48	Scraping	Soft Mat.	50
40.	C7 -100 48	Scraping	Soft Mat.	50
41.	C7 -100 48	Scraping	Soft Mat.	50
42.	C7 -110 8	Scraping	Fresh Hide	65
43.	C7 -110 8	Scraping	Fresh Hide	54
44.	C7 -114 10	Scraping	Dry Hide	67
45.	C7 -91 62	Scraping	Dry Hide	94
46.	C7 -91 62	Scraping	Dry Hide	100
47.	C7 -91 62	Scraping	Dry Hide	60
48.	C7 -91 62	Scraping	Dry Hide	75
49.	C8 -89 114	Scraping	Hide	85
50.	C8 -89 114	Scraping	Hide	56
51.	C8 -89 114	Scraping	Hide	80
52.	C8 -89 114	Scraping	Hide	62
53.	C8 -90 131	Scraping	Hide	80
54.	C8 -90 131	Scraping	Hide	90
55.	C9 -73 26	Cutting	Hard Mat.	45
56.	C9 -73 26	Scraping	Indeterm.	74

Piece num.	Activity	Worked Material	Edge Angle
57. D2 -80 52	Cutting	Hard Mat.	53
58. D2 -80 52	Cutting	Hard Mat.	35
59. D2 -80 52	Scraping	Indeterm.	56
60. D2 -83 93	No Traces	No Traces	73
61. D4 -78 13	Scraping	Hard Mat.	65
62. D4 -78 13	Scraping	Hard Mat.	39
63. D4 -78 13	Scraping	Hard Mat.	60
64. D4 -78 13	Scraping	Hard Mat.	90
65. D4 -78 74	Scraping	Dry Hide	65
66. D4 -83 46	Scraping	Dry Hide	68
67. D4 -83 46	Scraping	Dry Hide	40
68. D4 -83 46	Scraping	Dry Hide	73
69. D4 -83 46	Scraping	Dry Hide	69
70. D4 -89 131	Scraping	Hide	55
71. D4 -89 131	Scraping	Indeterm.	80
72. E2 -81 11	Scraping	Ant./Bone	72
73. E2 -81 11	Scraping	Ant./Bone	112
74. E2 -81 11	Scraping	Ant./Bone	72
75. E2 -81 11	Scraping	Ant./Bone	72
76. E2 -87 2	Scraping	Wood	84
77. E2 -87 2	Scraping	Wood	52
78. E2 -87 2	Scraping	Wood	84
79. E2 -87 2	Scraping	Wood	90
80. E4 -77 57	Cutting	Wood	55
81. E4 -77 57	Cutting	Wood	58
82. E4 -77 57	Scraping	Wood	90
83. E4 -77 57	Scraping	Wood	82
84. T11-100 11	Scraping	Indeterm.	60
85. T11-100 11	Cutting	Indeterm.	26
86. T11-100 11	Cutting	Indeterm.	26
87. T11-100 11	Scraping	Indeterm.	85
88. T11-101 17	Scraping	Hard Mat.	72
89. T11-101 17	Cutting	Hard Mat.	40
90. T11-101 17	Cutting	Hard Mat.	32
91. T11 -90 22	Scraping	Hard Mat.	64
92. T11 -93 47	Scraping	Dry Hide	115
93. T11 -93 47	Scraping	Dry Hide	80
94. T11 -93 47	Scraping	Dry Hide	72
95. T11 -93 47	Scraping	Dry Hide	60
96. T11 -96 15	Scraping	Indeterm.	60
97. T11 -96 15	Scraping	Indeterm.	60
98. T11 -96 15	Scraping	Indeterm.	85
99. T11 -99 8	Scraping	Hide	70
100.V15 -90 16	Scraping	Dry Hide	70
101.V15 -90 16	Scraping	Dry Hide	58
102.V15 -90 16	Scraping	Wood	70

Azilian level of Santa Catalina

Piece num.	Activity	Worked Material	Edge Angle
103.B6 42 14	Scraping	Hard Mat.	62
104.B6 110 64	Scraping	Wood	47
105.B6 110 64	Scraping	Wood	55
106.B6 110 64	Scraping	Wood	74
107.B8 26 11	Scraping	Fresh Hide	60
108.B8 26 11	Scraping	Fresh Hide	61
109.B8 26 11	Scraping	Fresh Hide	80
110.B8 26 11	Cutting	Fresh Hide	32
111.B8 57 23	Scraping	Fresh Hide	77
112.B8 57 23	Scraping	Fresh Hide	56
113.B8 60 12	Cutting	Dry Hide	35
114.B8 60 12	Scraping	Dry Hide	65
115.B8 60 12	Scraping	Dry Hide	68
116.B8 60 12	Cutting	Dry Hide	35
117.B8 112 229	Scraping	Dry Hide	65
118.C4 12 17	Scraping	Indeterm.	55
119.C4 14 29	Scraping	Fresh Hide	53
120.C4 14 29	Cutting	Fresh Hide	30
121.C4 14 29	Cutting	Fresh Hide	30
122.C4 14 29	Scraping	Fresh Hide	65
123.C4 40 21	Scraping	Indeterm.	52
124.C4 67 54	No Traces	No Traces	-
125.C4 70 29	Scraping	Hide	47
126.C6 11 17	Scraping	Fresh Hide	97
127.C6 11 17	Scraping	Fresh Hide	43
128.C6 11 17	Scraping	Fresh Hide	61
129.C6 40 17	Scraping	Fresh Hide	66
130.C6 67 51	Scraping	Fresh Hide	57
131.C6 67 51	Scraping	Fresh Hide	40
132.C6 68 35	No Traces	No Traces	-
133.C6 115 32	Scraping	Hide	70
134.C6 115 32	Scraping	Hide	40
135.C8 15 17	Scraping	Fresh Hide	69

Magdalenian level of Santa Catalina

Piece num.	Activity	Worked Material	Edge Angle
136.No sig.	Scraping	Hide	78
137.A4 59 132	No Traces	No Traces	-
138.B4 73 89	Scraping	Hide	71
139.B4 76 1	Cutting	Dry Hide	52
140.B4 76 1	Scraping	Dry Hide	67
141.B4 76 1	Scraping	Dry Hide	46
142.B4 82 172	Scraping	Bone	60
143.B4 82 172	Engraving	Bone	-
144.B4 82 172	Cutting	Bone	44
145.B4 82 172	Cutting	Bone	38
146.B4 82 172	Engraving	Bone	-
147.B4 82 174	Scraping	Antler	76
148.B4 82 174	Scraping	Antler	62
149.B4 82 174	Scraping	Antler	103
150.B4 85 291	Cutting	Fresh Hide	57
151.B4 85 291	Cutting	Fresh Hide	50
152.B4 85 291	Scraping	Fresh Hide	70
153.B4 88 57	Cutting	Antler	64
154.B4 88 57	Scraping	Antler	55
155.B4 96 85	Indeterm.	Indeterm.	66
156.B4 96 85	Scraping	Indeterm.	88
157.B4 117 81	Cutting	Butchering	39
158.B4 117 81	Cutting	Butchering	37
159.B4 117 81	Scraping	Indeterm.	59
160.B4 122 282	No Traces	No Traces	-
161.B4 126 26	Scraping	Hide	66
162.B4 126 26	Scraping	Hide	52

Piece num.	Activity	Worked Material	Edge Angle
163.B4 126 26	Scraping	Indeterm.	75
164.B4 126 26	Scraping	Indeterm.	75
165.B6 68 14	Scraping	Antler	38
166.B6 68 14	Scraping	Antler	65
167.B6 68 57	No Traces	No Traces	-
168.B6 72 3	Scraping	Wood	42
169.B6 72 3	Scraping	Wood	34
170.B6 72 3	Scraping	Wood	51
171.B6 89 67	Cutting	Antler	53
172.B6 89 67	Scraping	Antler	55
173.B6 89 67	Scraping	Antler	54
174.B6 89 67	Scraping	Antler	90
175.B6 120 119	No Traces	No Traces	-
176.B6 121 200	No Traces	No Traces	-
177.B6 123 174	Scraping	Antler	87
178.B6 123 174	Cutting	Antler	65
179.B6 123 174	Scraping	Antler	60
180.B6 123 174	Indeterm.	Indeterm.	60
181.B6 124 282	Scraping	Indeterm.	65
182.B6 124 282	Scraping	Indeterm.	70
183.B6 124 282	Cutting	Indeterm.	30
184.B6 126 42	Scraping	Antler	65
185.B6 126 42	Scraping	Ant./Bone	65
186.B6 127 103	Scraping	Hide	61
187.B6 127 103	Cutting	Hide	49
188.B6 127 103	Scraping	Hide	46
189.B6 127 103	Scraping	Dry Hide	75
190.B6 127 103	Scraping	Dry Hide	97
191.B6 127 103	Scraping	Ant./Bone	65
192.B8 69 31	Cutting	Soft Mat.	56
193.B8 69 31	Scraping	Indeterm.	65
194.B8 72 45	Scraping	Antler	101
195.B8 72 45	Scraping	Antler	56
196.B8 75 17	Cutting	Hide	35
197.B8 75 17	Scraping	Dry Hide	100
198.B8 75 17	Scraping	Dry Hide	59
199.B8 80 131	Cutting	Fresh Hide	38
200.B8 80 131	Cutting	Fresh Hide	42
201.B8 80 131	Scraping	Fresh Hide	89
202.B8 80 131	Scraping	Fresh Hide	83
203.B8 82 49	Scraping	Antler	38
204.B8 82 49	Scraping	Antler	85
205.B8 82 49	Scraping	Antler	95
206.B8 82 49	Scraping	Antler	88
207.B8 82 49	Scraping	Antler	66
208.B8 82 49	Scraping	Antler	92
209.B8 84 66	Scraping	Ant./Bone	70
210.B8 84 66	Scraping	Ant./Bone	66
211.B8 89 177	Scraping	Wood	76
212.B8 90 89	Scraping	Hide	90
213.B8 121 98	Scraping	Hide	58
214.C4 68 26	Scraping	Fresh Hide	68
215.C4 68 26	Scraping	Fresh Hide	64
216.C4 79 14	Scraping	Dry Hide	53
217.C4 82 40	Scraping	Dry Hide	61
218.C4 82 40	Scraping	Dry Hide	83
219.C4 86 2	Scraping	Dry Hide	104
220.C4 86 2	Scraping	Dry Hide	70
221.C4 86 2	Cutting	Dry Hide	38
222.C4 86 2	Scraping	Dry Hide	60
223.C4 86 2	Scraping	Dry Hide	76
224.C4 124 172	Scraping	Hide	72
225.C4 124 172	Scraping	Bone	80
226.C4 126 269	Scraping	Indeterm.	60
227.C4 127 107	Cutting	Dry Hide	39
228.C4 127 107	Cutting	Dry Hide	33
229.C4 127 107	Scraping	Dry Hide	58
230.C4 127 5	Scraping	Dry Hide	94
231.C4 127 5	Scraping	Dry Hide	71
232.C4 127 5	Cutting	Dry Hide	26
233.C4 127 5	Cutting	Dry Hide	29
234.C6 68 32	Cutting	Dry Hide	45
235.C6 68 32	Scraping	Dry Hide	59
236.C6 74 1	Cutting	Fresh Hide	30
237.C6 74 1	Scraping	Fresh Hide	58
238.C6 74 46	Scraping	Fresh Hide	60
239.C6 94 10	Indeterm.	Dry Hide	58
240.C6 94 10	Indeterm.	Dry Hide	72
241.C6 94 10	Scraping	Dry Hide	93
242.C6 94 10	Scraping	Dry Hide	44
243.C6 117 131	Scraping	Indeterm.	70
244.C6 120 93	Scraping	Fresh Hide	94
245.C6 120 93	Scraping	Fresh Hide	74
246.C6 127 224	Scraping	Antler	61
247.C6 127 224	Scraping	Antler	100
248.C6 127 224	Scraping	Antler	36
249.C6 127 224	Scraping	Antler	38
250.C6 127 29	Cutting	Fresh Hide	29
251.C6 127 29	Scraping	Fresh Hide	76
252.C6 127 29	Cutting	Fresh Hide	42
253.C8 67 28	Scraping	Fresh Hide	101
254.C8 67 28	Scraping	Fresh Hide	84
255.C8 67 28	Scraping	Fresh Hide	60
256.C8 67 28	Cutting	Dry Hide	55
257.C8 67 28	Scraping	Dry Hide	55
258.C8 82 11	Scraping	Ant./Bone	65
259.C8 82 11	Cutting	Ant./Bone	62
260.C8 120 145	Scraping	Hide	60
261.C8 120 155	No Traces	No Traces	-
262.C8 126 24	Scraping	Indeterm.	75
263.C8 126 24	Cutting	Indeterm.	35

END-SCRAPER+BURIN

Berniollo

Piece num.	Activity	Worked Material	Edge Angle
264.V11 -91 33	Scraping	Ant./Bone	52
265.V11 -91 33	Engraving	Ant./Bone	-
266.V11 -91 33	Scraping	Ant./Bone	70
267.V11 -91 33	Scraping	Ant./Bone	62

Piece num.	Activity	Worked Material	Edge Angle
Magdalenian level of Santa Catalina			
268.B4 71 8	Cutting	Ant./Bone	44
269.B4 71 8	Scraping	Indeterm.	68
270.C4 82 57	Cutting	Dry Hide	50
271.C4 82 57	Scraping	Dry Hide	38
272.C4 82 57	Scraping	Wood	50
273.C4 82 57	Engraving	Bone	-
274.C4 82 57	Scraping	Indeterm.	65
275.C4 127 126	Scraping	Dry Hide	65
276.C4 127 126	Scraping	Wood	95
277.C4 127 126	Engraving	Wood	-
278.C4 127 126	Scraping	Wood	99
279.C4 127 126	Engraving	Wood	-
280.C4 127 126	Scraping	Wood	74
RASPADOR-TRUNCADURA			
Magdalenian level of Santa Catalina			
281.C4 81 49	Scraping	Dry Hide	54
BURINS			
Berniollo			
282.A11 -88 40	Scraping	Ant./Bone	63
283.A11 -88 40	Scraping	Ant./Bone	97
284.A13 -84 9	No Traces	No Traces	-
285.A7 -83 64	Engraving	Antler	-
286.A7 -91 100	Scraping	Hard Mat.	72
287.A7 -91 100	Scraping	Hard Mat.	90
288.B7 -87 76	Scraping	Ant./Bone	105
289.B7 -87 76	Engraving	Ant./Bone	-
290.C7 -94 50	Cutting	Indeterm.	53
291.C7 -94 50	Cutting	Indeterm.	25
292.D10-86 149	Scraping	Antler	90
293.D10-86 149	Scraping	Antler	50
294.D10-86 149	Scraping	Antler	43
295.D10-86 149	Cutting	Antler	38
296.D10-86 149	Scraping	Antler	90
297.D10-86 149	Boring	Antler	-
298.D10-86 149	Scraping	Antler	90
299.D2 -81 3	Engraving	Antler	-
300.D2 -81 3	Engraving	Antler	-
301.D2 -81 3	Scraping	Antler	73
302.D2 -81 3	Scraping	Antler	73
303.E4 -75 43	Scraping	Antler	72
304.E4 -75 43	Engraving	Antler	-
305.E4 -75 43	Cutting	Antler	40
306.E4 -75 43	Scraping	Antler	26
307.E4 -75 43	Scraping	Antler	96
308.E4 -75 43	Engraving	Antler	-
309.T15 -87 12	Scraping	Antler	80
310.T15 -87 12	Scraping	Antler	106
311.T15 -87 12	Engraving	Antler	-
312.T15 -87 12	Cutting	Antler	53
313.T15 -87 12	Scraping	Antler	80
314.T15 -87 12	Scraping	Antler	90

Piece num.	Activity	Worked Material	Edge Angle
315.T15 -87 12	Scraping	Antler	53
316.T15 -87 12	Engraving	Antler	90
317.V13 -84 19	Scraping	Ant./Bone	92
318.V13 -84 19	Engraving	Ant./Bone	-
319.V19 -87 9	Engraving	Antler	-
320.V19 -87 9	Engraving	Antler	-
321.V19 -87 9	Engraving	Ant./Bone	-
Azilian level of Santa Catalina			
322.B4 37 1	Engraving	Ant./Bone	-
323.B4 37 1	Scraping	Ant./Bone	85
324.B4 37 1	Scraping	Ant./Bone	83
325.B4 37 1	Scraping	Ant./Bone	103
326.B4 42 42	Engraving	Antler	-
327.B4 42 42	Scraping	Antler	95
328.B4 42 42	Cutting	Indeterm.	29
329.B6 26 1	Scraping	Bone	60
330.B6 26 1	Scraping	Bone	65
331.B6 57 39	Engraving	Hide	-
332.B6 57 39	Cutting	Soft Mat.	53
333.B6 64 28	Scraping	Antler	72
334.B6 64 28	Engraving	Antler	-
335.B6 64 28	Engraving	Indeterm.	-
336.B6 76 16	Engraving	Indeterm.	98
337.B6 110 22	Scraping	Antler	90
338.B6 110 22	Engraving	Antler	-
339.B6 110 22	Scraping	Antler	34
340.B6 110 22	Scraping	Antler	86
341.B8 14 16	Scraping	Wood	95
342.B8 14 16	Scraping	Wood	68
343.B8 14 16	Engraving	Wood	-
344.B8 87 19	Scraping	Ant./Bone	70
345.B8 114 92	No Traces	No Traces	-
346.C4 75 57	Engraving	Indeterm.	-
347.C6 2 24	Scraping	Antler	71
348.C6 2 24	Scraping	Antler	98
349.C6 2 24	Scraping	Antler	65
350.C6 2 24	Cutting	Antler	48
351.C6 2 24	Engraving	Antler	-
Magdalenian level of Santa Catalina			
352.B4 125 5	Scraping	Hide	73
353.B4 125 5	Scraping	Antler	95
354.B4 125 5	Scraping	Bone	104
355.B4 125 5	Engraving	Bone	-
356.B4 127 45	Scraping	Ant./Bone	106
357.B4 127 45	Scraping	Ant./Bone	113
358.B4 127 45	Scraping	Ant./Bone	102
359.B4 127 45	Scraping	Ant./Bone	97
360.B6 67 9	Engraving	Antler	-
361.B6 67 9	Scraping	Antler	99
362.B6 67 9	Scraping	Stone	94
363.B6 74 74	Engraving	Antler	-
364.B6 74 74	Scraping	Antler	83

Piece num.	Activity	Worked Material	Edge Angle
365.B6 74 74	Indeterm.	Indeterm.	40
366.B6 74 74	Indeterm.	Indeterm.	88
367.B6 80 114	Engraving	Indeterm.	-
368.B6 84 79	Engraving	Antler	-
369.B6 87 135	Scraping	Antler	66
370.B6 87 135	Scraping	Antler	53
371.B6 87 135	Engraving	Hard Mat.	-
372.B8 78 104	Boring	Antler	-
373.B8 78 104	Scraping	Antler	90
374.B8 78 104	Scraping	Antler	74
375.B8 85 105	Scraping	Antler	62
376.B8 85 105	Scraping	Antler	100
377.B8 85 105	Scraping	Antler	90
378.B8 85 105	Scraping	Antler	90
379.B8 85 105	Engraving	Antler	-
380.B8 85 105	Cutting	Antler	42
381.B8 85 105	Cutting	Antler	50
382.B8 86 133	Engraving	Bone	-
383.B8 86 133	Scraping	Bone	106
384.B8 86 133	Scraping	Bone	55
385.B8 86 133	Scraping	Bone	116
386.B8 86 87	Scraping	Antler	68
387.B8 86 87	Scraping	Antler	50
388.B8 86 87	Scraping	Antler	55
389.B8 87 19	Scraping	Plant	47
390.B8 89 113	Indeterm.	Indeterm.	90
391.B8 127 52	Scraping	Antler	94
392.B8 127 52	Cutting	Antler	36
393.B8 127 52	Scraping	Antler	90
394.B8 127 52	Scraping	Antler	93
395.C4 75 42	No Traces	No Traces	-
396.C4 127 2	Scraping	Antler	78
397.C4 127 2	Scraping	Antler	78
398.C4 127 2	Scraping	Antler	80
399.C4 127 2	Scraping	Ant./Bone	102
400.C4 127 2	Scraping	Indeterm.	81
401.C6 77 8	Scraping	Indeterm.	74
402.C6 119 56	Engraving	Bone	-
403.C6 126 109	Scraping	Ant./Bone	93
404.C6 126 34	Cutting	Hide	26
405.C6 126 34	Engraving	Hide	-
406.C6 128 70	Scraping	Antler	127
407.C6 128 70	Engraving	Antler	-
408.C6 128 70	Scraping	Antler	119
409.C6 128 70	Scraping	Antler	90
410.C6 128 70	Engraving	Antler	-
411.C6 128 70	Scraping	Antler	89
412.C6 128 70	Scraping	Antler	80
413.C8 64 27	No Traces	No Traces	-
414.C8 126 3	Engraving	Antler	-
415.C8 126 3	Scraping	Antler	52
416.C8 126 3	Scraping	Antler	67
417.C8 126 3	Scraping	Antler	81

Piece num.	Activity	Worked Material	Edge Angle
DENTICULATES			
Azilian level of Santa Catalina			
418.B8 63 27	Cutting	Butchering	51
419.C8 116b 21	Boring	Hide	-
Magdalenian level of Santa Catalina			
420.B6 72 28	Scraping	Bone	107
421.B6 86 166	Scraping	Antler	56
422.B6 86 166	Scraping	Antler	90
423.B6 86 166	Scraping	Antler	50
424.B6 86 166	Scraping	Antler	107
425.B6 86 166	Scraping	Antler	92
426.B6 121 214	No Traces	No Traces	-
427.B6 123 279	Cutting	Ant./Bone	33
428.B6 123 279	Scraping	Ant./Bone	59
429.B8 66 2	Scraping	Fresh Hide	53
430.B8 79 145	Cutting	Dry Hide	34
431.B8 79 145	Cutting	Dry Hide	37
432.C4 128 27	Scraping	Soft Mat.	51
433.C8 85 12	Cutting	Butchering	33
WEDGES			
Berniollo			
434.B3 -77 28	Wedge	Hard Mat.	-
Magdalenian level of Santa Catalina			
435.B8 79 130	Chopping	Indeterm.	65
DENTICULATED BLADELETS			
Berniollo			
436.C4 -87 60	Cutting	Wood/Plant	32
437.D2 -81 55	No Traces	No Traces	-
438.D2 -84 57	Scraping	Plant	30
439.D2 -84 57	Scraping	Plant	30
440.D4 -79 7	No Traces	No Traces	-
Azilian level of Santa Catalina			
441.B6 16 20	No Traces	No Traces	-
Magdalenian level of Santa Catalina			
442.B4 87 4	No Traces	No Traces	-
443.B6 127 104	Scraping	Soft Mat.	47
BACKED BLADELETS			
Berniolllo			
444.A17 -75 79	No Traces	No Traces	-
445.A2 -88 119	Cutting	Indeterm.	21
446.A2 -88 90	No Traces	No Traces	-
447.A7 -90 61	Cutting	Hard Mat.	38
448.B1 -90 24	No Traces	No Traces	-
449.B2 -81 16	No Traces	No Traces	-

Piece num.	Activity	Worked Material	Edge Angle
450.B3 43	No Traces	No Traces	-
451.B6 -95 92	No Traces	No Traces	-
452.C6 -88 120	Indeterm.	Indeterm.	-
453.C6 42	Cutting	Butchering	-
454.C8 -93 116	Boring	Wood	-
455.C8 -98 187	No Traces	No Traces	-
456.C9 -67 95	No Traces	No Traces	-
457.D4 -78 10	No Traces	No Traces	-
458.D4 -80 12	Cutting	Butchering	30
459.D4 77	No Traces	No Traces	-
460.D6 -86 213	Indeterm.	Indeterm.	-
461.V17 -75 29	Cutting	Butchering	-
462.V17 -79 17	No Traces	No Traces	-
463.V17 -86 16	Cutting	Butchering	-

Azilian level of Santa Catalina

Piece num.	Activity	Worked Material	Edge Angle
464.B4 24 67	Cutting	Soft Mat.	35
465.B4 25 13	No Traces	No Traces	-
466.B4 32 24	Cutting	Butchering	44
467.B4 43 23	No Traces	No Traces	-
468.B4 113 77	Scraping	Hard Mat.	65
469.B6 20 18	Cutting	Butchering	35
470.B6 23 16	Cutting	Butchering	24
471.B6 97 15	Cutting	Butchering	30
472.B6 110 22	No Traces	No Traces	-
473.B6 110 30	Cutting	Butchering	42
474.B6 112 36	No Traces	No Traces	-
475.B6 112 66	No Traces	No Traces	-
476.B6 116 130	No Traces	No Traces	-
477.B8 22 18	No Traces	No Traces	-
478.B8 4 15	No Traces	No Traces	-
479.B8 7 13	Cutting	Butchering	52
480.B8 7 13	Boring	Hard Mat.	-
481.B8 8 10	Cutting	Butchering	28
482.B8 110 237	No Traces	No Traces	-
483.B8 110 47	Cutting	Hide/Meat	23
484.B8 113 273	No Traces	No Traces	-
485.B8 115 74	No Traces	No Traces	-
486.C4 48 48	No Traces	No Traces	-
487.C4 52 22	No Traces	No Traces	-
488.C4 66 53	Cutting	Butchering	33
489.C4 111 28	No Traces	No Traces	-
490.C4 115 233	Cutting	Butchering	39
491.C6 19 14	Cutting	Butchering	27
492.C6 114 65	Projectile	Hunting	-
493.C6 115 124	No Traces	No Traces	-
494.C8 10 14	Scraping	Wood	37
495.C8 74 12	No Traces	No Traces	-
496.C8 91 11	Cutting	Butchering	38
497.C8 115 65	Cutting	Butchering	30
498.C8 116 78	Cutting	Butchering	30

Magdalenian level of Santa Catalina

Piece num.	Activity	Worked Material	Edge Angle
499.B4 86 179	Cutting	Indeterm.	33
500.B4 119 73	No Traces	No Traces	-
501.B4 125 206	No Traces	No Traces	-
502.B4 127 298	No Traces	No Traces	-
503.B6 69 51	Projectile	Hunting	-
504.B6 73 4	Scraping	Fresh Hide	60
505.B6 73 4	Scraping	Fresh Hide	70
506.B6 121 233	No Traces	No Traces	-
507.B6 122 250	Cutting	Butchering	25
508.B6 123 275	No Traces	No Traces	-
509.B6 123 276	No Traces	No Traces	-
510.B6 124 440	No Traces	No Traces	-
511.B6 124 497	No Traces	No Traces	-
512.B6 125 347	Scraping	Hide	34
513.B6 126 138	No Traces	No Traces	-
514.B6 126 335	No Traces	No Traces	-
515.B8 75 64	Projectile	Hunting	-
516.B8 80 124	No Traces	No Traces	-
517.B8 121 200	Cutting	Butchering	45
518.B8 121 353	Cutting	Butchering	20
519.B8 121 353	Cutting	Butchering	19
520.B8 122 611	No Traces	No Traces	-
521.B8 123 388	Cutting	Butchering	35
522.B8 124 230	Cutting	Butchering	44
523.B8 126 389	Cutting	Butchering	21
524.C4 94 91	No Traces	No Traces	-
525.C4 121 187	No Traces	No Traces	-
526.C4 127 222	Cutting	Butchering	46
527.C4 128 64	Cutting	Butchering	26
528.C6 71 21	No Traces	No Traces	-
529.C6 117 98	No Traces	No Traces	-
530.C6 127 52	Cutting	Butchering	30
531.C6 128 433	No Traces	No Traces	-
532.C8 60 2	Cutting	Indeterm.	36
533.C8 64 8	Cutting	Butchering	27
534.C8 124 171	Cutting	Butchering	24
535.C8 127 454	No Traces	No Traces	-

TRUNCATED BACKED BLADELETS

Azilian level of Santa Catalina

Piece num.	Activity	Worked Material	Edge Angle
536.B4 24 41	No Traces	No Traces	-

Magdalenian level of Santa Catalina

Piece num.	Activity	Worked Material	Edge Angle
537.B6 121 278	Projectile	Hunting	-
538.C4 83 25	Cutting	Butchering	40

DUFOUR BLADELET

Azilian level of Santa Catalina

Piece num.	Activity	Worked Material	Edge Angle
539.C4 19 38	Cutting	Hide	40

RETOUCHED FLAKES AND BLADES

Berniollo

Piece num.	Activity	Worked Material	Edge Angle
540.A11 -82 13	Projectile	Indeterm.	-

Piece num.	Activity	Worked Material	Edge Angle
541.A4 -93 158	Cutting	Indeterm.	54
542.B3 -86 42	Indeterm.	Ant./Bone	48
543.B3 -86 42	Scraping	Ant./Bone	70
544.B7 -103 34	Scraping	Indeterm.	58
545.B7 -103 34	Cutting	Indeterm.	38
546.B7 -78 37	Cutting	Wood	33
547.B7 -78 37	Cutting	Wood	48
548.B7 -78 37	Scraping	Wood	48
549.B7 -79 18	Projectile	Indeterm.	-
550.B7 -79 60	Cutting	Indeterm.	45
551.B7 -79 60	Cutting	Indeterm.	25
552.B7 -90 26	Scraping	Dry Hide	53
553.B7 -90 26	Scraping	Hard Mat.	44
554.B7 -96 25	Cutting	Butchering	42
555.B7 -96 25	Cutting	Butchering	34
556.C7 -89 26	Scraping	Indeterm.	43
557.C7 -89 26	Cutting	Indeterm.	54
558.C7 -89 26	Scraping	Indeterm.	54
559.C7 -89 26	Cutting	Indeterm.	43
560.D8 -88 73	Scraping	Ant./Bone	109
561.D8 -88 73	Scraping	Hard Mat.	65
562.D8 -88 73	Scraping	Hard Mat.	65
Azilian level of Santa Catalina			
563.B6 110 20	Cutting	Hard Mat.	36
564.B6 110 20	Scraping	Hard Mat.	62
565.B8 59 45	Scraping	Antler	90
566.B8 59 45	Scraping	Hard Mat.	73
567.B8 61 31	Cutting	Dry Hide	39
568.B8 61 31	Cutting	Dry Hide	49
569.C4 17 16	No Traces	No Traces	-
570.C4 22 13	Scraping	Fresh Hide	66
571.C4 22 13	Cutting	Fresh Hide	66
572.C4 22 13	Scraping	Antler	60
573.C4 22 13	Scraping	Antler	123
574.C4 22 13	Boring	Antler	-
575.C4 32 34	Scraping	Ant./Bone	66
576.C4 32 34	Scraping	Ant./Bone	69
577.C6 21 14	No Traces	No Traces	-
578.C6 23 10	Cutting	Butchering	52
579.C6 38 19	Cutting	Butchering	52
580.C8 18 40	No Traces	No Traces	-
Magdalenian level of Santa Catalina			
581.B4 80 40	Cutting	Butchering	56
582.B4 126 55	Cutting	Ant./Bone	31
583.B4 126 55	Cutting	Ant./Bone	72
584.B4 126 55	Scraping	Ant./Bone	78
585.B6 87 127	Scraping	Wood	97
586.B6 87 127	Scraping	Wood	29
587.B6 87 127	Scraping	Wood	53
588.B6 126 45	Cutting	Wood	58
589.B8 126 174	Indeterm.	Indeterm.	29
590.B8 126 32	No Traces	No Traces	-
591.C4 127 124	Scraping	Wood	79

Piece num.	Activity	Worked Material	Edge Angle
592.C4 127 124	Cutting	Wood	53
593.C4 127 124	Cutting	Wood	62
594.C6 94 42	Cutting	Dry Hide	40
595.C6 94 42	Cutting	Dry Hide	48
596.C8 85 10	Chopping	Wood	47
TRUNCATED BLADELET			
Azilian level of Santa Catalina			
597.B6 65 30	Scraping	Antler	85
BACKED BLADES			
Magdalenian level of Santa Catalina			
598.B6 78 125	Scraping	Hide	85
599.B6 78 125	Scraping	Indeterm.	32
600.B8 84 5	Scraping	Hide	65
601.C4 83 25	Scraping	Soft Mat.	54
602.C4 128 12	Cutting	Butchering	30
603.C8 122 121	No Traces	No Traces	-
BORERS			
Berniollo			
604.B4 -90 37	Boring	Wood	-
605.B4 -90 37	Boring	Indeterm.	-
Azilian level of Santa Catalina			
606.C6 20 14	Boring	Hard Mat.	-
Magdalenian level of Santa Catalina			
607.B8 79 61	Scraping	Antler	60
608.B8 79 61	Scraping	Antler	74
609.B8 79 61	Boring	Antler	-
610.B8 79 61	Cutting	Antler	42
611.C4 89 1	Boring	Bone	-
612.C4 89 1	Cutting	Bone	50
613.C4 89 1	Cutting	Bone	58
614.C4 93 89	Boring	Wood	-
615.C4 93 89	Cutting	Wood	44
616.C4 93 89	Cutting	Wood	43
617.C6 78 48	Boring	Bone	-
618.C6 78 48	Scraping	Bone	40
619.C6 78 48	Cutting	Bone	36
620.C6 78 48	Scraping	Bone	34
621.C8 90 7	Scraping	Antler	86
BACKED POINTS			
Berniollo			
622.A13 -87 5	Projectile	Hunting	-
623.A2 -88 99	Projectile	Hunting	-
624.A2 -91 49	Projectile	Hunting	-
625.A4 -94 18	Projectile	Hunting	-
626.B6 -93 75	No Traces	No Traces	-

Piece num.	Activity	Worked Material	Edge Angle
627.C2 -84 1	Projectile	Hunting	-
628.C6 -91 34	No Traces	No Traces	-
629.C8 -89 149	No Traces	No Traces	-
630.C8 70	No Traces	No Traces	-
631.D6 -81 58	Projectile	Hunting	-
632.E4 -80 19	Projectile	Hunting	-
Azilian level of Santa Catalina			
633.B4 45 33	No Traces	No Traces	-
634.B4 72 5	No Traces	No Traces	-
635.B6 63 4	No Traces	No Traces	-
636.B6 113 73	Projectile	Hunting	-
637.B8 115 131	Projectile	Hunting	-
638.C4 1 10	Cutting	Butchering	35
639.C4 13 17	No Traces	No Traces	-
640.C4 53 23	Projectile	Hunting	-
641.C4 67 58	No Traces	No Traces	-
642.C4 113 105	Projectile	Hunting	-
643.C4 114 108	Projectile	Hunting	-
644.C6 21 21	No Traces	No Traces	-
645.C6 25 18	Projectile	Hunting	-
646.C8 116b 18	No Traces	No Traces	-
Magdalenian level of Santa Catalina			
647.B4 79 3	No Traces	No Traces	-
648.B4 82 81	No Traces	No Traces	-
649.B4 126 237	Projectile	Hunting	-
650.B4 127 1	Projectile	Hunting	-
651.B4 127 1	Cutting	Butchering	38
652.B6 70 56	Projectile	Hunting	-
653.B6 77 25	Projectile	Hunting	37
654.B6 123 278	No Traces	No Traces	-
655.B6 126 13	Boring	Hide	-
656.B6 127 73	Projectile	Hunting	-
657.B8 73 71	No Traces	No Traces	-
658.B8 86 57	No Traces	No Traces	-
659.B8 89 24	Projectile	Hunting	-
660.B8 122 281	No Traces	No Traces	-
661.B8 126 168	Cutting	Butchering	24
662.C4 119 143	No Traces	No Traces	-
663.C4 119 173	No Traces	No Traces	-
664.C4 120 58	Projectile	Hunting	-
665.C4 125 203	No Traces	No Traces	-
666.C6 71 10	No Traces	No Traces	-
667.C6 73 64	No Traces	No Traces	-
668.C6 88 19	Cutting	Butchering	32
669.C6 119 138	Projectile	Hunting	-
670.C6 122 265	Cutting	Butchering	35
671.C6 122 98	Boring	Dry Hide	-
672.C6 122 98	Scraping	Dry Hide	80
673.C6 124 127	No Traces	No Traces	-
674.C6 126 32	Cutting	Butchering	24
675.C8 125 194	Cutting	Hide	25
676.C8 127 141	No Traces	No Traces	-
677.C8 127 41	Indeterm.	Soft Mat.	47

Piece num.	Activity	Worked Material	Edge Angle
TRUNCATED TOOLS			
Berniollo			
678.A4 -93 78	Cutting	Butchering	48
679.A4 -93 78	Cutting	Butchering	65
680.B1 -79 53	Scraping	Wood	75
681.B1 -79 53	Scraping	Wood	85
682.B1 -79 53	Scraping	Wood	57
683.C11 -69 53	Cutting	Butchering	38
684.C11 -69 53	Cutting	Indeterm.	35
685.C6 -84 5	Scraping	Antler	79
686.C6 -84 5	Cutting	Antler	46
687.C6 -84 5	Scraping	Antler	104
688.C6 -84 5	Scraping	Antler	90
689.V15 -86 20	Cutting	Butchering	40
Azilian level of Santa Catalina			
690.C8 19 11	Scraping	Wood	99
691.C8 19 11	Cutting	Wood	39
692.C8 19 11	Scraping	Soft Mat.	40
693.C8 19 11	Scraping	Hard Mat.	86
Magdalenian level of Santa Catalina			
694.B4 90 25	Scraping	Antler	63
695.B6 73 84	No Traces	No Traces	-
696.B6 121 203	No Traces	No Traces	-
697.B6 127 233	Cutting	Butchering	40
698.B6 127 233	Cutting	Butchering	34
699.B8 67 49	Cutting	Antler	66
700.B8 67 49	Scraping	Antler	103
701.B8 67 49	Cutting	Antler	35
702.B8 67 49	Scraping	Antler	81
703.B8 67 49	Scraping	Antler	80
704.B8 83 123	Indeterm.	Hide	59
705.B8 83 123	Scraping	Hide	76
706.B8 83 64	Boring	Indeterm.	-
707.B8 126 150	Engraving	Antler	91
708.B8 126 150	Scraping	Antler	91
709.B8 126 150	Scraping	Antler	92
710.B8 126 150	Scraping	Antler	70
711.B8 126 150	Scraping	Antler	66
712.B8 126 40	Scraping	Antler	92
713.B8 126 40	Scraping	Antler	103
714.B8 126 40	Engraving	Hard Mat.	-
715.B8 126 40	Scraping	Hard Mat.	69
716.C4 89 34	Cutting	Dry Hide	26
717.C4 89 34	Cutting	Dry Hide	39
718.C4 127 172	Cutting	Butchering	49
719.C4 127 172	Cutting	Butchering	28
720.C4 128 234	Scraping	Soft Mat.	84
721.C6 77 2	Cutting	Antler	59
722.C6 77 2	Cutting	Antler	29
723.C6 77 2	Scraping	Antler	89
724.C6 77 2	Scraping	Antler	59
725.C6 77 2	Scraping	Antler	110

Piece num.	Activity	Worked Material	Edge Angle
726.C6 127 22	Cutting	Antler	36
727.C6 127 22	Cutting	Antler	47
728.C6 127 22	Scraping	Antler	90
729.C6 127 22	Scraping	Antler	69
730.C8 77 13	Scraping	Hard Mat.	105
731.C8 86 1	Scraping	Fresh Hide	67
732.C8 86 1	Cutting	Fresh Hide	24
733.C8 86 1	Scraping	Fresh Hide	86
734.C8 123 128	Scraping	Antler	103
735.C8 123 128	Scraping	Antler	80
736.C8 126 70	Cutting	Butchering	33

DIVERSE (92 TYPE)

Azilian level of Santa Catalina

Piece num.	Activity	Worked Material	Edge Angle
737.B4 52 20	Indeterm.	Indeterm.	-
738.B4 52 20	Indeterm.	Indeterm.	-
739.B4 68 12	No Traces	No Traces	-
740.C6 39 17	No Traces	No Traces	-
741.C6 4 10	Scraping	Fresh Hide	85
742.C6 4 10	Scraping	Fresh Hide	90
743.C6 4 10	Scraping	Fresh Hide	76
744.C6 4 10	Scraping	Fresh Hide	80
745.C6 4 10	Scraping	Fresh Hide	81
746.C6 4 10	Scraping	Fresh Hide	80
747.C6 9 15	Scraping	Wood	37

Magdalenian level of Santa Catalina

Piece num.	Activity	Worked Material	Edge Angle
748.B4 126 36	Boring	Wood	-
749.B6 81 55	Scraping	Fresh Hide	75
750.B6 81 55	Scraping	Hide	112
751.B6 81 88	Scraping	Wood	87
752.B6 81 88	Scraping	Wood	84
753.B6 81 88	Scraping	Indeterm.	63
754.B6 87 125	Scraping	Fresh Hide	56
755.B6 87 125	Scraping	Fresh Hide	97
756.B6 87 125	Scraping	Fresh Hide	55
757.B6 125 253	Scraping	Ant./Bone	93
758.B6 125 253	Scraping	Indeterm.	61
759.B8 120 68	Scraping	Indeterm.	77
760.C4 126 228	No Traces	No Traces	-
761.C4 127 125	Cutting	Butchering	25
762.C4 127 125	Cutting	Butchering	46
763.C4 127 165	Chopping	Indeterm.	5

UNRETOUCHED BLADES

Berniollo

Piece num.	Activity	Worked Material	Edge Angle
764.A2 -94	No Traces	No Traces	-
765.B7 -84 224	Scraping	Wood	85
766.B7 -84 224	Scraping	Wood	80
767.B7 -84 224	Cutting	Wood	28
768.B7 -84 224	Cutting	Wood	30
769.B7 -89 175	No Traces	No Traces	-

Piece num.	Activity	Worked Material	Edge Angle
770.B7 -89 178	Cutting	Butchering	29
771.B7 -89 178	Cutting	Butchering	37
772.B7 -89 211	Cutting	Butchering	43
773.B7 -89 211	Cutting	Butchering	42
774.B7 -89 222	Scraping	Wood	72
775.B7 -89 222	Scraping	Wood	90
776.B7 -89 222	Cutting	Wood	30
777.B7 -89 227	No Traces	No Traces	-
778.B7 -89 294	No Traces	No Traces	-
779.C1 -76 11	No Traces	No Traces	-
780.C11-112 44	Scraping	Wood	22
781.C11-112 44	Cutting	Wood	22
782.C11-112 44	Cutting	Wood	46
783.C3 -82 69	No Traces	No Traces	-
784.C7 -104 79	Cutting	Butchering	47
785.C7-114 115	Indeterm.	Indeterm.	23
786.C7-114 115	Indeterm.	Indeterm.	30
787.C7 -78 36	No Traces	No Traces	-
788.C7 -95 68	Cutting	Butchering	44
789.C8 -91 95	No Traces	No Traces	-
790.C9 -78 10	No Traces	No Traces	-
791.C9 -88 111	No Traces	No Traces	-
792.D2 -84 62	No Traces	No Traces	-
793.D2 -85 23	No Traces	No Traces	-
794.E4 -76 1	Cutting	Soft Mat.	41
795.V15 -73 5	No Traces	No Traces	-

Azilian level of Santa Catalina

Piece num.	Activity	Worked Material	Edge Angle
796.B4 19 33	Cutting	Butchering	27
797.B4 19 33	Cutting	Butchering	46
798.B4 25 27	No Traces	No Traces	-
799.B4 3 14	Scraping	Hard Mat.	62
800.B4 33 23	No Traces	No Traces	-
801.B4 40 14	No Traces	No Traces	-
802.B6 30 24	Cutting	Butchering	21
803.B8 6 12	Cutting	Butchering	45
804.B8 6 12	Scraping	Hide	71
805.B8 6 12	Scraping	Indeterm.	26
806.B8 110 267	No Traces	No Traces	-
807.C4 10 20	No Traces	No Traces	-
808.C4 14 67	Cutting	Butchering	38
809.C4 27 1	Scraping	Hide	91
810.C4 37 9	No Traces	No Traces	-
811.C4 116 235	No Traces	No Traces	-
812.C4 18 63	Cutting	Butchering	21
813.C4 18 63	Scraping	Antler	98
814.C6 13 19	Scraping	Indeterm.	100
815.C6 23 6	No Traces	No Traces	-
816.C6 3 15	No Traces	No Traces	-
817.C6 3 22	Cutting	Butchering	21
818.C6 116 111	Cutting	Butchering	31
819.C6 116 111	Cutting	Butchering	64
820.C8 21 10	Scraping	Plant	39

Piece num.	Activity	Worked Material	Edge Angle
Magdalenian level of Santa Catalina			
821.A8 77 377	No Traces	No Traces	-
822.B4 77 12	No Traces	No Traces	-
823.B4 82 54	No Traces	No Traces	-
824.B4 93 19	No Traces	No Traces	-
825.B6 88 64	No Traces	No Traces	-
826.B6 126 114	No Traces	No Traces	-
827.B6 126 130	No Traces	No Traces	-
828.B6 128 199	Scraping	Indeterm.	73
829.B8 67 42	No Traces	No Traces	-
830.B8 68 30	Cutting	Butchering	21
831.B8 68 30	Cutting	Butchering	54
832.B8 71 51	No Traces	No Traces	-
833.B8 79 151	No Traces	No Traces	-
834.B8 82 22	No Traces	No Traces	-
835.B8 84 19	No Traces	No Traces	-
836.B8 86 56	No Traces	No Traces	-
837.B8 121 100	No Traces	No Traces	-
838.B8 121 187	No Traces	No Traces	-
839.B8 121 356	Cutting	Dry Hide	18
840.B8 126 4	No Traces	No Traces	-
841.B8 126 40	Scraping	Hard Mat.	43
842.B8 126 54	Scraping	Antler	103
843.B8 126 56	No Traces	No Traces	-
844.C4 72 28	Cutting	Butchering	41
845.C4 72 28	Scraping	Hard Mat.	86
846.C4 72 56	No Traces	No Traces	-
847.C4 75 3	Scraping	Wood	34
848.C4 76 50	Indeterm.	Indeterm.	39
849.C4 79 15	No Traces	No Traces	-
850.C4 84 8	No Traces	No Traces	-
851.C4 95 61	Cutting	Hide/Meat	29
852.C4 95 61	Scraping	Ant./Bone	89
853.C4 95 61	Scraping	Ant./Bone	90
854.C4 97 48	No Traces	No Traces	-
855.C4 97 94	No Traces	No Traces	-
856.C4 97 95	Cutting	Butchering	20
857.C4 97 95	Engraving	Wood	-
858.C4 127 100	No Traces	No Traces	-
859.C4 127 127	Scraping	Ant./Bone	95
860.C4 127 294	No Traces	No Traces	-
861.C4 127 73	Cutting	Butchering	49
862.C4 127 73	Cutting	Butchering	41
863.C4 127 86	Boring	Antler	-
864.C4 127 87	No Traces	No Traces	-
865.C6 69 15	No Traces	No Traces	-
866.C6 74 45	Scraping	Wood	46
867.C6 128 240	Cutting	Butchering	37
868.C6 128 240	Cutting	Butchering	59
869.C8 52 115	No Traces	No Traces	-
870.C8 67 18	No Traces	No Traces	-
871.C8 81 10	No Traces	No Traces	-

Piece num.	Activity	Worked Material	Edge Angle
UNRETOUCHED BLADELETS			
Berniollo			
872.A2 -92 10	Projectile	Indeterm.	-
873.B2 -92 120	Boring	Indeterm.	-
874.B7 -87 74	No Traces	No Traces	-
875.C4 -85 65	No Traces	No Traces	-
876.C6 -87 100	No Traces	No Traces	-
877.C7-100 113	No Traces	No Traces	-
878.C7-101 153	No Traces	No Traces	-
879.C9 -93 73	No Traces	No Traces	-
880.D2 -82 76	No Traces	No Traces	-
881.D4 -83 21	No Traces	No Traces	-
882.D4 92	No Traces	No Traces	-
883.D4 92	No Traces	No Traces	-
884.D6 138	No Traces	No Traces	-
885.E4 -72 24	No Traces	No Traces	-
Azilian level of Santa Catalina			
886.B4 19 28	No Traces	No Traces	-
887.B4 2 17	No Traces	No Traces	-
888.B4 30 20	No Traces	No Traces	-
889.B4 40 17	No Traces	No Traces	-
890.B4 60 17	No Traces	No Traces	-
891.B4 7 17	No Traces	No Traces	-
892.B4 72 11	No Traces	No Traces	-
893.B6 10 24	No Traces	No Traces	-
894.B6 24 10	No Traces	No Traces	-
895.B6 24 15	No Traces	No Traces	-
896.B8 5 12	No Traces	No Traces	-
897.B8 112 101	No Traces	No Traces	-
898.C4 1 16	No Traces	No Traces	-
899.C4 25 22	No Traces	No Traces	-
900.C4 116 180	No Traces	No Traces	-
901.C6 24 27	No Traces	No Traces	-
902.C6 33 16	No Traces	No Traces	-
903.C6 67 10	No Traces	No Traces	-
904.C6 113 74	No Traces	No Traces	-
905.C6 18 53	No Traces	No Traces	-
906.C8 111 51	Scraping	Indeterm.	70
907.C8 111 51	Cutting	Indeterm.	48
908.C8 116 57	No Traces	No Traces	-
909.C8 18 26	No Traces	No Traces	-
Magdalenian level of Santa Catalina			
910.B4 121 249	No Traces	No Traces	-
911.B6 126 274	Cutting	Hide/Meat	36
912.B8 73 18	Cutting	Soft Mat.	32
913.B8 73 18	Cutting	Soft Mat.	52
914.B8 79 16	No Traces	No Traces	-
915.B8 86 47	No Traces	No Traces	-
916.B8 122 299	No Traces	No Traces	-
917.C4 89 16	No Traces	No Traces	-
918.C4 127 74	No Traces	No Traces	-
919.C6 69 16	No Traces	No Traces	-

Piece num.	Activity	Worked Material	Edge Angle
920.C6 117 179	Scraping	Ant./Bone	83
921.C6 126 689	No Traces	No Traces	-
922.C8 89 62	No Traces	No Traces	-
923.C8 127 56	No Traces	No Traces	-

UNRETOUCHED FLAKES

Berniollo

Piece num.	Activity	Worked Material	Edge Angle
924.A11 -36 76	No Traces	No Traces	-
925.A17 -80 10	No Traces	No Traces	-
926.A2 -93 59	No Traces	No Traces	-
927.A4 -93 104	No Traces	No Traces	-
928.A5 -88 53	No Traces	No Traces	-
929.A9 -89 70	No Traces	No Traces	-
930.B10 -99 65	No Traces	No Traces	-
931.B2 -82 126	No Traces	No Traces	-
932.B2 -86 2	Scraping	Indeterm.	70
933.B4 -86 4	No Traces	No Traces	-
934.B6 -91 96	No Traces	No Traces	-
935.B7 -80 36	No Traces	No Traces	-
936.B7 -83 62	Indeterm.	Indeterm.	82
937.B7 -83 62	Indeterm.	Indeterm.	62
938.B7 -84 71	No Traces	No Traces	-
939.B7 -89 189	Cutting	Wood/Plant	23
940.B7 -89 189	Scraping	Wood	108
941.B7 -89 189	Scraping	Wood	72
942.B7 -89 197	No Traces	No Traces	-
943.B7 -89 217	Scraping	Bone	61
944.B9 -86 5	No Traces	No Traces	-
945.C1 68	No Traces	No Traces	-
946.C10-99 150	No Traces	No Traces	-
947.C3 -80 50	No Traces	No Traces	-
948.C5 60	No Traces	No Traces	-
949.C7-101 153	No Traces	No Traces	-
950.C7 -109 43	Scraping	Indeterm.	23
951.C7 -109 43	Scraping	Indeterm.	18
952.C7 -109 43	Scraping	Indeterm.	92
953.C7 -109 43	Scraping	Indeterm.	80
954.C8 -89 162	No Traces	No Traces	-
955.C8 -91 46	No Traces	No Traces	-
956.C9-115 106	Engraving	Wood	-
957.C9-115 106	Scraping	Wood	95
958.C9-115 106	Scraping	Wood	82
959.C9-115 106	Scraping	Wood	23
960.C9-115 106	Scraping	Wood	85
961.C9 -80 96	Scraping	Wood	89
962.C9 -80 96	Scraping	Bone	56
963.C9 -80 96	Cutting	Indeterm.	33
964.D10 -88 65	No Traces	No Traces	-
965.D2 -74 31	No Traces	No Traces	-
966.D2 -75 66	No Traces	No Traces	-
967.D2 -80 53	No Traces	No Traces	-
968.D2 -88 92	No Traces	No Traces	-
969.D4 -80 50	No Traces	No Traces	-
970.D4 -90 136	No Traces	No Traces	-
971.D6 -80 54	No Traces	No Traces	-
972.D6 -82 137	Scraping	Wood	85
973.D6 -82 137	Scraping	Wood	55
974.D6 -82 137	Scraping	Wood	83
975.D6 -82 137	Scraping	Wood	26
976.D8 -83 42	No Traces	No Traces	-
977.D8 -83 88	No Traces	No Traces	-
978.D8 -85 83	No Traces	No Traces	-
979.D8 -86 145	Scraping	Ant./Bone	93
980.D8 -86 145	Scraping	Ant./Bone	98
981.D8 -86 145	Scraping	Ant./Bone	97
982.D8 -86 145	Cutting	Indeterm.	35
983.E4 -76 37	No Traces	No Traces	-
984.T11 -90 49	No Traces	No Traces	-
985.V17 -83 38	No Traces	No Traces	-
986.V19 -69 25	Scraping	Ant./Bone	92
987.V19 -69 25	Cutting	Indeterm.	56

Azilian level of Santa Catalina

Piece num.	Activity	Worked Material	Edge Angle
988.B4 11 14	Scraping	Wood	106
989.B4 20 13	No Traces	No Traces	-
990.B4 20 7	Scraping	Bone	44
991.B4 20 7	Scraping	Hard Mat.	66
992.B4 22 83	No Traces	No Traces	-
993.B4 36 3	No Traces	No Traces	-
994.B4 40 26	Engraving	Hard Mat.	-
995.B4 43 39	No Traces	No Traces	-
996.B4 43 40	No Traces	No Traces	-
997.B6 23 22	No Traces	No Traces	-
998.B6 30 41	Indeterm.	Indeterm.	24
999.B6 42 12	Cutting	Butchering	34
1000.B6 42 12	Scraping	Hard Mat.	82
1001.B6 71 11	No Traces	No Traces	-
1002.B6 72 15	No Traces	No Traces	-
1003.B6 80 16	Boring	Hide	-
1004.B6 80 16	Scraping	Hide	65
1005.B6 111 57	Cutting	Fresh Hide	51
1006.B8 18 11	No Traces	No Traces	-
1007.B8 23 1	Scraping	Bone	92
1008.B8 23 1	Scraping	Bone	74
1009.B8 23 1	Scraping	Bone	94
1010.B8 23 1	Scraping	Bone	105
1011.B8 23 1	Scraping	Bone	58
1012.B8 61 9	No Traces	No Traces	-
1013.B8 17 41	No Traces	No Traces	-
1014.C4 34 22	Cutting	Hide	21
1015.C4 34 22	Scraping	Hard Mat.	50
1016.C4 8 25	Scraping	Hide	80
1017.C4 8 25	Scraping	Hide	74
1018.C6 1 30	Cutting	Fresh Hide	40
1019.C6 17 23	Scraping	Fresh Hide	42
1020.C6 25 28	Cutting	Hide	19
1021.C6 25 28	Engraving	Indeterm.	-
1022.C6 39 18	No Traces	No Traces	-
1023.C6 39 5	Cutting	Butchering	33

Piece num.	Activity	Worked Material	Edge Angle
1024.C6 116 119	No Traces	No Traces	-
1025.C6 18 35	No Traces	No Traces	-
1026.C8 19 36	No Traces	No Traces	-
Magdalenian level of Santa Catalina			
1027.B4 84 75	No Traces	No Traces	-
1028.B4 87 5	Scraping	Ant./Bone	94
1029.B4 87 5	Scraping	Ant./Bone	44
1030.B4 87 5	Scraping	Indeterm.	104
1031.B6 121 205	Scraping	Hide	50
1032.B6 121 205	Scraping	Hide	70
1033.B6 126 117	No Traces	No Traces	-
1034.B6 126 137	No Traces	No Traces	-
1035.B6 126 262	No Traces	No Traces	-
1036.B6 126 63	No Traces	No Traces	-
1037.B6 128 16	Cutting	Plant	28
1038.B8 72 43	No Traces	No Traces	-
1039.B8 73 22	No Traces	No Traces	-
1040.B8 75 70	Scraping	Fresh Hide	71
1041.B8 75 70	Cutting	Hide	28
1042.B8 79 155	Scraping	Hard Mat.	95
1043.B8 79 159	No Traces	No Traces	-
1044.B8 79 8	Scraping	Plant	90
1045.B8 79 8	Scraping	Plant	90
1046.B8 80 160	Scraping	Soft Mat.	40
1047.B8 82 93	No Traces	No Traces	-
1048.B8 83 18	No Traces	No Traces	-
1049.B8 83 79	No Traces	No Traces	-
1050.B8 122 169	No Traces	No Traces	-
1051.B8 122 282	No Traces	No Traces	-
1052.B8 124 219	No Traces	No Traces	-
1053.B8 124 220	No Traces	No Traces	-
1054.B8 126 42	No Traces	No Traces	-
1055.C4 75 39	Scraping	Antler	40
1056.C4 75 39	Scraping	Antler	64
1057.C4 79 1	Scraping	Antler	84
1058.C4 87 19	No Traces	No Traces	-
1059.C4 87 20	Scraping	Plant	17
1060.C4 96 58	No Traces	No Traces	-
1061.C4 127 134	Scraping	Ant./Bone	115
1062.C4 127 208	Scraping	Antler	69
1096.			

Piece num.	Activity	Worked Material	Edge Angle
1063.C4 127 65	No Traces	No Traces	-
1064.C4 127 734	No Traces	No Traces	-
1065.C6 68 26	No Traces	No Traces	-
1066.C6 72 34	No Traces	No Traces	-
1067.C6 82 12	No Traces	No Traces	-
1068.C8 90 15	Scraping	Bone	93
1069.C8 90 15	Scraping	Bone	59
1070.C8 90 15	Scraping	Bone	79
1071.C8 120 133	Scraping	Hide	80
1072.C8 127 341	Cutting	Butchering	37
CORES			
Berniollo			
1073.A4 -94 196	No Traces	No Traces	-
1074.C2 -81 87	Scraping	Wood	104
1075.C2 -81 87	Scraping	Wood	120
1076.C2 -81 87	Scraping	Wood	78
1077.C2 -81 87	Scraping	Wood	83
1078.C8 -91 185	No Traces	No Traces	-
1079.D8 -84 40	No Traces	No Traces	-
1080.D8 -86 150	No Traces	No Traces	-
1081.D8 -87 110	No Traces	No Traces	-
1082.V17 -87 8	No Traces	No Traces	-
Azilian level of Santa Catalina			
1083.B6 26 41	Scraping	Antler	70
1084.B6 26 41	Scraping	Antler	83
1085.B6 26 41	Scraping	Antler	76
1086.B6 61 12	Scraping	Dry Hide	71
1087.B6 61 12	Engraving	Dry Hide	-
1088.B6 61 12	Scraping	Dry Hide	90
1089.B6 61 12	Scraping	Dry Hide	92
Magdalenian level of Santa Catalina			
1090.B4 87 14	No Traces	No Traces	-
1091.B4 87 72	Scraping	Hide	89
1092.B4 125 205	No Traces	No Traces	-
1093.B6 83 85	No Traces	No Traces	-
1094.B6 86 176	No Traces	No Traces	-
1095.C6 128 15	No Traces	No Traces	-

APPENDIX 6

PHOTOGRAPHS OF ARCHAEOLOGICAL USE TRACES

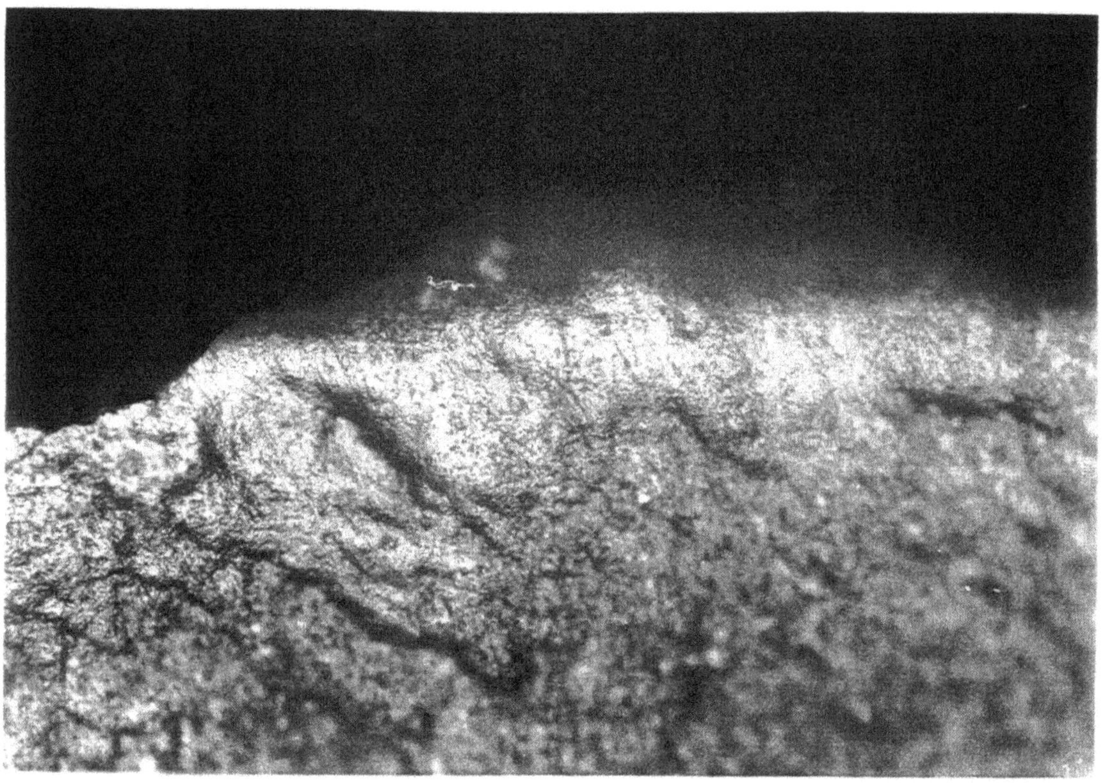

Photo 1. Dry hide polish. SC B8 75 15. 100 X.
Photo 2. The same zone at 200 X.

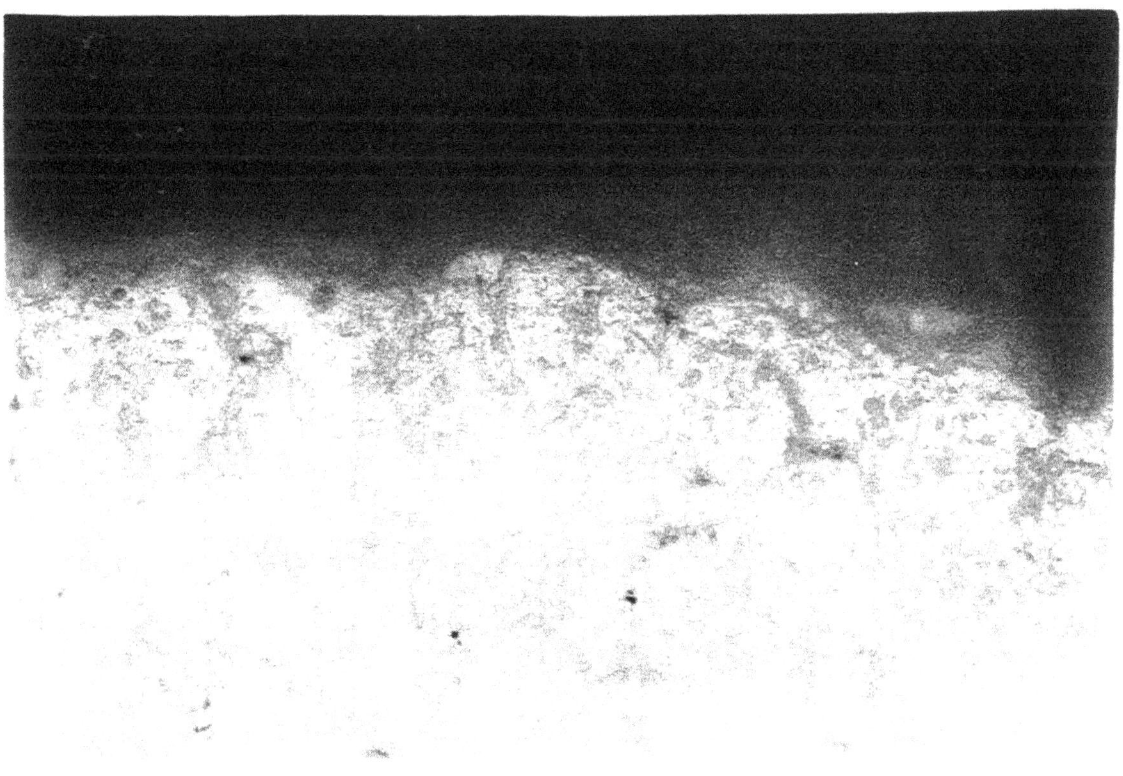

Photo 3. Dry hide polish. SC B8 75 17. 200 X.
Photo 4.. Dry hide polish. BN T11 -93 47. 200 X.

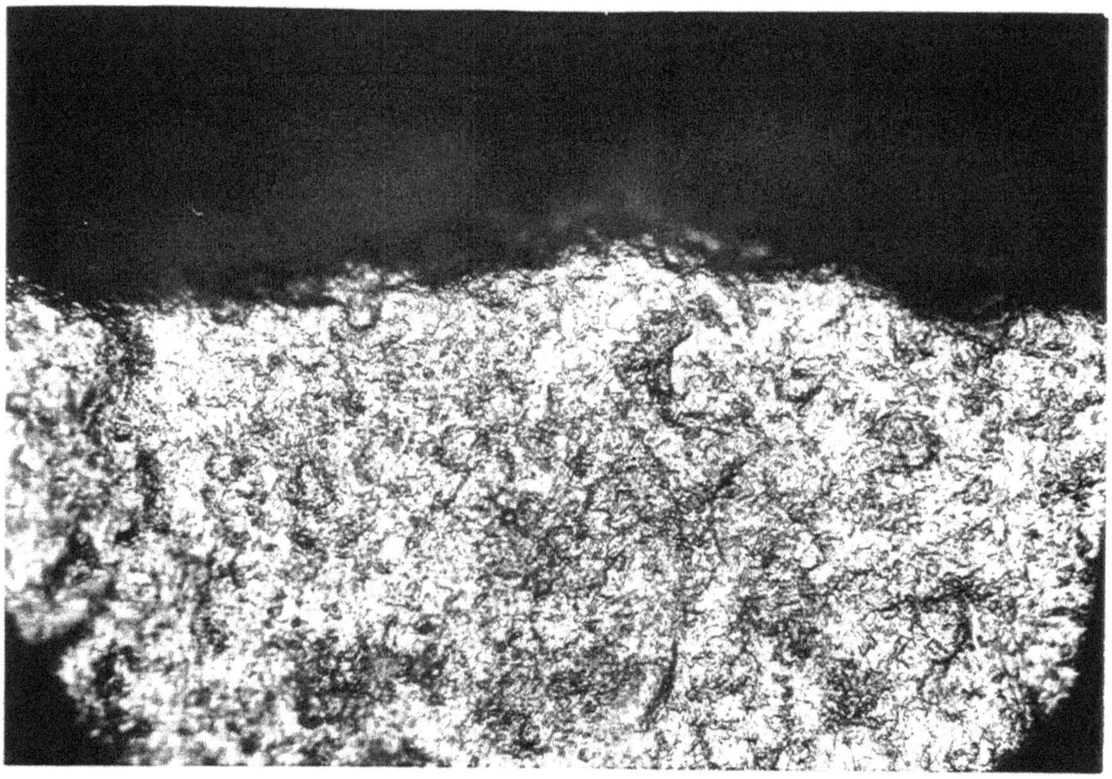

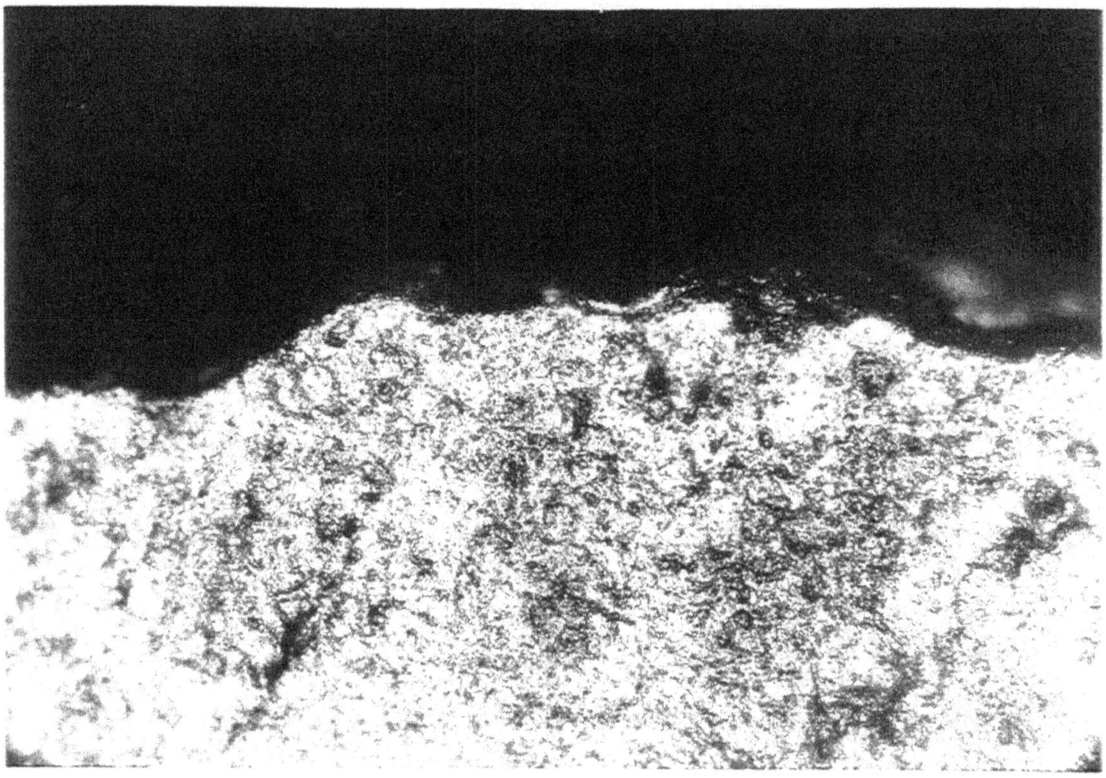

Photo 5. Fresh hide polish. SC C4 22 13. 200 X
Photo 6 Fresh hide polish. SC B8 57 23. 100 X.

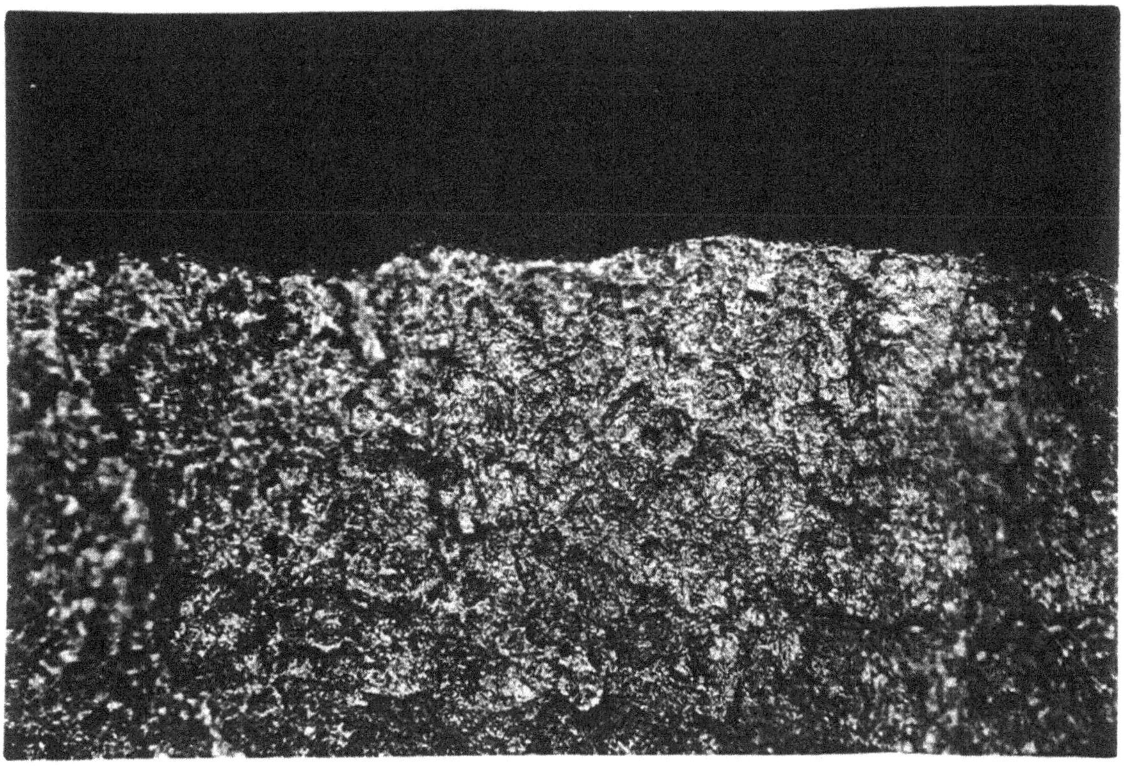

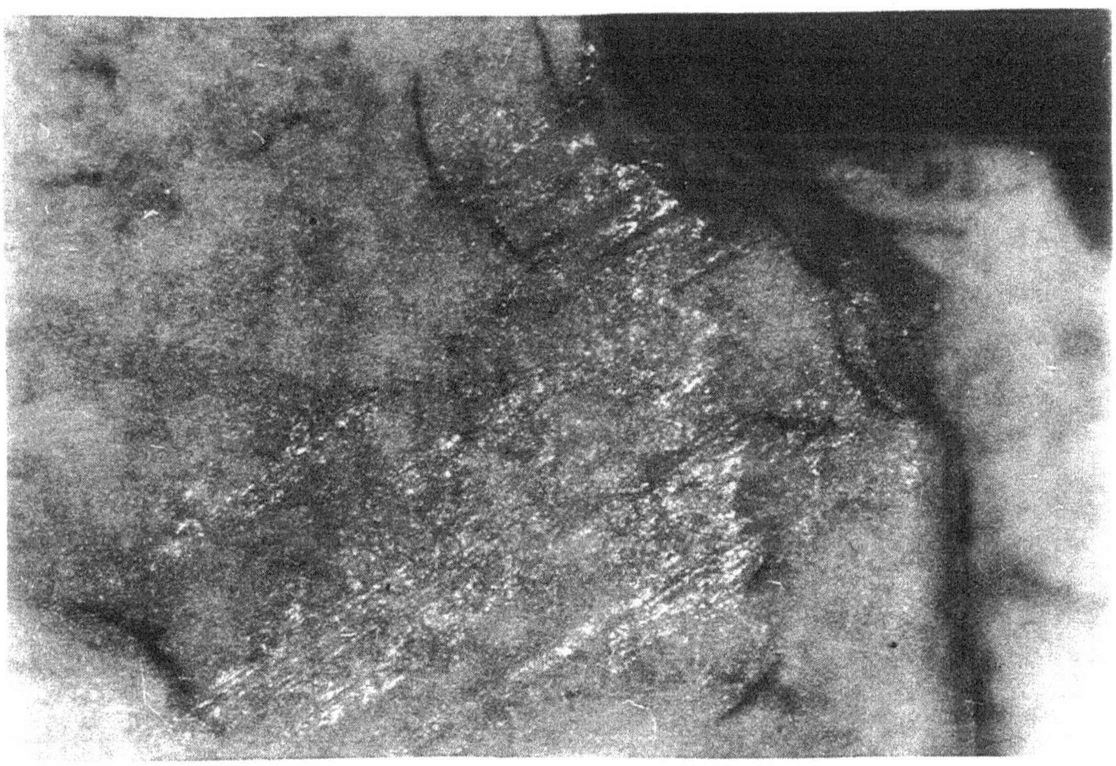

Photo 7. Fresh hide polish. SC B8 L21 98. 200X.
Photo 8. Striations from flaking. BN D8 -86 150. 50X.

Photo 9. Bone polish from a butchering activity. SC B8 L21 353. 200 X
Photo 10 Bone polish from a butchering activity. SC C6 23 10. 200 X.

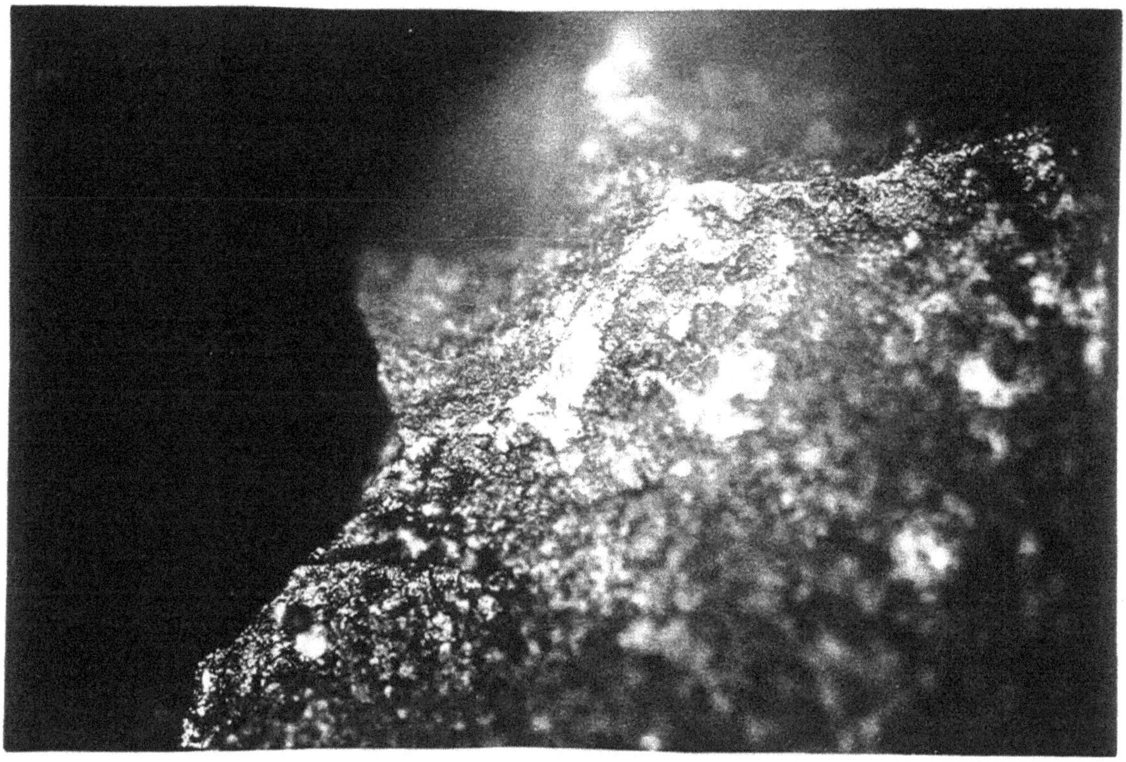

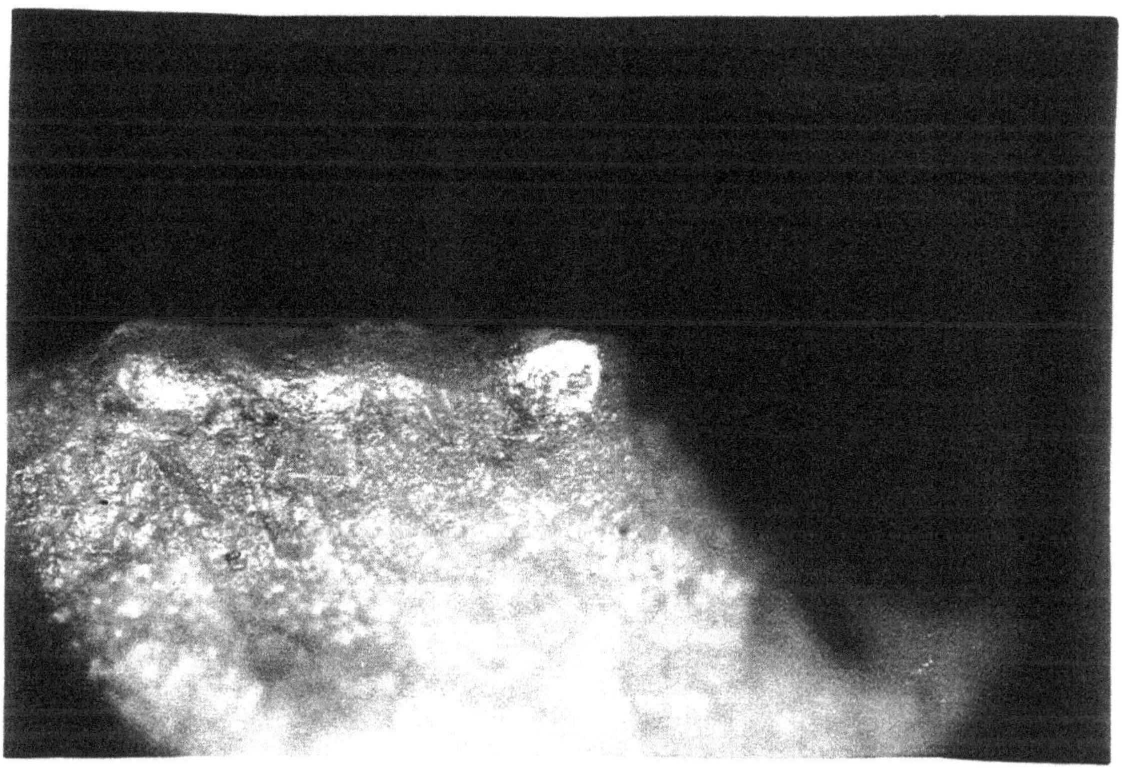

Photo 11. Bone polish from a butchering activity. SC B6 23 L6. 200 X.
Photo 12. Bone polish from a butchering activity. BN D4 -80 12. 200 X.

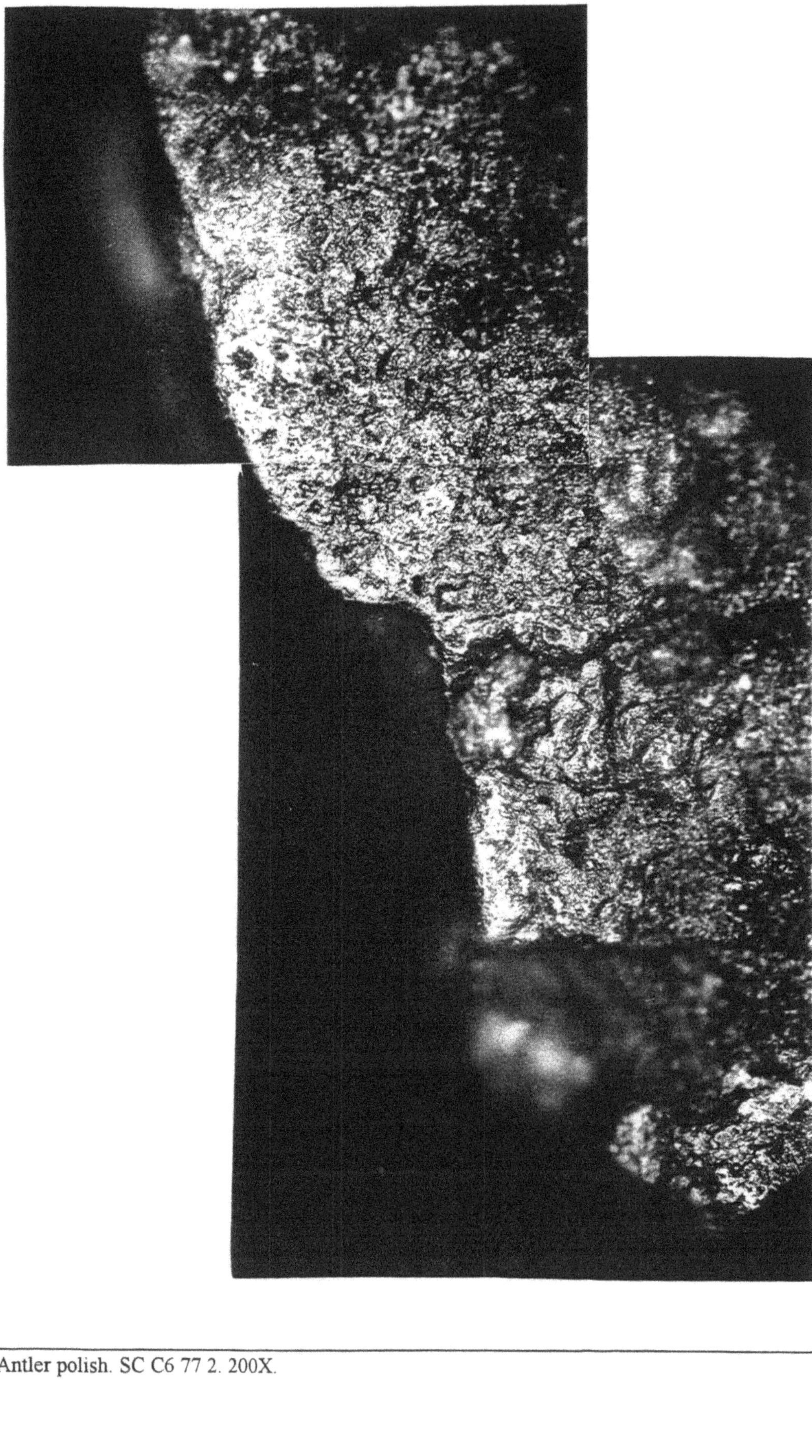

Photo 13. Antler polish. SC C6 77 2. 200X.

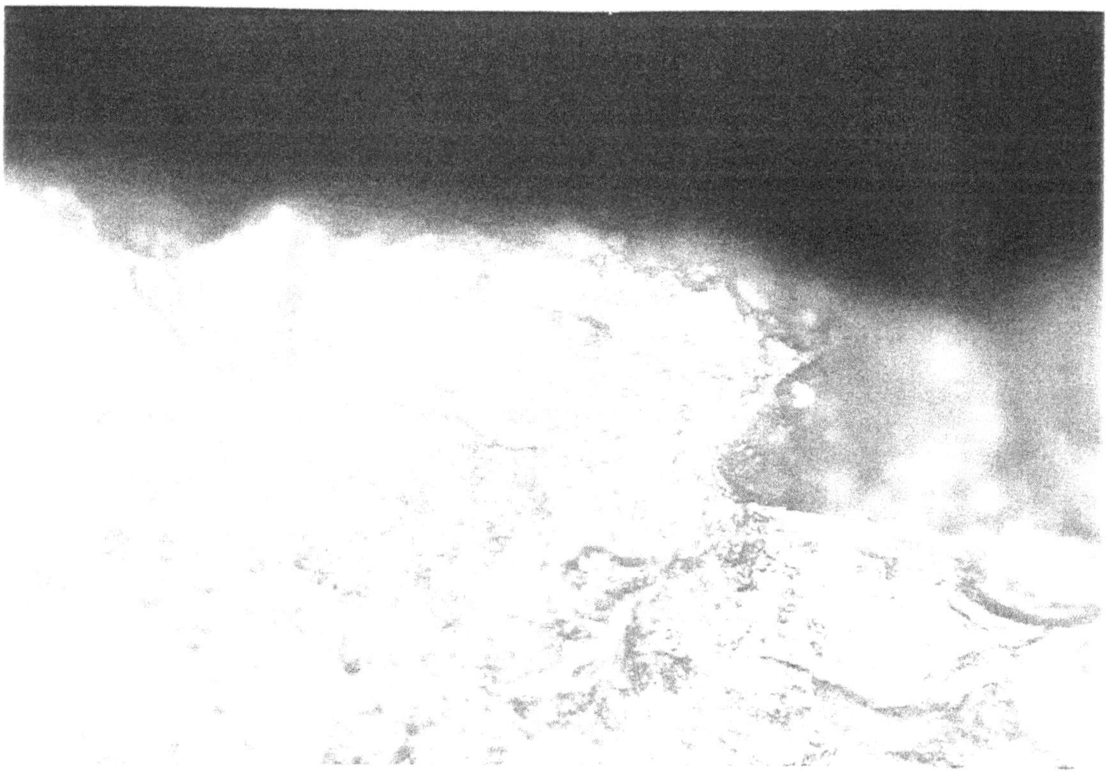

Photo 14. Antler polish. SC C4 22 13. 200 X.
Photo 15. Antler polish. SC C6 2 24. 200 X. .

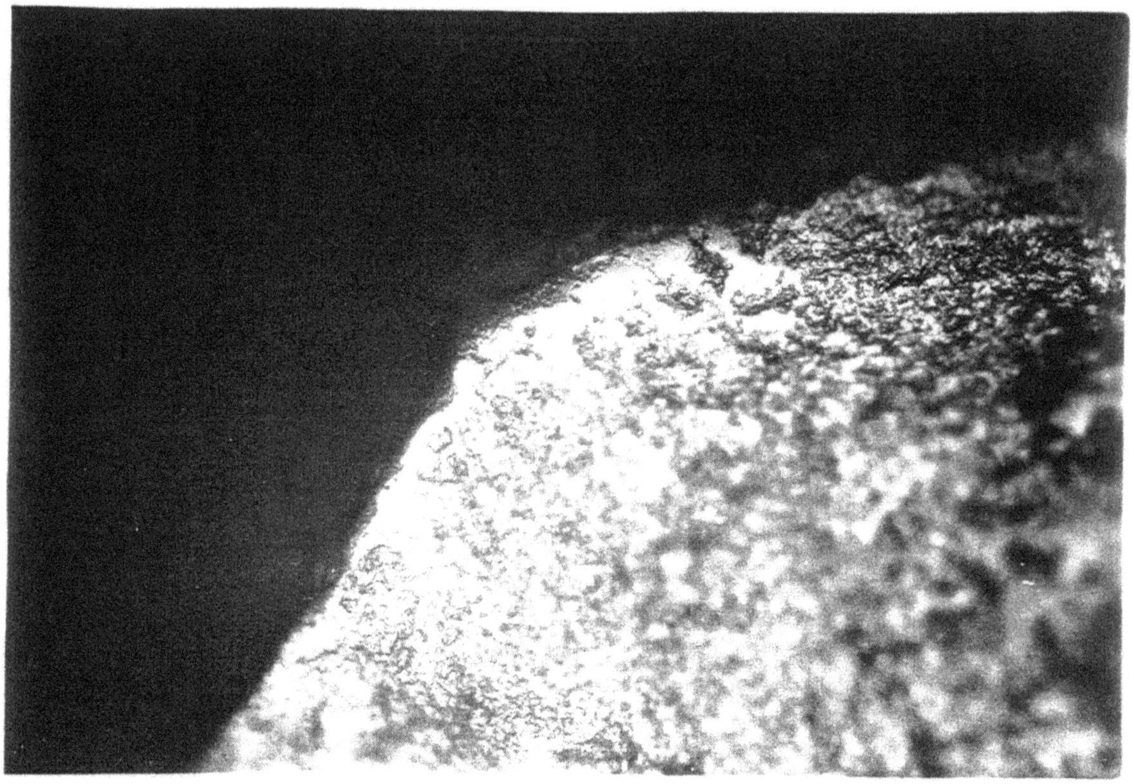

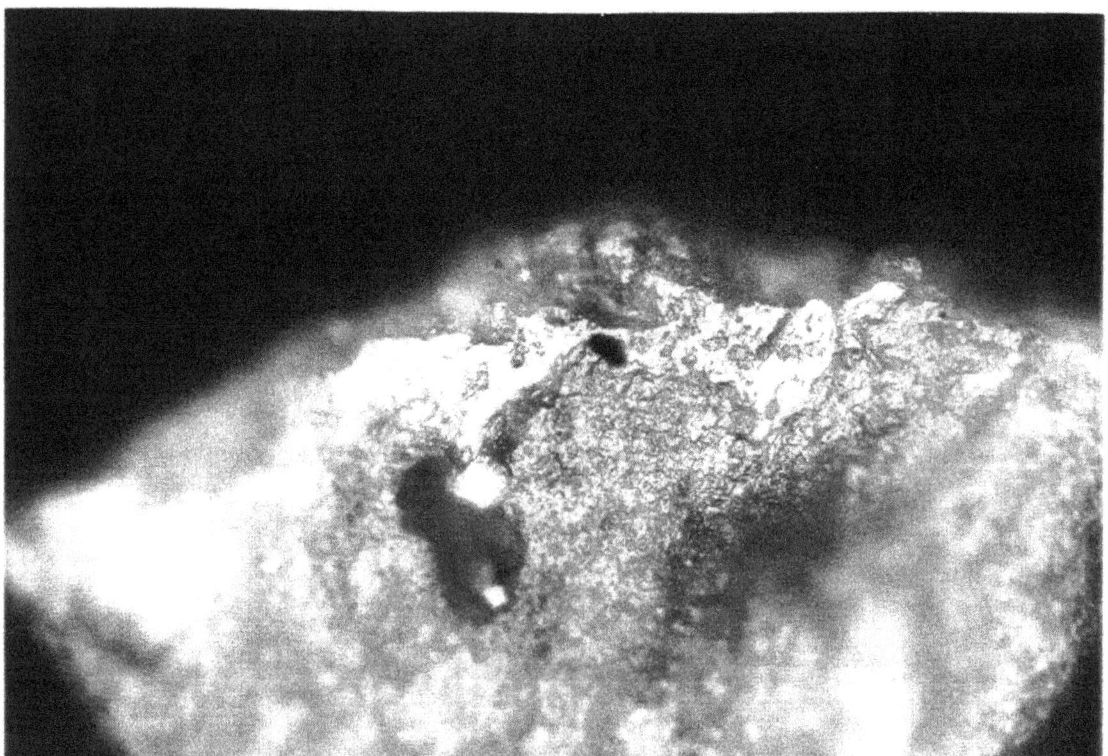

Photo 16. Bone polish. SC C6 78 48. 200 X
Photo 17. Bone polish. BN C9 -80 46. 200 X.

Photo 18. Plant polish. SC B8 87 19. 200X.

BIBLIOGRAPHY

Abreu Galindo,J. 1602?/1977 *Historia de la conquista de las siete islas de Canaria*, Goya Ediciones, Santa Cruz de Tenerife.

Akoshima,K. 1987 "Microflaking quantification", G. de G. Sieveking and M.H. Newcomer (eds.) *The human uses of flint and chert*, 71-80, Cambridge University Press.

Albright,S.L. 1984 *Tahltan Ethnoarchaeology*, Simon Fraser University, n. 15, Vancouver.

Allain,J., Fritchs,R., Rigaud,A. and Trotignon,F. 1974 "Le débitage du bois de renne dans les niveaux à raclettes du Badegoulien de l'Abri Fritsch et sa signification", *L'industrie d'os dans la Préhistoire*, 67-71.

Allain,J. and Rigaud,A. 1986 "Décor et fonction. Quelques exemples tirés du magdalénien", *L'Anthropologie*, 90, 4, 713-738.

Anderson,P.C. 1981. *Contribution méthodologique a l'analyse des microtraces d'utilisation sur les outils préhistoriques.* Thèse de 3ème. cycle, Université de Bordeaux I.

Anderson,p.C. 1987 "L'enmanchement au mousterien", in D.Stordeur (dir.) *La main et l'outil. Manches et enmanchements préhistoriques.* Travaux de la Maison de l'Orient, 15: 37-54. G.S. Maison de l'Orient, Lyon.

Anderson,P.C. 1988 "Using prehistoric stone tools to harvest cultivated wild cereals: preliminary observations of traces and impact", in S. Beyries (ed.) *Industries Lithiques. Tracéologie et Technologie*, vol..I, 175-196, B.A.R. International Series, 411, Oxford.

Anderson,P.C. 1992 "Experimental cultivation, harvesting and thresing of wild cereals and their relevance for interpreting the use of Epipapalaeolithic and Neolithic artefacts", in P.C. Anderson (ed.) *Préhistoire de la Agriculture: Nouvelles approches expérimentales et ethnographiques.* Monographies du CRA 6, 179-210, Eds. du C.N.R.S., Paris.

Anderson,P.C.(ed.) 1992 *Préhistoire de la Agriculture: Nouvelles approches expérimentales et ethnographiques.* Monographies du CRA 6, Paris, Eds. du C.N.R.S.

Anderson,P.C., S. Beyries, M. Otte and H. Plisson (eds.) 1993 *Traces et fonction: les gestes rétrouvés*, E.R.A.U.L., 50.

Anderson,P., Inizan,M.L., Lechevallier,M., Pelegrin,J., Pernot,P. 1989 "Des lames de silex dans un atelier de potier harapéen: interaction des domaines techniques" *Comptes rendues de l'Académie des Sciences de Paris*, 308, Série II, 443-449.

Apellániz,J.M. 1991 *Modelo de Análisis de la Autoría en el Arte Figurativo del Paleolítico.* Cuadernos de Arqueología de Deusto, 13, Universidad de Deusto, Bilbao.

Audouze,F., Cahen,D., Keeley,L.H. and Schmider,B. 1981 "Le site magdalénien du Buisson Campin à Verberie (Oise)", *Gallia Préhistoire*, 24, 1, 99-143.

Audouin,F. and Plisson,H. 1982 "Les ocres et leurs témoins au Paléolithique en France: enquête et expériences sur leur validité archéologique", *Cahiers du Centre de Recherches Préhistoriques*, 8, 33-80.

Bailey,G. (ed.) 1983 *Hunter-Gatherer Economy in Prehistory: A European Perspective*, Cambridge University Press, Cambridge.

Balfet,H. (ed) 1991 *Observer l'action technique. Des chaînes opératoires, pour quoi faire?*, Ed. C.N.R.S., Paris.

Balicki,A. 1968 "The Netsilik Eskimo: Adaptive Processes", in R.B. Lee and I. De Vore (eds.) *Man the Hunter*, 78-82, Aldine De Gruyter, New York.

Balicki,A. 1970 *The Netsilik Eskimo*, Natural History Press, New York.

Bamforth,D.B. 1986 "Technological Efficiency and Tool Curation", *American Antiquity*, 51, 38-50.

Barber,E.J.W. 1990 "Neolithic Textiles in Europe and the Near East", *Archeomaterials*, 4, 1, 63-69.

Beaune,S.A.de 1989 "Exemple ethnographique de l'usage pluri-fonctionnelle d'un gallet de quartz", *Bulletin de la Société Préhistorique Francaise*, 86, 2, 61-64.

Belitz,B. 1973 *Brain Tanning, the Sioux Way*, Pine Ridge Reservation Publication, South Dakota.

Bergman,C.A. and Newcomer,M.H. (1983) "Flint arrowhead breakage: examples from Ksar Akil, Lebanon", *Journal of Field Archaeology*, 10: 238-243.

Bergman,C.A.; Barton,R.N.E.; Collcutt,S.N. and Morris,G. (1987) "Intentional breakage in a Late Upper Palaeolithic assemblage from Southern England" In G.de G. Sieveking and M.H.Newcomer (ed.) *The human uses of flint and chert*. Cambridge University Press, Cambridge.

Berke,H. 1977 "Déchets de harpons magdaleniens découverts à Gonnersdorf (Neuwied)", *Bulletin de la Société Préhistorique Française"*, 74, 3, 86-88.

Beyries,S. 1987 *Variabilité de l'industrie lithique au Mousterien. Approche fonctionnelle sur quelques gisements françaises.* B.A.R. International Series, 328, Oxford.

Beyries,S.(ed.) 1988 *Industries lithiques. Tracéologie et Technologie*. 2 vol.. B.A.R. International Series 411 (I and II), Oxford.

Beyries,S., Delamare,F. and Quantin,J.-C. 1988 "Tracéologie et rugosimétrie tridimensionnelle ingeneering use-wear models", in S. Beyries (ed.) *Industries Lithiques. Tracéologie et Technologie*, 115-132, B.A.R. International Series, 411, Oxford.

Beyries,S. and Roche,H. 1982 "Technologie et traces d'utilisation: application a des industries acheuléennes (Carrières Thomas, Casablanca, Maroc)", *Studia Praehistorica Belgica*, 2, 267-278.

Bienenfeld,P. 1985 "Preliminary Results from a Lithic Use-wear Study of the Swifterbant Sites, S-51, S-4, and S-2 (Swifterbant Contribution 13)", *Helinium*, 25, 194-211.

Bienenfeld,P. 1988 "Stone tool use and the organization of technology in the dutch Neolithic", in S. Beyries (ed.) *Industries Lithiques. Tracéologie et Technologie*, vol. 1, 219-230, B.A.R. Internacional Series, 411, Oxford.

Billamboz,A. 1977 "L'industrie du bois de cerf en Franche Comte au Néolithique et au début de l'Age du Bronze", *Gallia Préhistoire*, 20, 1, 7-176.

Binder,D. and Gassin,B. 1988 "Le débitage laminaire chasséen après chauffe: technologie et traces d'utilisation", in S. Beyries (ed.) *Industries Lithiques. Tracéologie et Technologie*, vol. 1, 93-126, B.A.R. International Series, 411, Oxford.

Binder,D. and Perlès,C. 1990 "Stratégies de gestion des outillages lithiques au Néolithique", *Paléo*, 2, 257-284.

Binford,L.R. (ed.) 1977 *For Theory Building in Archaeology*, Academic Press, New York.

Binford,L.R. 1978 *Nunamiut ethnoarchaeology*, Academic Press, New York.

Binford,L.R. 1979 "Organization and Formation Processes: Looking at Curated Technologies", *Journal of Anthropological Research*, 35, 255-273.

Binford,L.R. 1980 "Willow smoke and dog's tail: Hunter-gatherer systems and archaeological site formation", *American Antiquity*, 45, 4-22.

Binford,L.R. 1981 *Bones. Ancient Men and Modern Myths*, Studies in Archaeology, Academic Press, New York.

Binford,L.R. 1983 *In pursuit of the past*, Thames and Hudson, London.

Binford,L.R. 1986 "An Alyawara day: making men's knives and beyond". *American Antiquity*, 51, 547-562.

Binford,L.R. and Bertram,J.B. 1977 "Bone frecuencies and attritional processes", in L.R.Binford (ed.) *For Theory Building in Archaeology*, 75-153, Academic Press, New York.

Binford,L.R. and Binford,S.R. 1966 "A preliminary analysis of the functional variability in the Mousterian and Levallois facies", *American Anthropologist*, 68 (2), 238-295.

Bird,C. and Beek,C. 1980 "Bone points and spatulae: salvage ethnography in Southwest Australia". *Archaeol. Phys. Anthropol. Oceania*, 15, 168-171.

Blanc,J. and Gury,A. 1989 "Techniques de vannerie d'après les réliefs Gallo-romains", in *Tissage, corderie, vannerie*, 193-207, éd. APDCA, Juan-les-Pins, France.

Bocek,B.R. 1987 *Hunter-gatherer ecology and settlement mobility along San Francisquito Creek*, UMI Dissertation Services.

Bocquet,A. 1989 "Le travail des fibres textiles au Néolithique récent à Charavines (Isère)" in *Tissage, corderie, vannerie*, 113-128, éd. APDCA, Juan-les-Pins, France.

Bonnichsen,R. and Will,R. 1980 "Cultural Modification of Bone: the Experimental Approach in Faunal Analysis", in *Mamalian Osteology*, 7-30.

Bordes,F. 1950 "Du poli particulier de certains silex taillés", *L'Anthropologie*, 54, 161-163.

Bordes,F. 1952 "A propos des outils a bord abattu: quelques remarques", *Bulletin de la Societé Préhistorique Française*, 49, 645-647.

Bordes,F. 1965 "Utilisation possible des côtes des burins". *Fundberichte aus Schwaben, neue folge*, 17: 3-4.

Bordes,F. 1973 "Position des traces d'usure sur les grattoirs simples du Perigordien supérieur evolué de Corbiac (Dordogne)", *Estudios dedicados al profesor D. Luis Pericot*, 55-60. Universidad de Barcelona.

Brezillon,M. 1983 *La dénomination des objets de pierre taillée*. Éd.C.N.R.S., Paris.

Brown,A.G. and M.Edmonds,M. (eds.) *Lithic Analysis and Later British Prehistory. Some Problems and approches*, B.A.R. International Series, 162, Oxford.

Cahen,D.; Caspar,J.P. and Otte,M. 1986 *Industries lithiques danubiénnes de Belgique*. E.R.A.U.L., 21: 89 p.. Université de Liège.

Cahen,D. and Gysels,J. 1983 "Techniques et fonctions dans l'industrie lithique du groupe de Blicquy (Belgique)", in M.-C. Cauvin (ed.) *Traces d'utilisation sur les outils néolithiques du Proche Orient*, 37-52. Maison de l'Orient, Lyon.

Cahen,D., Keeley,L.H., Karlin,C. and Van Noten,F.L. 1980 "Méthodes d'analyse technique, spatiale et fonctionnelle d'ensembles lithiques". *Helinium*, 20, 209-259.

Cahen,D.; Keeley,L.H. and Van Noten,F. 1979 "Stone tools, toolkits and human behaviour in Prehistory". *Current Anthropology*, 20 (4): 661-684.

Camilli,E. 1989 "The occupational history of sites and the interpretation of prehistoric technological systems: an example from Cedar Mesa, Utah", in R. Torrence (ed.) *Time, energy and stone tools*, Cambridge University Press.

Campana,D.V. 1989 *Natufian and protoneolithic bone tools. The manufacture and use of bone implements in the Zagros and the Levant*, B.A.R. International Series, 494, Oxford.

Camps-Fabrer,H. and D'Anna,A. 1976 Fabrication experimental d'outils à partir de métapodes de mouton et de tibias de lapin", *Méthodologie apliquée à l'industrie de l'os préhistorique*, 311-323, Colloques Intern. du C.N.R.S..

Carneiro,R. 1979 "Tree felling with the stone axe: An experiment carried out among the Yanomano indians of Southern Venezuela", in C. Kramer (ed.) *Etnoarchaeology: Implications of Ethnography for Archaeology*, Columbia University Press, New York.

Caspar,J.P. 1985 "Etude tracéologique de l'industrie de silex du village rubané de Darion: données preliminaires". *Bulletin de la Societé royale belge d'Anthropologie et de Préhistoire*, 96, 49-79.

Caspar,J.P. and Cahen,D. 1987 "Enmanchement des outils danubiens de Belgique: données techniques et tracéologiques", in D.Stordeur (dir.) *La main et l'outil. Manches et enmanchements préhistoriques.* Travaux de la Maison de l'Orient, 15: 185-196. G.S. Maison de l'Orient, Lyon.

Castro Curiel,Z. 1990 "Información gráfica de plaquetas de Parpalló: consideraciones sobre el inicio de tecnologías vegetales", *Cypsela*, 8, 15-20.

Cauvin,M.-C. (ed.) 1983 *Traces d'utilisation sur les outils néolithiques du Proche Orient*, Maison de l'Orient, Lyon.

Chahine,C. 1991 "Le cuir: structure de la peu, composition chimique", in *Autour du cuir*, 61-75, Musée Archéologique Departamental du Val D'Oise.

Celerier and Moss,E.H. 1983 "L'abri sous-roche de Pont d'Ambon a Bourdeilles (Dordogne). Un gisement Magdalénien- Azilien: microtraces et analyse fonctionnelle de l'industrie lithique." *Gallia Préhistoire*, 26, 2, 81-109.

Clark,J.G.D. 1981 "Ethnoarchaeology in Ethiopia and its relevance for archaeological interpretation", *Préhistoire Africaine*, 6, 69-79.

Clark,J.G.D. and Kurashina,H. 1981 "A study of the work of a modern tanner in Ethiopia and its relevance for archaeological interpretation", in R.Gould and M.Schiffer (eds.) *Modern material culture: the archaeology of us*, 303-321, Academic Press, New York.

Clark,J.G.D. and Thompson,M.V. 1953 "The groove and splinter technique of working antler in Upper Paleolithic and Mesolithic Europe with special reference to the materials from Star Carr", *Proceedings of the Prehistoric Society*, 19, 148-160.

Clarke,D.L. (ed.) 1972 *Models in Archaeology*, Methuen, London.

Clarke,D.L. (1978) Analytical Archaeologist: Collected Papers of David L. Clarke. Edited and Introduced by His Colleagues. Academic Press, New York.

Clemente,I. 1995 *Instrumentos de trabajo líticos de los Yámanas (canoeros-nómadas de la Tierra del Fuego): una perspectiva desde el análisis funcional.* Tesis doctoral, Universitat Autònoma de Barcelona.

Clemente,I. (in press) "Thermal alteration on flint surfaces. The influence for the determination of use-wear traces". *VI Flint International Symposium*, Madrid, 1991.

Coles,J.M.; Heal,V.E. and Orme,B.J. 1978 "The Use and Character of Wood in Prehistoric Britain and Ireland" *Proceedings of the Prehistoric Society*, 44: 1-45.

Coles,J. 1979 *Experimental Archaeology*, Academic Press, London.

Collins,M.B. 1975 "Lithic Technology as a Means of Processual Inference", in E.Swanson (ed.), *Lithic Technology*, 15-34, Mouton, The Hague.

Coqueugniot,E. 1983 "Analyse tracéologique d'une serie de grattoirs et herminettes de Mureybet, Syrie (9eme-7eme millenaires)", in M.-C.Cauvin (ed.) *Traces d'utilisation sur les outils néolithiques du Proche Orient*, 163-172, Maison de l'Orient, Lyon.

Curwen, E.C. (1930) "Prehistoric flint sickles". *Antiquity*, 4,

Damas,D. 1968 "The diversity of Eskimo societies", in R.B. Lee and I. De Vore (eds.) *Man the Hunter*, 111-117, Aldine de Gruyter, New York.

Delluc,B. and Delluc,G. 1979 "L'acces aux parois", in Arl. Leroi-Gourhan and J. Allain (dir.) *Lascaux Inconnu*, 175-184, éd. C.N.R.S., Paris.

Delporte,H. and Mons,L. 1992 "Sagaie à base fourchue", *Fiches typologiques de l'industrie osseuse préhistorique*, Cahier I: Sagaies, Publications de l'Université de Provence.

Délaporte,Y. and Roué,M. 1978 "La preparation de la peau de renne chez les Lapons de Kautokeino", *Journal d'Agriculture traditionnelle et de Botanie*, 25, 4, 219-244.

Diamond,G. 1979 "The nature of so-called polish surface on stone artifacts", in B. Hayden (ed.) *Lithic Use-Wear Analysis*, 159-166. Academic Press, New York.

Digard,J.-P. 1981 *Techniques des nomades Baxtyari d'Iran*, Cambridge University Press. Editions de la Maison des Sciences du Homme.

Donahue,R.E. 1985 "Landscape and site function: A preliminary report on Plagicci Cave, level 4a", in S. Stoddart and C. Malone (eds.), *Papers in Italian Archaeology IV: The Cambridge Conference*, 28-49, B.A.R. International Series 244, Oxford.

Donahue,R.E. 1988 "Microwear analysis and site function of Paglicci Cave, level 4A", *World Archaeology*, 19, 3, 357-375.

Driskell,B.N. 1986 *The Chipped Stone Tool Production/Use Cycles. Its Potential in Activity Analysis of Disturbed Sites*, B.A.R. International Series, 305. Oxford.

Dumont,J.V. 1982 "The quantification of microwear traces: a new use for interferometry". *World Archaeology*, 14, 2, 206-217.

Dumont,J.V. 1983 "An Interim Report of the Star Carr Microwear Study", *Oxford Journal of Archaeology*, 2 (2) 127-145.

Dumont,J.V. 1988 *A Microwear Analysis of Selected Artefact Types from the Mesolithic Sites of Star Carr and Mount Sandel*, B.A.R. British Series 187, Oxford.

Duvois,M. 1974 "Industrie osseuse préhistorique et experimentation", in H. Camps-Fabrer (dir.) *L'industrie de l'os préhistorique*, 73-84, Publications de la Université de Provence.

Ebert,J.I. 1979 "An Etnoarchaeological Approach to reassessing the meaning of variability in stone tools assemblages", in C. Kramer (ed.) *Ethnoarchaeology: implications of Etnography for Archaeology*, 59-74, Columbia University Press, New York.

Edmonds,M. 1987 "Rocks and Risk: Problems with Lithic Procurement Strategies", in A.G. Brown and M. Edmonds,M. (eds.) *Lithic Analysis and Later British Prehistory. Some Problems and approches*, B.A.R. International Series, 162, Oxford.

Elorza,J.J. and Bustillo,M.A. 1989 "Early and late diagenetic chert in carbonate turbidities of the senonian flysch, northeast Bilbao, Spain" In J.R.Hein and J.Obradovic (ed.) *Siliceous deposits of the Tethys and Pacific regions*, 93-105. Springer-Verlag.

Elorza,J.J. and Garcia-Garmilla,F. 1993 "Chert appearance in the Cueva-Bedon carbonate platform (upper Cretaceous, northern Spain)", *Geological Magazine*, 130 (6) 805-816

Fischer,A., Hansen,P.V. and Ramussen,P. 1984 "Macro and Micro Wear Traces on Lithic Projectile Points. Experimental Results and Prehistoric Examples". *Journal of Danish Archaeology*, 3, 19-46.

Forbes,R. 1966 *Studies in ancient technology*, vol.V, E.J. Brill, Leiden.

Fujimoto,T. 1988 "Early cereal utilization? Sickle polish on microliths from the Upper- and Epi-Paleolithic assemblages from Palmyra basin, Syria", in S. Beyries (éd.) *Industries Lithiques. Tracéologie et Technologie*, vol.1, 165-174. B.A.R. International Series, 411, Oxford.

Gallagher,J.P. 1977 "Contemporary stone tools in Ethiopia: implications for archaeology", *Journal of Field Archaeology*, 4, 4, 407-414.

Gansser,A. 1951 "Les éléments du tannage primitif", *Cahiers du CIBA*, 34, 1154-1179.

Gassin,B. 1993 *Approche fonctionnelle des industries lithiques du Neolithique provençal*, Thèse de doctorat, Université Paris X.

Gibaja,J.F. 1994 *Análisis funcional del material lítico de las sepulturas de la Bóbila Madurell*. Tesina, Universitat Autònoma de Barcelona.

Geneste,J.M. 1991 "Systèmes techniques de production lithique: variations techno-économiques dans les processus de realisation des outillages paleolithiques", *Techniques et Culture*, 17-18, 1-35.

Godeau,G. 1989 "Villaines-les-Rocheurs, cité troglodytique de la vannerie", in *Tissage, corderie, vannerie*, 283-293, éd APDCA, Juan-les-Pins, France.

González Sainz,C. 1989 *El magdaleniense superior final de la región cantábrica*, ed.Tantin, Santander.

González, J.E. 1993 *Tecnología lítica y organización económica*. Tesis doctoral, Universidad de Deusto.

González,J.E. and Ibáñez,J.J. 1991 "La tecnología de talla laminar en la ocupación epipaleolítica de Berniollo (Subijana- Morillas, Alava)", in R.Mora, X.Terradas, A.Parpal and C.Plana (eds.) *Tecnología y cadenas operativas líticas*, 201-222, Universitat Autònoma de Barcelona.

González,J.E. and Ibáñez,J.J. 1994a *Metodología de análisis funcional de instrumentos tallados en sílex*, Cuadernos de Arqueología, n°14, Universidad de Deusto, Bilbao.

González,J.E. and Ibáñez,J.J. 1994b "Análisis funcional del utillaje en sílex en el yacimiento de Laminak II", *Kobie*, 21, 111-129.

Goodyear,A.C. 1979 *A Hypothesis for the Use of Cryptocrystalline Raw Materials Among Paleo-Indian Groups of North Columbia*, University of South Carolina, U.S.A.

Gould,R.A. 1970 "Spears and Spear-Throwers of the Western Desert Aborigines of Australia", *American Museum Novitat*, 2403, 1-42.

Gould,R.A. 1977 "Ethno-archaeology; or, where do models come from? A close look at Australian Aboriginal lithic technology", in R.V.S. Wright (ed.), *Stone tools as cultural markers. Change, evolution and complexity*, 163-168, Humanities Press, New Jersey.

Gould,R.A. 1980 *Living Archaeology*, Cambridge University Press, Cambridge.

Gould,R.A., Koster,D.A. and Sontz,A.H. 1971 "The lithic assemblages of the Western Desert Aborigines of Australia", *American Antiquity*, 36, 2, 149-169.

Gould,R. and M.Schiffer (eds.) 1981 *Modern material culture: the archaeology of us*, Academic Press, New York.

Grace,R. 1989 *Interpreting the Function of Stone Tools*, B.A.R. International Series, 474, Oxford.

Grace,R., Ataman,K., Fábregas,R. and Hagren,C.M.B. 1988 "A multivariate approach to the functional analysis of stone tools", in S. Beyries (edt.) *Industries lithiques. Technologie et Typologie*, 217-230, B.A.R. International Series, 411, Oxford.

Grace,R., Graham,I.D.G. and Newcomer,M.H. 1985 "The quantification of microwear polishes", *World Archaeology*, 17, 2, 112-120.

Gräslund,B. (ed.) 1990 *The interpretative possibilities of microwear studies*, A.U.N., 2, Uppsala.

Gubser, N.J. 1965 *The Nunamiut Eskimo: Hunters of Caribou*, Yale University Press, New Haven.

Gusinde,M. 1930/1982 *Los indios de Tierra de Fuego*, tomo 1, volumen 1, ed. CAEA, Buenos Aires.

Guthrie,R. 1983 "Osseous Projectile Point: Biological Considerations Affecting Raw Material Selection and Desing Among Paleolithic and Paleoindian Peoples", in *Animals and Archaeology 1. Hunters and their Prey*, 274-294, B.A.R. International Series, 163, Oxford.

Gutiérrez,C. 1993 "L'identification des activités à travers la tracéologie", in P.C.Anderson, S.Beyries, M.Otte and H.Plisson (dir.) *Traces et fonction: les gestes retrouvés*, 477-488, éd. E.R.A.U.L., 50

Gutiérrez,C., González,J.E. and Ibáñez,J.J. 1988 "Alteraciones microscópicas en el tratamiento convencional del material lítico: su incidencia en las huellas de uso", *Munibe*, suplemento 6, 83-89, Sociedad de Ciencias Aranzadi, San Sebastian.

Halasz-Csiba,E. 1991 "Transformation de la peau en cuir", in *Autour du cuir*, 27-50, Musee Archeologique Departamental du Val D'Oise.

Hardouin,J.-P. "Là où le saule devient osier", in *Tissage, corderie, vannerie*, 275-282, éd. APDCA, Juan-les-Pins, France.

Harlan,J.R. 1992 "Wild grass seed harvesting and implications for domestication", in P.C. Anderson (ed.) *Préhistoire de la Agriculture: Nouvelles approches expérimentales et Ethnographiques*. Monographies du CRA 6, Paris, Eds. du C.N.R.S.

Hatt,G. 1969 "Arctic clothing in Eurasia and America, an ethnographic study", *Arctic Anthropology*, 5, 2.

Hayden,B. 1977 "Stone tool functions in the western desert", in R.V.S. Wright (ed.) *Stone Tools as Cultural Markers. Change, Evolution and Complexity*, 178-188, Humanities Press, New Jersey.

Hayden,B. 1979 *Palaeolithic reflections: Lithic Technology and Ethnographic Excavation among Australian Aborigines*, Humanities Press, New York.

Hayden,B. (ed.) 1979 *Lithic Use-Wear Analysis*, Academic Press, New York.

Hayden,B. 1987 "Past and Present Uses of Stone Tools in the Maya Highlands", in B. Hayden (ed.) *Lithic Studies Among the Contemporary Highland Maya*, The University of Arizona Press, Tucson.

Hayden,B. (de.) 1987 *Lithic Studies Among the Contemporary Highland Maya*, The University of Arizona Press, Tucson.

Hayden,B. 1989 "From chopper to celt: the evolution of resharpening techniques", in R. Torrence (ed.) *Time, energy and stone tools*, Cambridge University Press.

Hayden,B. 1990 "The right rub: hide working in high ranking households", in B. Gräslund (ed.) *The interpretative possibilities of microwear studies*, 89-102, Uppsala.

Hayden,B. 1993 "Investigating status with hideworking use-wear: a preliminary assessment", in P.C. Anderson, S. Beyries, M. Otte and H. Plisson (eds.) *Traces et fonction: les gestes retrouvés*, E.R.A.U.L. 50, 119-130.

Hayden,B., Franco,N. and Spafford,J. 1996 "Evaluating Lithic Strategies and Design Criteria", in G. Odell (ed.) *Stone Tools Theoretical Insights into Human Prehistory*, 9-45, Plenum Press, New York and London.

Hein,J.R. and J.Obradovic (ed.) 1989 *Siliceous deposits of the Tethys and Pacific regions*, Springer-Verlag.

Heitzer,R.F. 1951 "The sickle in North America", *American Antiquity*, 3, 247-252.

Helmer,D., Villa,P. and Courtin,J. 1987 "Quelques exemples de découpe dans le Neolithique du Sud-est de la France", *Anthropozoologica*, premier numéro spécial.

Hillman,G.C. and Stuart Davies,M. 1992 "Domestication rate in wild wheats and barley under primitive cultivation", in P.C. Anderson (ed.) *Préhistoire de la Agriculture: Nouvelles approches expérimentales et Ethnographiques*, 113-158, Monographies du CRA 6, Paris, Eds. du C.N.R.S.

Hitchcock,R.K. 1982 *The ethnoarchaeology of sedentism: mobility strategies and site structure among foraging and food producing populations in the Eastern Kalahari Desert,Botswana*, UMI Dissertation Services.

Hodder,I. 1982 *Symbols in Action: Ethnoarchaeological Studies of Material Culture*, Cambridge University Press, Cambridge.

Hurcombe,L.M. 1992 *Use-wear analysis and obsidian: theory, experiments and results*, Sheffield Archaeological Monographs 4, J.R. Collis Publications, Univesity of Sheffield.

Ibáñez,J.J., González,J.E., Ruiz,R. and Berganza,E. 1993 "Huellas de uso en sílex en el yacimiento de Santa Catalina. Consideraciones sobre la manufactura del utillaje óseo y la funcionalidad dcl asentamiento", in P.C.Anderson, S.Beyries, M.Otte and H.Plisson (dir.) *Traces et fonction: les gestes retrouvés*, 225-234, éd. E.R.A.U.L., 50.

Ibáñez,J.J. and González,J.E. 1994 "Utilización de algunos cantos rodados en Laminak II", *Kobie*, 21, 132-155

Jeske,R.J. 1989 "Economies in raw material use by prehistoric hunter-gatherers" in R. Torrence (ed.) *Time, energy and stone tools*, Cambridge University Press.

Johnson,J.K. and C.A. Morrow (eds.) *The Organization of Core Technology*, Westview Press, Boulder.

Juel Jensen,H. 1982 "A preliminary analysis of blade scrapers from Ringkloster, a Danish late Mesolithic site", *Studia Praehistorica Belgica*, 2, 323-327.

Juel Jensen,H. 1994 *Flint tools and plant working*, Aarhus University Press, Denmark.

Juel Jensen,H. and Brinch Petersen,E. 1986 "A Functional Study of Lithics from Vaenget Nord, a Mesolithic Site at Vedbaek, N.E. Sjaelland", *Journal of Danish Archaeology*, 4, 40-51.

Juel-Jensen,H. 1988 "Plant harvesting and processing with flint implements in the danish stone age", *Acta Archaelogica*, 59, 131-141.

Julien,M. 1978-80 "La industria ósea de Telarmachay", *Revista del Museo Nacional de Lima*, 44, 69-93.

Julien,M. 1982 *Les harpons magdaléniens*, Gallia Préhistoire, XVIIe. supplement.

Kamminga,J. 1978 *Journey into the Microcosms: A Functional Analysis of Certain Classes of Prehistoric Australian Stone Tools*, Ph.D. thesis, University of Sydney.

Kamminga,J. 1979 "The nature of use-polish and abrasive smoothing on stone tools", in B.Hayden (ed.) *Lithic Use-Wear Analysis*, 143-157, Academic Press, New York.

Kamminga,J. 1982 *Over the Edge*, Occasional Papers in Anthropology, 12, University of Queensland, Australia.

Kamminga,J. 1988 "Wood artefacts: A checklist of plant species utilised by Australian Aborigines", *Australian Aboriginal Studies*, 2, 26-58.

Karlin,C. 1992 "Analyse d'un processus technique: le débitage laminaire des magdaleniens de Pincevent (Seine et Marne), in R.Mora, X.Terradas, A.Parpal and C.Plana (ed.) *Tecnología y cadenas operativas líticas*, 125-162, Universidad Autónoma de Barcelona.

Karlin,C., Bodu,P. and Pelegrin,J. 1991 "Processus techniques et chaînes opératoires. Comment les préhistoriens s'approprient un concept élaboré par les ethnologues", H. Balfet (ed) *Observer l'action technique. Des chaînes opératoires, pour quoi faire?*, 101-117, C.N.R.S., Paris.

Keeley,L.H. 1980 *Experimental determination of stone tool uses. A microwear analysis*. Prehistory, Archaeology and Ecology Series. The University of Chicago Press.

Keeley,L.H. 1982 "Hafting and retooling: effects on the archaeological record", *American Antiquity*, 47, 4, 798-809.

Keeley,L.H. 1983 "Neolithic novelties: the view from ethnography and microwear analysis", in M.-C. Cauvin (ed.) *Traces d'utilisation sur les outils néolithiques du Proche Orient*, 251-256, Maison de l'Orient. Lyon.

Keeley,L.H. 1987 "Hafting and 'Retooling' at Verberie". D. Stordeur (dir.) *La main et l'outil. Manches et enmanchements préhistoriques*, Travaux de la Maison de l'Orient, 15, 89-98, Maison de l'Orient, Lyon.

Keeley,L.H. 1988 "Lithic Economy, Style and Use: A Comparison of Three Late Magdalenian Sites", *Lithic Technology*, 17, 1, 19-25.

Keeley,L.H. 1992 "The use of plant food among hunter-gatherers: a cross cultural survey", in P.C. Anderson (ed.) *Préhistoire de la Agriculture: Nouvelles approches expérimentales et Ethnographiques*, 29-38, monographies du CRA 6, éds. du C.N.R.S.

Keeley,L.H. and Newcomer,M.H. 1977 "Microwear analysis of experimental flint tools: a test case", *Journal of Anthropological Science*, 4, 29-62.

Keeley,L.H. and Toth,N. 1981 "Microwear polishes on early stone tools from Koobi-Fora, Kenia". *Nature*, 293 (5832), 464-465.

Kelly,R.L. 1985 *Hunter-gatherer mobility and sedentism: a Great Basin study*, UMI Dissertation Services.

Kew,D. and Goddard,P.E. 1993 *Indian Art and Culture*, Hancock House Publishers.

Kimball,L.R. 1989 *Planning and Functional Variability in the Upper Palaeolithic: Microwear Analysis of Upper Perigordian Tools from Le Flageolet I (Dordogne)*, Ph. D. Thesis, Northwestern University, Illinois.

Kimball,L.R., Kimball,J.F. and Allen,P.E. 1995 "Microwear polishes as viewed through the atomic force microscope", *Lithic Technology*, 20, 1, 6-28.

Knutsson,K. 1988 *Making and using stone tools. The analysis of the lithic assemblages from sites with flint in Uästerbotten, northern Sweden*, AUN II, Uppsala.

Knutsson,K. 1989 "Analyse tracéologique des outillages de quartz: Les enseignements du site Néolithique moyen-tardif de Bjurselet, Suède Septentrionale", *L'Anthropologie*, 93, 3, 705-738.

Kramer ,C. (ed.) 1979 *Ethnoarchaeology: implications of Etnography for Archaeology*, 59-74, Columbia University Press, New York.

Kuoni,B. 1981 *Cestería tradicional ibérica*, Ediciones del Serbal, Barcelona.

Le Brun,J. (ed.) 1984 *Fouilles récentes à Khirokitia (Chypre), 1977-1981.*

Le Mouël,J.F. 1973 "Preparation et utilisation des peaux de phoques chez les Eskimo Naujamiut", in *L'Homme, hier et aujourd'hui.*

Lee,R.B. and De Vore,I. (eds.) 1968 *Man, the Hunter*, Aldine De Gruyter, New York.

Lee,R.B. 1979 *The !Kung San: Men, Women, and Work in a Foraging Society*, Cambridge University Press.

Le Mouël, J.F. 1973 "Preparation et utilisation des peaux de phoques chez les esquimo Naujâmiut", in *L'homme hier et aujourd'hui, recueil d'études en hommage à A. Leroi-Gourhan*, Cujas, Paris.

Lemonier,P. 1976 "La description des chaînes opératoires: contribution à l'analyse des systèmes techniques", *Techniques et Culture*, 1, 100-151.

Lemonier,P. 1983 "L'étude des systèmes techniques, une urgence en technologie culturelle", *Techniques et Culture*, 1, 11-20.

Leroi-Gourhan,A. 1964 *Le geste et la parole I. Techniques et langage*, Albin Michel, Paris.

Levi-Sala,I. 1988 "Processes of polish formation on flint tool surface", in S.Beyries (éd.) *Industries Lithiques.*

Tracéologie et Technologie, vol.II, 83-98, B.A.R. International Series, 411, Oxford.

Lobo,P.J. (1986) "El yacimiento al aire libre de 'ElAlbardón' (Tobera, Alava)". *Estudios de Arqueologia Alavesa*, 13, p.63-111.

Mansur-Franchomme,M.E. 1983 *Traces d'utilisation et technologie lithique: exemples de la Patagonie*. Thèse de 3ème. cycle, Université de Bordeaux I.

Mansur-Franchomme,M.E. 1986 *Microscopie du materiel lithique: Traces d'utilisation, alterations naturelles, accidentelles et technologiques. Exemples de Patagonie*, Cahiers du Quaternaire, 9, C.N.R.S.

Mariezkurrena,C. 1990 "Dataciones absolutas para la arqueologia vasca", *Munibe*, 42.

Mason,O.T. 1889 "Aboriginal skin-dressing", *Report of the U.S. National Museum*, 553-589.

McComb,P. (1989) *Upper palaeolithic osseous artifacts from Britain and Belgium. An inventory and technological description*. B.A.R. International Series, 481, Oxford.

McGrath,K. 1970 *A model for the use of ethnographic data in the analysis of prehistoric activities*, Master Thesis, University of Pennsylvania.

McGregor,A. 1985 *Bone, Antler, Ivory and Horn. The technology of Skeletal Materials Since the Roman Period*,

Meeks,N.D., Sieveking,G., Tite,M.S., Cook,J. 1982 "Gloss and use-wear traces on flint sickles and similar phenomena", *Journal of Archaeological Science*, 9, 317-340.

Miles,C. 1963/1986 *Indian and eskimo artifacts of North America*, Bonanza Books, New York.

Mons,L. 1972 "Notes de technologie de l'art Paléolithique (I et II), *Bulletin des Antiquités Nationales*, 4, 15-21.

Mons,L. 1992 "Sagaie à base raccourci", *Fiches typologiques de l'industrie osseuse préhistorique*, Cahier sagaies, Publications de l'Université de Provence.

Mons,L. and Stordeur,D. 1977 "Des objets només `lissoirs' de la Grotte du Placard (Charente)", *Bulletin des Antiquités Nationales*, 9, 15-25.

Mora,R., X.Terradas, A.Parpal and C.Plana (ed.) 1991 *Tecnologia y cadenas operativas líticas*, Universidad Autónoma de Barcelona.

Moss,E.H. 1983 *The functional analysis of flint implements. Pincevent and Pont d'Ambon: two case studies from the French final Palaeolithic*, B.A.R. International Series, 177, Oxford.

Moss,E.H. 1987 "A review of "Investigating microwear polishes with blind tests". *Journal of Archaeological Science*, 14 (5), 473-482, Academic Press, London.

Moss,E.H. and Newcomer,M.H. 1982 "Reconstruction of stone tool use at Pincevent: microwear and experiments", *Studia Praehistorica Belgica*, 2: 289-312.

Múgica,J.A. 1983 "Industria de hueso en la prehistoria de Guipúzcoa", *Munibe*, 35 (3-4), 451-631.

Murray,C. 1979 "Les techniques de débitage de métapodes de petits rumiants a Auvernier-Port", in H. Camps-Fabrer (dir.) *L'industrie en os et bois de cervidé durant de Néolithique et l'Age des métaux*, 27-32, éd. C.N.R.S., Paris.

Myers,A. 1987 "All shot to pieces? Interassemblage variability, lithic analysis and Mesolithic assemblage types; some preliminary observations", in A.G.Brown and M. Edmonds (eds.) *Lithic Analysis and Later British Prehistory. Some Problems and approches*, B.A.R. British Series, 162, Oxford.

Myers,A. 1989 "Reliable and maintainable technological strategies in the Mesolithic of mainland Britain", in R. Torrence (ed.) *Time, energy and stone tools*, 57-66, Cambridge University Press.

Nash,S.E. 1996 "Is Curation a Useful Heuristic?", in G. Odell (ed.) *Stone Tools Theoretical Insights into Human Prehistory*, 81-99, Plenum Press, New York and London.

Nations,J.D. and Clark,J.E. 1983 "The Bows and Arrows of the Lacandon Maya" *Archaeology*, 36, 1, 36-43.

Nelson,R.K. 1973 Hunters of the Northern Forest, 303 p., The University of Chicago Press

Nelson,M.C. 1981 *Chipped Stone Analysis in the Reconstruction of Prehistoric Subsistence Practices: An Example from Southwestern New Mexico*, Ph.D. dissertation, University of California, Santa Barbara.

Nelson,M. 1991 "The Study of Technological Organization", in M. Schiffer (ed.) *Archaeological Method and Theory*, vol. 3, 57-100, University of Arizona Press, Tucson.

Newcomer,M.H. 1974 "Study and replication of bone tools from Ksar Akil", *World Archaeology*, 6, 138-153.

Newcomer,M. 1976 "Experiments in Upper Palaeolithic Bone Work", in *Methodologie apliquée à l'industrie de l'os préhistorique*, 293-301, Colloques Intern. du C.N.R.S..

Nissen,K. and Dittemore,M. 1974 "Ethnographic data and wear pattern analysis: a study of socketed eskimo scrapers", *Tebiwa*, 17, 67-88.

O'Connell,J.F. and Hawkes,K. 1988 "Hadza hunting, Butchering and Bone Transport and their Archaeological Implications", *Journal of Anthropological Research*, 44, 2, 113-161.

Odell,G.H. 1977 *The application of microwear analysis to the lithic component of an entire prehistoric settlement: methods, problems and functional reconstructions*, Ph.D.Dissertation, University of Harvard.

Odell,G.H. 1980 "Toward a more behavioral approach to archaeological lithic concentrations", *American Antiquity*, 45, 3, 401-431.

Odell,G.H. 1981 "The mechanism of use-breakage of stone tools: some testable hypothesis", *Journal of Field Archaeology*, 8, 197-209.

Odell,G.H. 1996 "Economizing Behavior and the Concept of Curation", in G. Odell (ed.) *Stone Tools Theoretical*

Insights into Human Prehistory, 51-80, Plenum Press, New York and London.

Odell,G. (ed.) 1996 *Stone Tools Theoretical Insights into Human Prehistory*, 9-45, Plenum Press, New York and London.

Odell,G.H. and Odell-Vereecken,F. 1980 "Verifying the reliability of lithic use-wear assessments by 'blind-tests': the low-power approach", *Journal of Field Archaeology*, 7, 87-120.

Olsen,S.L. 1979 "A Study of Bone Artifacts from Grasshopper Pueblo", *The Kiva*, 44, 4, 341-373.

Ortiz,L.; Vivanco,J.J.; Ferreira,A.; Lobo,P.; Pinillos,R.; Tarriño,J.M.; Tarriño,A. and Muñoz,M.D. (1986) *El habitat en la prehistoria en el valle del rio Rojo (Alava)*, Cuadernos de Seccion Prehistoria-Arqueología. Sociedad de Estudios Vascos.

Osgood,C. 1940 *Ingalik material culture*, Yale University Press.

Parry,W.J. and Kelly,R.L. 1987 "Expedient Core Technology and Sedentism", in J.K. Johnson. and C.A. Morrow (eds.) *The Organization of Core Technology*, 285-304, Westview Press, Boulder.

Pecheux,H. 1922 *Le Cuir. Les Os, l'Ivore, la Corne, l'Ecaille, la Nacre et les Perles, le Corail.*

Pelegrin,J., Karlin,C. and Bodu,P. 1988 "Chaines operatoires": un outil pour le préhistorien", in Tixier,J. (dir.) *Notes et Monographies Techniques*, 25, 55-62, C.N.R.S.

Peltier,A. and Plisson,H. 1986 "Micro-tracéologie fonctionnelle sur l'os: quelques résultats expérimentaux", in E. Aguirre and Patou,M. (dir.) *Outillage peu elaboré en os et en bois de cervidé*, 69-80, Centre d'Études et de Documentation Archéologique, Viroinval, Belgique.

Perlès,C. 1992 "In Search of Lithic Strategies. A Cognitive Approach to Prehistoric Chipped Stone Assemblages", in J.C. Gardin and C. Peebles *Representation in Archaeology*, 223-247, Indiana University Press.

Perlés,C. and Vaughan,P.C. 1983 "Pièces lustrées, travail des plantes et moissons à Franchthi (Grèce) (Xème-IVème mill. B.C.)" in M.-C. Cauvin (éd.) *Traces d'utilisation sur les outils néolithiques du Proche Orient*, 209-224. Lyon.

Peyrony,D. 1933 "De l'usage présumé des lamelles droites ou courbées à bord abattu du Paléolithique Supérieur et du Mésolithique". *A.F.A.S.*, Congrès de Chambéry (57th. session).

Philibert,S. 1993 "Quelle interprétation fonctionelle pour les grattoirs ocrés de la Balma Margineda (Andorre)?", in P.C.Anderson, S.Beyries, M.Otte and H.Plisson (dir.) *Traces et fonction: les gestes retrouvés*, ed. E.R.A.U.L., 50, 131-137.

Pigeot,N. 1987 *Magdaléniens d'Etiolles: Economie du débitage et organisation sociale (L'unité d'habitation U5)*, Gallia Préhistoire, Supplément 25, éd. C.N.R.S..

Plisson,H. 1985 *Etude fonctionnelle d'outillages lithiques préhistoriques par l'analyse des micro-usures: recherche méthodologique et archéologique*, Thèse de 3ème. cycle, Université de Paris I.

Plisson,H. 1987 "L'emmanchement dans l'habitation n 1 de Pincevent", in D. Stordeur (dir.) *La main et l'outil. Manches et enmanchements préhistoriques*, 75-88, Travaux de la Maison de l'Orient, 15, Lyon.

Plisson,H. 1988 "Technologie et tracéologie des outils lithiques mousteriens en Union Sovietique: les travaux de V.E.Shchelinskii", in *L'Homme de Néandertal*, vol.4, La Technique, 121-168, ERAUL 35, Liège.

Plisson,H. 1991 "Le cuir au Paleolithique", in *Autour du cuir*, 7-19, Musee Archeologique Departamental du Val D'Oise.

Ploux,S. 1990 "Technologie, technicité, techniciens: Méthode de determination d'auteurs et comportements techniques individuels", in *25 ans d'études technologiques en préhistoire*, 201-214, XI Rencontres Internacionales d'Archéologie et d'Histoire d'Antibes, éd. APDCA, Juan-les-Pins, France.

Poplin,F. 1974 "Deux cas particuliers de débitage par usure" in *L'industrie de l'os dans la Préhistoire*

Reed,R. 1972 *Ancient Skins Parchments and Leathers*, Seminar Press, London and New York.

Richardson,N. 1988 "Selected filmography on Aboriginal woodworking", *Australian Aboriginal Studies*, 2, 56-59.

Rivera,D. and Obón,C. 1991 *La guía de INCAFO de las plantas útiles y venenosas de la Península Ibérica y Baleares (excluidas medicinales)*, INCAFO S.A., Madrid.

Rodríguez,A. 1993 "Analyse fonctionelle d'outils lithiques en basalte de l'île de La Palma (îles Canaries): premiers résultats", in P.C.Anderson, S.Beyries, M.Otte and H.Plisson (dir.) *Traces et fonction: les gestes retrouvés*, ed. E.R.A.U.L., 50, 295-302.

Rodríguez,A. 1993 "L'analyse fonctionnelle de l'industrie lithique du gisement epipaléolithique/ mesolithique d'El Roc de Migdia (Catalogne). *Prehistoire européenne*, 4, p. 63-84

Rodríguez Santana,C. 1989 "Tejidos, cestería y cordelería en la Prehistoria de Gran Canaria", in *Tissage, corderie, vannerie*, 81-94, ed. AP.D.C.A., Juan-les-Pins, France.

Roux,V. 1985 *Le matériel de broyage à Tichitt (Mauritanie). Etude ethno-archéologique*, Editions Recherches sur les Civilisations, 58, Paris.

Rueda,J.M. 1985 "El treball de les matèries dures animals al Paleolític Superior del Reclau Viver", *Cypsela*, 5, 7-20. Centre d'Investigacions Arqueologiques de Girona, Girona.

Rueda,J.M. 1993 *L'acció antròpica sobre les matéries dures animals durant el Pleistocè del Nord-est de Catalunya*, Tesis doctoral, Univeritat de Girona.

Sahlins,M.D. 1968 *Tribesmen*, Prentice-Hall, New Jersey.

Schaefer,C. 1973 *The Indian Art of Tanning Buckskin*.

Schick,T. 1989 "Early Neolithic twined Basketry and Fabrics from the Hemar Cave, Israel", in *Tissage, corderie et vannerie*, 41-52, APDCA, Juan-les-Pins, France.

Schiffer,M.B. 1976 *Behavioral Archaeology*, Academic Press, New York.

Schiffer,M.B. (ed.) 1991 *Archaeological Method and Theory*, vol. 3, University of Arizona Press, Tucson.

Semenov,S.A. 1964 *Prehistoric technology. An experimental study of the oldest tools and artifacts from traces of manufacture and wear*. Cory, Adams and Mackay, London.

Senépart,I. 1992 *Les industries en matière dure animale de l'Epipaléolithique au Néolithique Final dans le Sud-est de la France*, Thèse de 3ème. cycle, Université de Paris X.

Shea,J.J. 1988 "Metodological considerations affecting the choice of analytical techniques in lithic use-wear analysis: test, result and applications", in S. Beyries (ed.) *Industries Lithiques. Tracéologie et Technologie*, vol. 2, 65-82, B.A.R. International Series, 411, Oxford.

Shennan,S.J. (ed.) 1989 *Archaeological approaches to cultural identity*, Unwin Hyman, London.

Shott,M.J. 1986 "Technological Organization and Settlement Mobility: an Ethnographic Examination", *Journal of Anthropological Research*, 42, 15-51.

Sidera,I. 1993 *Les assemblages osseux en Bassin Parisien et Rhenan du VIe. Millenaire B.C. Histoire, techno-economie et culture*, Thèse de doctorat, Université de Paris I.

Sieveking,G.de G. and M.H. Newcomer (eds.) 1987 *The human uses of flint and chert*, Cambridge University Press.

Sigaut,F. 1985 "Moisson et fenaison", *Nouvelles de l'Archeologie*, 19, 28-38.

Smith,A.B. and Poggenpoel,C. 1988 "The technology of bone tool fabrication in the South-Western Cape, South Africa", *World Archaeology*, 20, 1, 103-114.

Spencer,B. and Gillen, F.J. (1927) *The Arunta*. Mcmillan and Co., Londres.

Spiess,A.E. 1979 *Reindeer and Caribou Hunters: An Archaeological Study*, Academic Press, London.

Steinbring,S. 1966 "The manufacture and use of bone defleshing tools", *American Antiquity*, 31, 4, 575-581.

Stewart,H. 1973 *Artifacts of the Northwest Coast Indians*, Hancock House Publishers.

Stoddart,S. and C. Malone (eds.) 1985 *Papers in Italian Archaeology IV: The Cambridge Conference*, B.A.R. International Series 244, Oxford.

Stordeur,D. 1976a "Les poinçons d'os à poulie articulaire: observations techniques d'après quelques exemples syriénnes", *Bulletin de la Société Préhistorique Française*, 73, 2, 39-42.

Stordeur,D. 1976b "La fabrication des aiguilles à chas. Observation et expérimentation", *Méthodologie apliquée à l'industrie de l'os préhistorique*, 251-256, Colloques Intern. du C.N.R.S..

Stordeur-Yedid,D. 1979 *Les aiguilles à chas au Paléolithique. 13 supplément à Gallia Préhistoire*, C.N.R.S..

Stordeur,D. 1983 "Quelques remarques pour attirer l'attention sur l'intérêt d'une recherche commune entre tracéologues du silex et technologues de l'os", in M.-C. Cauvin (ed.) *Traces d'utilisation sur les outils néolithiques du Proche Orient*, 231-240, Maison de l'Orient, Lyon.

Stordeur,D. 1984 "L'industrie osseuse de Khirokitia", in J.Le Brun (ed.) *Fouilles récentes à Khirokitia (Chypre), 1977-1981.*

Stordeur,D. (dir.) 1987 *La main et l'outil. Manches et emmanchements préhistoriques*, Travaux de la Maison de l'Orient, 15, Lyon.

Stordeur,D. 1989 "Vannerie et tissage au Proche Orient néolithique: IX-V millenaire", in *Tissage, corderie, vannerie*, 19-40, éd APDCA, Juan-les-Pins, France.

Stordeur,D.; Anderson-Gerfaud,P.C.; Deraprahmian,G.; Reynolds,P.; Le Ribault,L. 1989 "Les omoplates encochées néolithiques de Ganj-Dareh (Iran): étude morphologique et fonctionnelle" *Cahiers de l'Euphrate*, 4

Sussman,C. 1985 "Microwear on quartz: fact or fiction?". *World Archaeology*, 17 (1), 101-111.

Sussman,C. 1987 "Résultats d'un étude des microtraces d'usure sur un echantillon d'artefacts d'Olduvai (Tanzanie)", *L'Anthropologie*, 91, 2, 375-380.

Sussman,C. 1988 "Aspects of microwear as applied to quartz", in Beyries,S. (ed.) *Industries Lithiques. Traceologie et Technologie*. vol.II, p.3-28, B.A.R. International Series, 411, Oxford.

Swanson,E. (de.) 1975 *Lithic Technology*, Mouton, The Hague.

Symens,N. 1986 "A Functional Analysis of Selected Stone Artefacts from the Magdalenian Site at Verberie, France", *Journal of Field Archaeology*, 13, 213-222.

Tepper,L.H. 1994 *Earth Line and Morning Star. Nlaka'pamux Clothing Traditions*, Canadian Museum of Civilization, Quebec.

Thompson,D.F. 1964 "Some wood and stone implements of the Bindibu tribe of Central Western Australia", *Proceedings of the Prehistoric Society*, 30, 400-422.

Torrence,R. 1983 "Time budgeting and hunting-gathering economy", in Bailey G. (ed.) *Hunter-Gatherer Economy in Prehistory: A European Perspective*, 11-22, Cambridge University Press.

Torrence,R. 1986 *Production and Exchange of Stone Tools*, Cambridge University Press.

Torrence,R. 1989 "Re-tooling: towards a behavioral theory of stone tools", in R. Torrence,R. (ed.) *Time, energy and stone tools*, Cambridge University Press.

Torrence,R. (ed.) 1989 *Time, Energy and Stone Tools*, Cambridge University Press.

Tringham,R., Cooper,G., Odell,G.H., Voytek,B. And Whitman,A. 1974 "Experimentation in the formation of edge-damage: a new approach to lithic analysis". *Journal of Field Archaeology*, 1, 171-196.

Unger-Hamilton,R. 1988 *Method in microwear analysis. Prehistoric sickles and other tools from Arjoune, Syria*, B.A.R. International Series, 435, Oxford.

Unger-Hamilton,R. 1989 "Epipaleolithic Palestine plant cultivation: the evidence from harvesting experiments and microwear study", *Current Anthropology*, 30, 1, 88-103.

Van Gijn,A.L. 1989 *The wear and tear of flint*, Analecta Praehistorica Leidensia, 22, University of Leiden.

Van Gijn,A.L. in press "The representativity of microwear traces: implication for archaeological inferences". *VI Flint International Symposium*, Madrid, 1991.

Van Noten,F., Cahen,D., Keeley,L.H. and Moeyersons,J. 1978 *Les chasseurs de Meer*. Dissertationes Archaeologicae Gandenses XVIII, 2 vols. De Tempel, Brugge.

Vaughan,P.C. 1985 *Use-wear analysis of flaked stone tools*, The University of Arizona Press.

Vaughan,P.C. 1986 "A sampling method for use-wear analysis of large flint assemblages", *Early Man News*, 9/10/11, 183-186.

Vaughan,P.C. and Bocquet,A. 1987 "Première étude fonctionnelle d'outils lithiques néolithiques du village de Charavines, Isère", *L'Anthropologie*, 91, 2, 399-410, Masson, Paris.

Velter,A. and Lamothe,M.J. 1976 *Le livre de l'outil, 1. La terre, le bois*, éd Hier et Demain, Paris.

Victor,P.E. and Robert-Lamblin,J. 1989 *La civilisation du phoque*, A. Colin and R. Chabaud, Paris.

Vila,A. 1987 "L'assentament paleolitic del Castell (Vilanova de Sau)", *Cypsela*, 6, 111-123. Centre d'Investigacions Arqueologiques de Girona, Girona.

Vila,A. and Gallart,F. 1991 "Caracterización de los micropulidos de uso: ejemplo de de aplicación del análisis de imágenes digitalizadas", in P.C. Anderson, S. Beyries, M. Otte and H. Plisson (eds.) *Traces et fonction: les gestes rétrouvés*, E.R.A.U.L. 50, 459-466.

Walker,P.L. 1978 "Butchering and stone tool function", *American Antiquity*, 43, 710-715.

Watanabe,H. 1968 "Subsistence and ecology of northern food gatherers with special reference to the Ainu", in R.B. Lee and I. De Vore (eds.) *Man the Hunter*, 49-67.

White,J. And Thomas,D. 1972 "What Mean these Stones?: Ethno-taxonomic Models in Archaeological Interpretations in The New Guinea Highlands", in D.L. Clarke (ed.) *Models in Archaeology*, 275-308, Methuen, London.

Wilke,P.J. 1988 "Bow Staves Harvested from Juniper Trees by Indians of Nevada", *Journal of California and Great Basin Anthropology*, 10, 1, 3-31.

Winiarska-Kabacinska,M. 1988 "Examination of hafting traces on end-scrapers concentration III/75 from the site of Wojnovo 'a', (Western Poland)", in S. Beyries (éd.), *Industries lithiques. Tracéologie et technologie*, vol.I, 47-52, B.A.R. International Series, 411, Oxford.

Wright,R.V.S. (ed.), 1977 *Stone tools as cultural markers. Change, evolution and complexity*, Humanities Press, New Jersey.

Yamada,S. 1993 "The formation process of use-wear polishes", in P.C.Anderson, S.Beyries, M.Otte, and H.Plisson (eds.) *Traces et fonction: les gestes retrouvés*, E.R.A.U.L., 50, 433-446.

Yamada,S. and Sawada,A. "The method of description for polished surfaces", in P.C.Anderson, S.Beyries, M.Otte and H.Plisson (dir.) *Traces et fonction: les gestes retrouvés*, 447-458, E.R.A.U.L., 50.

Yellen,J.E. 1977 *Archaeological approaches to the present. Models for reconstructing the past*, Academic Press, New York.

RESUMEN. SPANISH SUMMARY.

Este trabajo está orientado hacia la reconstrucción de las estrategias económicas de los grupos de cazadores-recolectores que vivieron en torno al final del paleolítico en el País Vasco. El estudio se basa en el material recuperado en los niveles magdaleniense y aziliense de la cueva de Santa Catalina (Lekeitio, Bizkaia) y en el nivel epipaleolítico del yacimiento al aire libre de Berniollo (Subijana, Álava). Estas ocupaciones se produjeron entre hace 12.000 y hace 10.000 años.

La información empleada procede del estudio del uso de los útiles de piedra tallada: cuáles eran los útiles y cómo se usaron, qué materiales trabajaron y de qué forma. Los criterios empleados en el análisis de las huellas de uso se exponen de forma explícita en el primer capítulo.

Esta información funcional es relevante para la cuestión pero no es suficiente. Dos limitaciones son las más claras. Por una parte, no disponemos de datos sobre otros comportamientos técnicos y económicos que podrían deducirse de los restos materiales. Por otra, las actividades de los grupos de cazadores-recolectores son variables en el tiempo y en el espacio. Para comprender la organización económica de estos grupos es preciso conocer esta variabilidad de comportamientos en un número amplio de ocupaciones, lo que implica el estudio de una mayor cantidad de yacimientos.

Debido a estas dos restricciones hemos decidido centrar el estudio en la siguiente cuestión: la información sobre el uso de los útiles de piedra tallada, ¿cómo puede ayudarnos a comprender la organización económica de los grupos de cazadores-recolectores? Lo que proponemos por tanto es una estrategia para "leer" los resultados del análisis funcional desde esta perspectiva.

Para ello hemos establecido tres niveles de interpretación. En el **primer nivel** relacionamos las características del utillaje -la materia prima en la que se fabrica, el soporte elegido, el tipo de retoque y de morfotipo- con las características del uso en el que participa el útil: la materia trabajada, la actividad, la intensidad y el contexto de uso. Este nivel corresponde, al menos en parte, con los enfoques interesados en la economía de materias primas y de productos y en las estrategias de aprovechamiento del utillaje.

Algunos de los rasgos observados en este nivel de interpretación son comunes para los tres conjuntos estudiados. En prácticamente todos los casos se emplea el frente del raspador, para raspar, sobre todo piel, tanto cuando la materia está fresca como seca. También se emplean a menudo otras zonas de la pieza, casi siempre en la misma materia y estado que la trabajada con el frente en lo que parecen ser labores complementarias a la labor principal de raspado. Los buriles se emplean sobre todo para raspar materias óseas -asta y hueso- con los fuertes ángulos de las facetas creadas por el golpe de buril; menos habituales son las labores de ranurado, realizadas con el bisel.

En las piezas truncadas se ha podido establecer una distinción: unas fueron usadas por la parte truncada para el raspado de materias duras, sobre todo óseas; otras, con truncadura cóncava y sobre lámina, fueron empleadas en carnicería y en el corte de materias blandas. En estas últimas, por la disposición de las huellas, deducimos que la truncadura es una adaptación para favorecer el enmangue.

El utillaje de dorso, tanto puntas como laminillas, también muestra una división funcional. Una parte de estos útiles son elementos de proyectil mientras otras piezas participan en el procesado de materias blandas, sobre todo carnicería y corte de piel. En las puntas esta diferencia se establece entre puntas regulares y finas, tipo "microgravette" por una parte y puntas de tipo aziliense por otro. En las laminillas de dorso la diferencia se asocia al tamaño: las más pequeñas funcionan a menudo como barbas de proyectil mientras las mayores se emplean más como cuchillos de carnicería o en el corte y raspado de la piel.

El análisis funcional de las piezas no retocadas ha proporcionado asimismo interesantes informaciones. Fue usada una pequeña parte de las lascas brutas, elegidas entre las de mayor tamaño. La materia trabajada con estas piezas fue precisamente aquella que más se trabajó en cada conjunto: las materias óseas en el nivel magdaleniense de Santa Catalina; la carne y la piel fresca en el nivel aziliense y la madera en Berniollo.

Las láminas brutas fueron empleadas más a menudo que las lascas; en el nivel aziliense de Santa Catalina se llegó a usar algo más de la mitad de estas láminas no retocadas. En los tres conjuntos, la actividad dominante entre estas piezas fueron las labores de carnicería y el corte de piel, aunque también participaron en otros trabajos variados como el raspado de diferentes materias duras. También en este caso el tamaño de los soportes ha influido en la selección de los soportes para el uso, en favor de las más grandes.

Las laminillas brutas, por el contrario, se emplearon de forma marginal. Tan sólo en el nivel aziliense de Santa Catalina se recurrió a algunas de las mayores quizá debido a la relativa escasez de láminas en este conjunto.

Hemos observado otras elecciones técnicas generales, relacionadas esta vez con la materia prima y el tipo de soporte en el que se fabricaron los útiles. Entre las más pronunciadas se encuentra el tratamiento de los diferentes tipos de sílex en Berniollo. Allí se aprecia que el sílex de mayor calidad, de origen probablemente más lejano, se explota con una talla más cuidada que produce las láminas de mayor tamaño. Los útiles retocados a partir de estas láminas, por ejemplo los raspadores, son los que se enmangan de forma preferente. Una parte de las láminas producidas en estos núcleos se han conservado enteras y sin usar; otra parte además ha sido probablemente aportada para ser usada fuera del asentamiento.

El tipo de soporte -lasca, lámina,...- en el que se fabrican los útiles influye también en el destino que reciben. En general, se observa que los útiles sobre lámina se destinan a las labores que tienen más importancia en cada conjunto. En el nivel aziliense de Santa Catalina los raspadores sobre lámina se destinan al raspado de la piel fresca, una de las labores principales llevadas a cabo en este nivel; en el nivel magdaleniense, por el contrario, los raspadores sobre lámina participan sobre todo en el trabajo de la piel seca, que es una de las fases de trabajo más importantes en este conjunto.

En un **segundo nivel** pretendemos reconstruir las actividades llevadas a cabo con los útiles de piedra: qué materiales fueron trabajados, de qué forma y con qué objetivo. Para ello hemos hecho uso de los conceptos de cadena operativa y de proceso técnico y hemos recurrido de manera extensiva a las referencias arqueológicas y etnográficas sobre las técnicas de trabajo de los grupos de cazadores-recolectores. La documentación y estructuración de los procesos técnicos generales en materias tales como las óseas, la piel o la madera crea un marco donde es posible comprender el sentido que tienen los trabajos reconocidos en los instrumentos de piedra. Para ello hemos establecido las actividades que tienen lugar en cada una de las fases de captación, de transformación o de mantenimiento de los productos elaborados en cada materia. Así es posible relacionar más tarde el trabajo reconocido en un útil de piedra con una fase más o menos concreta del proceso técnico general. En este punto se emplean criterios basados en el tipo de materia trabajada, el tipo de actividad, la intensidad de uso, la variedad de actividades entre el conjunto de útiles, la asociación de actividades en el mismo útil, la longitud de las zonas activas, el tamaño de los útiles y el uso de instrumentos o zonas activas específicas.

A partir de los resultados obtenidos en el conjunto del utillaje es posible entrever los rasgos que caracterizan el trabajo de cada una de las materias. La información sobre el trabajo de la piel proporciona algunas de las referencias más interesantes. Así, en el nivel aziliense de Santa Catalina domina el trabajo de la piel aún fresca, es decir la limpieza de la piel más inmediata a la captura del animal. Sin embargo, el trabajo sobre piel que se lleva a cabo en Berniollo es radicalmente distinto; la fase de limpieza de la piel está casi ausente y lo que domina es el raspado de piel seca, para adelgazarla o flexibilizarla, la fase más costosa en términos de tiempo, energía y utillaje en el procesado de este material.

El trabajo de materias óseas en los tres conjuntos refleja una transformación bastante compleja del asta en todos ellos, en la que participan sobre todo útiles específicos como los buriles. Estos útiles son empleados intensamente, por varias zonas activas distintas y con una gran variedad de actividades, desde el ranurado y la perforación hasta el corte y el raspado. Mayores diferencias se han visto en el trabajo del hueso. Por ejemplo, en el conjunto aziliense de Santa Catalina este material se trabaja de forma sencilla y poco intensa -quizá en labores de mantenimiento y reparación del utillaje. Por el contrario, en el nivel magdaleniense los útiles muestran evidencias de casi todas las fases del proceso de fabricación. Los restos óseos conservados en

Santa Catalina ofrecen un panorama coherente con estas deducciones.

En general, las óseas y la piel son las materias más trabajadas con los útiles de piedra. El trabajo de madera sólo es importante en Berniollo donde, a pesar de realizarse con útiles poco específicos -dominan las lascas no retocadas- debe tratarse de la fabricación de productos bastante elaborados como prueba la variedad e intensidad de las actividades sobre esta materia. Las labores de carnicería están presentes en los tres conjuntos, aunque destaca la importancia que cobran entre el utillaje del nivel aziliense de Santa Catalina. Como es habitual en contextos de cronología similar, el trabajo de vegetales no leñosos está poco extendido.

En un **tercer nivel** intentamos reconstruir las estrategias económicas de los grupos estudiados. Nos centramos sobre todo en el grado y en el tipo de planificación de las actividades de captación y transformación de los diferentes recursos.

Para ello empleamos la información reunida en los dos niveles anteriores. En el primero habíamos visto cómo se gestionaban los útiles, con qué criterios, para atender a qué necesidades. En el segundo hemos puesto de manifiesto, para cada conjunto estudiado, qué fases de transformación de las materias trabajadas están ausentes o presentes, y con qué intensidad. Estas elecciones tienen una lógica que puede establecerse gracias a la información etnográfica sobre las estrategias económicas de los grupos de cazadores-recolectores. "Estrategias económicas" quiere decir en este contexto cómo se organizan en el tiempo y en el espacio la captación, la transformación y el consumo de los recursos y cómo se organiza el propio grupo para realizar estas labores.

Las limitaciones para alcanzar este nivel de interpretación son aún mayores. Algunas son teóricas y aquí se discuten diferentes enfoques sobre la organización de los cazadores-recolectores, subrayándose la variabilidad entre diferentes grupos humanos y la adaptabilidad a circunstancias cambiantes que son propias de estos sistemas. Otras se refieren al limitado número de yacimientos estudiados o a la ausencia en este estudio de otras informaciones relevantes que proporciona el registro arqueológico -el aprovisionamiento de materias primas, el tipo de fauna cazada y la forma de su procesamiento o las distribuciones macro- y microespaciales por no citar más que algunas de ellas.

Sin embargo, es posible una primera síntesis que explica desde este punto de vista los comportamientos observados en el uso de los útiles.

En el nivel aziliense de Santa Catalina, son especialmente abundantes las evidencias de proyectiles usados, de carnicería y del trabajo de la piel fresca, es decir, el conjunto de las labores de captación y la primera fase de procesado de las materias animales. Estas actividades parecen reflejar el principal motivo de la ocupación del asentamiento. A ellas se dedica un utillaje comparativamente más cuidado. El resto de los trabajos sobre otros materiales, sobre todo hueso y madera, se reduce al mantenimiento para no interferir las labores principales durante la ocupación. La imagen que ofrece es la que se genera en un asentamiento especializado dentro de una estrategia planificada, en la que se segmentan las actividades. Algunas se aplazan o reservan para otro tiempo o lugar para no interferir el trabajo principal que motiva la ocupación.

En Berniollo, por el contrario, las actividades de captación y primera transformación, aunque presentes, están poco desarrolladas. Dominan los trabajos destinados a la fabricación de productos más o menos complejos en madera y en materias óseas; junto a ellos, también tiene importancia la fase de trabajo de la piel más costosa, el acondicionamiento de la piel seca por raspado. El estudio paralelo de la fabricación del utillaje lítico mostró la producción de utillaje abundante y variado incluyendo soportes de calidad y útiles destinado a un uso diferido o en el exterior del asentamiento. En conjunto, el asentamiento de Berniollo parece haber funcionado como un lugar donde se centralizan las actividades de manufactura más complejas de una gran variedad de productos, todo ello en detrimento de las labores de captación de recursos.

El nivel magdaleniense de Santa Catalina proporciona una imagen diferente a la vez de la de los otros dos conjuntos. Están presentes todas las fases de transformación de las diferentes materias, desde las primeras hasta las finales -se raspa tanto la piel fresca como seca, por ejemplo- pero los trabajos de captación o de carnicería también son importantes. El conjunto de actividades realizadas por los útiles de piedra reflejaría un asentamiento sin función definida. Esto puede ser debido al tipo de organización del grupo, con escasa planificación y segmentación de las actividades pero también cabe la alternativa de que se trate de una indefinición producida por la superposición de ocupaciones con finalidades diferentes, un efecto palimpsesto.

LABURPENA. BASQUE SUMMARY.

Paleolitikoaren amaieran Euskal Herrian bizi izan ziren ehiztari-fruitubiltzaile taldeen estrategia ekonomikoak ezagutzea da lan honen xedea. Santa Katalina (Lekeitio, Bizkaia) kobako magdalen-aro eta azil-aroko mailetatik atera den materialean oinarritzen da azterketa hau, baita Arabako Subijana udalerrian haize agerian dagoen Berniollo aztarnategiko epipaleolitiko mailako materialean ere, hau da, orain 12.000-10.000 urte okupatuta zeuden lurzoruetan aurkitutako aztarnetan.

Landutako harrizko tresnen erabilera estudiatu ondoren eskuratutakoa da agertzen den informazioa: zeintzuk ziren tresnak eta nola erabili ziren, zernolako materialak landu zituzten eta nola. Erabilera-aztarnen analisian erabiltzen diren irizpideak beren beregi azaltzen dira lehenengo kapituluan.

Funtzioari buruzko informazio hori garrantzitsua da, ez ordea nahikoa, gai honetarako. Bi dira mugarik argienak. Alde batetik, ez daukagula daturik aztarna material hauetatik ondoriozta litezkeen beste portaera tekniko eta ekonomiko batzuri buruz. Beste alde batetik, ehiztari-fruitubiltzaile taldeen jardunak aldatu egiten direla denboran eta espazioan zehar. Talde hauen antolakuntza ekonomikoa ezagutu nahi izanez gero iharduerak nola aldatzen diren ikusi beharko dugu ainitz okupaziotan, eta horrek aztarnategi kopuru handiagoa estudiatzera behartzen gaitu.

Bi muga hauek direla eta, estudio honetan honako gai honi ekitea erabaki dugu: landutako harrizko tresnen erabilerari buruzko informazioak, nola lagun liezaiguke ehiztari-fruitubiltzaile taldeen antolakuntza ekonomikoa ulertzera? Proposatzen duguna, beraz, analisi funtzionalaren emaitzak ikuspuntu honetatik "irakurtzeko" estrategia da.

Horretarako hiru interpretazio-maila ezarri ditugu. **Lehen mailan** uztartu egin ditugu tresneriaren ezaugarriak (fabrikazioan erabilitako lehengaia, aukeratutako euskarria, ukitu-mota eta morfotipoa) eta lanabesak eduki litzakeen erabileraren ezaugarriak (landu duen materia, burutu duen iharduera, erabilera-neurria eta erabileraren testuingurua). Maila hau, hein batean, lehengaien eta ekoizkinen ekonomia eta tresnerien aprobetxamendurako estrategiak behatzen dituzten ikuspuntuei dagokie.

Interpretazio-maila honetan behatzen diren zenbait ezaugarri amankomunekoak dira estudiatutako hiru aztarnategientzat. Harraskaren muturra harraskatzeko erabiltzen da ia kasu guztietan, eta batez ere larrua harraskatzeko, bai fresko bai lehor dagoenean material hori. Sarritan harraskaren beste gune batzuk ere erabiltzen dira harraskaketa lana osatzen omen duten beste eginbehar batzutarako, harraska-muturra erabili izan den material eta egoera berean. Burilak hezur-materiak, hau da, adarrak eta hezurrak harraskatzeko erabili ohi dira batez ere, buril-kopeez sortutako angelu zorrozdun fazetak dituztelako; urriagoak dira, ordea, burilen biselez egindako zirrikituak.

Pieza moztuetan bereizketa bat egin da: batzuk, materia gogorrak - hezurrezko materiak, batik bat- harraskatzeko erabili izan ziren beren gune moztuez baliatuz; beste batzuk, berriz, mozketa konkabodunak, laminetan eginak, aragia eta materia bigunak ebakitzeko erabili izan ziren. Azken hauetan agertzen diren aztarnen finkalekua kontuan hartzen badugu, mozketa girtena jarri ahal izateko egokitzapena dela ondoriozta dezakegu.

Ukitu bortitzez ertza landuta daukaten tresnek ere, ("dorsodunak") bai puntak eta bai laminatxoak, bereiztuta daude funtzioari dagokionean. Tresna hauetako batzuk jaurtikigaiak dira, beste batzuk, aldiz, materia bigunen prestaketarako erabiltzen dira, aragia eta larrua ebakitzeko, batik bat. Punten arteko bereizketari dagokienean, alde batetik punta erregular finak, "microgravette" motakoak daude, eta beste alde batetik azil-aroko puntak. Dorsodun laminatxoetan tamainak ezartzen du diferentzia: txikienak jaurtikigaien alboetako hortz modura erabiltzen ziren; handienak, berriz, aragia ebakitzeko edota larrua harraskatzeko ganibet gisa.

Era berean, ukiturik gabeko piezen funtzioaren analisiak informazio interesgarriak eman dizkigu. Landu gabeko zenbait harri-printza erabili ziren, handienetatik aukeratuak, hain zuzen ere. Pieza hauek landu zuten materia aztarnategi bakoitzean gehien landu zen materia izan zen, alegia: hezur-materiak Santa Katalinako magdalen-aroko mailan; aragia eta larru freskoa azildar mailan eta zura Berniollon.

Landu gabeko laminak printzak baino sarriago erabili izan ziren; Santa Katalinako azildar mailan ukiturik gabeko lamina hauen erdia baino zeozer gehiago erabili zen. Hiru aztarnategietan, pieza hauen egiteko nagusia aragia eta larrua ebakitzea izan zen, nahiz eta beste zenbait lanetan ere parte hartu, adibidez materia gogor desberdinen harraskaketa-lanetan. Kasu honetan ere euskarrien neurria hartu da kontuan hautapena egiterakoan, handienak aukeraru direlarik.

Landu gabeko laminatxoak, aitzitik, noiz behinka baino ez ziren erabili. Santa Katalinako azildar mailan baino ez ziren haietako handienez baliatu, agian koba honetan lamina eskazia zegoelako.

Beste hautaketa tekniko orokor batzuk ere behatu ditugu, lanabesak fabrikatzeko hartu zen lehengaia eta euskarriarekin lotuta daudenak, alegia. Nabarmenenen artean, Berniollon sukarri-mota desberdinei egokitzen zaien xede desberdina aipa genezake. Kalitaterik handieneko sukarria -seguraski urrundik ekarria- taila txukunagoaz lantzen da eta lamina handiagoak ateratzen dira bertatik. Lamina hauek euskarri izanik ukitzen diren lanabesei, harraskei adibidez, lehentasunez jartzen zaie girtena. Nukleo hauetatik ateratako lamina-multzo bat bat oso-osorik eta erabili gabe kontserbatu da; beste multzo bat, seguraski, aztarnategitik at erabiltzeko eramango zen kanpora.

Lanabesak egiteko aukeratzen diren euskarri mota desberdinak ere -printzak, laminak, e.a.- lanabes horiek edukiko dituzten betebeharrarekin lotuta egongo dira. Orohar esan genezake, aztarnategi bakoitzean garrantzirik handiena duten lanetarako erabiltzen direla laminetan egindako tresnak. Santa Katalinako azildar mailan laminetan egindako harraskak larru freskoa harraskatzeko erabiltzen dira, maila honetan burutzen den lan garrantzitsuenetakoa bat baita; magdalendar mailan, aldiz, laminetan egindako harraskak larru lehorra prestatzeko lanetan hartzen dute parte, garai hartako lan-faserik hedatuenean, alegia.

Bigarren mailan harrizko tresnekin burutu ziren iharduketak ezagutu nahi ditugu: zeintzuk izan ziren landutako materialak, nola landu ziren eta zertarako. Horretarako kate operatibo eta prozesu tekniko kontzeptuez baliatu gara eta ehiztari-fruitubiltzaile taldeen lan-teknikei buruzko hainbat adibide arkeologiko eta antropologikoetara jo dugu. Hezur-materietan, larruan edo zurean egindako prozesu tekniko orokorren dokumentazioari eta egituraketari esker, harrizko tresnetan antzeman diren lanen zentsua ulertzeko markoa eskuratu dugu. Horretarako, materia bakoitzez egindako ekoizkinen hartze-, transformazio- eta mantenimendu-fase bakoitzeko iharduerak ezarri ditugu. Horrela, geroago, harrizko tresna batean antzeman den lana prozesu tekniko orokorreko fase zehatz batekin lotu daiteke. Puntu honetan hauexek dira erabiltzen diren irizpideak: landutako materia-mota, iharduera-mota, erabilera-neurria, tresna-multzoaren iharduera aniztasuna, tresna bakoitzaren iharduera desberdinak, gune aktiboen luzera, tresnen neurria eta tresna edo gune aktibo espezifikoen erabilera.

Tresna-multzoaren azterketatik ateratako emaitzetan oinarrituz materia bakoitzean egindako lanaren ezaugarriak erdikus daitezke. Larrugintzari buruzko informazioak erreferentzia interesgarriak eskaintzen ditu. Izan ere, Santa Katalinako azildar mailan larrua oraindik freskua dagoenean egiten zaizkion lanak dira nabarienak, hau da, animalia ehizatu bezain laster haren larrua garbitzea. Hala ere, Berniollon burutzen den larrugintza erabat desberdina da; ia ez da larru-garbitze fasea isladatzen eta gehienen antzematen dena, ordea, larru lehorra mehetzeko edo malgutzeko harraskaketa-lana da, izan ere, faserik neketsuena, material hau prestatzeko behar den denbora, energia eta tresneria kontuan hartzen bada.

Hezur-materietan egindako lanak adarraren transformazio konplexu samarra isladatzen du hiru aztarnategietan, eta horretarako tresna bereziak erabili izan direla, burilak adibidez, askotan eta biziki erabiliak gune aktibo desberdinetatik gainera, hainbat iharduera burutzeko: zirrikituak eta zuloak egiteko, ebakitzeko, harraskatzeko, etabar.

Bereiztasun handiagoak ikusi izan dira hezurgintzan. Adibidez, Santa Katalinako multzo azildarrean hezurra era soil eta indar gutxiz lantzen da, agian tresneria mantentzeko eta konpontzeko. Magdalen aroko mailean, ordea, fabrikazio-prozesuaren ia fase guztietako arrastoak erakusten dituzte tresnek. Santa Katalinan kontserbatu diren hezur-hondakinak bat datoz ondorio hauekin.

Orohar, hezur-materiak eta larrua dira harrizko tresnez gehien landutako materiak. Zurgintza Berniollon baino ez da aipagarria. Aipatu aztarnategiko zur-ekoizkinak, gutxi landutako tresnez (ukitu gabeko tresnez) eginda badaude ere, ekoizkin nahikoa onak izan behar zuten, materia honetan iharduera ugari eta kementsuak egin zirela kontuan hartzen bada. Aragigintza hiru aztarnategietan ageri da, baina azpimarratzekoa da Santa Katalinako azildar mailako tresneriaren artean daukan garrantzia. Kronologia beretsuko testuinguruetan ohi denez, gutxi zabalduta dago landare ez egurtsuen langintza.

Aztertu ditugun taldeen estrategia ekonomikoak ezagutzen saiatuko gara **hirugarren maila** batean. Batez ere baliabide desberdinak eskuratzeko eta transformatzeko ihardueren maila eta planifikazio-mota izango dugu kontuan. Horretarako, aurreko bi mailetan bildutako informazioaz baliatuko gara. Lehenengoan ikusi genuen nola kudeatzen ziren lanabesak, zeintzuk ziren horretarako jarraitzen ziren irizpideak, eta zernolako premiak asetzen zituzten. Bigarrenean, landutako materiaren transformazio-fase desberdinak agertzen diren ala ez adierazi dugu eta agertzekotan zein neurritan agertzen diren. Transformazio-faseen presentzia/ausentziak eta trinkotasunak logika bat dute eta logika hori ehiztari-fruitubiltzaile taldeen estrategia ekonomikoei buruzko informazio etnografikoei esker ikus daiteke: nola antolatzen duten baliabideen lorpena, transformazioa eta kontsumoa denboran eta espazioan zehar eta nola antolatzen duten taldea bera iharduera hauek burutzeko.

Interpretazio maila honetara iristeko mugak askoz ere handiagoak dira. Muga batzuk teorikoak dira eta ehiztari-fruitubiltzaileen antolaketari buruzko ikuspegi desberdinez eztabaidatzen da hemen, giza-talde desberdinen arteko bariabilitatea eta era berean sistema hauek inguruabar aldakorretara moldatzeko duten gaitasuna azpimarratuz. Hauexek dira beste muga batzuk: estudiaturiko aztarnategi kopurua urria dela eta estudio honetan ez daudela arkeologiak eskaintzen dituen garrantzi handiko zenbait informazio: lehengaia non eta nola eskuratzen den, ehizaturiko animaliak zeintzuk diren eta nola prestatzen diren premia desberdinak estaltzeko, edo zeintzuk diren banaketa micro-macroespazialak, informazio horietako gutxi batzuk baino ez aipatzearren.

Hala ere, lanabesen erabileran antzeman diren portaerak azaltzen dituen lehen laburpen bat egin daiteke ikuspegi honetatik.
Santa Katalinako azildar mailan, ugariak dira, batik bat, erabilitako jaurtikigaien hondakinak, eta aragigintzarako eta larru freskua lantzeko tresnen hondakinak, hau da, animaliak harrapatzeko lanetan eta materia hori prozesatzeko lehen fasean parte hartzen duten tresneriaren aztarnak. Iharduera hauek omen dira haitzuloa okupatzeko arrazoirik funtzezkoena eta horretarako gordetzen dira, hain zuzen ere, tresnarik hoberenak. Beste material batzutan egiten den lana, batez ere, hezurrean edo zurean, mantenimendu-lana baino ez da, haitzuloan egingo den egonaldiko lan garrantzitsuenei enbarazurik ez egiteko. Aztarnategi honek finkaleku espezializatua dirudi ihardueral beharrizanen arabera geroratzen edo zatikatzen baitira, estrategia planifikatu baten barruan.

Berniollon, aitzitik, materia hartzeko eta lehen transformaziorako ihardueral egon dauden arren, oso gutxi garatuta daude. Zurezko edo hezur-materiazko ekoizkin konplexuagoak fabrikatzeko lanak dira nagusi; hauekin batera, larrugintzaren atalik neketsuenak ere garrantzi nabaria du, hau da, larru lehorra harraskaketaren bidez egokitzeak. Aldi berean egin zen harrizko tresnen fabrikazioari buruzko estudioak agerian jarri du tresneria oparoa eta ainitza egiten zela kalitatezko euskarrietan, beranduago edota orduko finkalekutik kanpo erabiltzeko. Orohar, Berniolloko finkalekuan produktu ugari prestatzeko burutu behar diren ihardeurik konplexuenak zentralizatzen zituen leku modura funtzionatzen omen zuen, beraz, askoz ere gutxiago ziren baliabideak eskuratzeko ihardueral.

Santa Katalinako magdalen-aroako maila oso bestelakoa da beste bi multzoekin erkatuz gero. Bertan materia desberdinen transformazio-aldi guztiak agertzen dira, lehenengoetatik hasi eta bukaerakoetaraino -larru freskua zein lehorra harraskatzen da, adibidez; baina baliabideak eskuratzeko eta aragigintzako lanak ere garrantzitsuak dira. Betebehar zehatz bako finkalekua dela isladatzen du harrizko tresnek burututako iharduera-multzoak. Agian, taldea antolatzeko modua izan daiteke horren arrazoia, ihardueral gutxi planifikatzen eta zatikatzen baitira. Baina, hala ere, bestelako zio bat egon liteke: helburu desberdinetarako egindako zenbait egonaldi gainjarrita egotea, eta horren ondorioz itxura zehazgabea ematea finkalekuari, palinpsesto efektua lortuz.

www.ingramcontent.com/pod-product-compliance
Lightning Source LLC
LaVergne TN
LVHW071208210926
848482LV00015B/757